Full Field Optical Metrology and Applications

Online at: https://doi.org/10.1088/978-0-7503-3027-5

IOP Series in Advances in Optics, Photonics and Optoelectronics

SERIES EDITOR

Professor Rajpal S Sirohi Consultant Scientist

About the Editor

Rajpal S Sirohi is currently working as a faculty member in the Department of Physics, Alabama A&M University, Huntsville, AL, USA. Prior to this, he was a consultant scientist at the Indian Institute of Science, Bangalore, and before that he was Chair Professor in the Department of Physics, Tezpur University, Assam. During 2000–11 he was an academic administrator, being vice-chancellor to a couple of universities and the director of the Indian Institute of Technology, Delhi. He is the recipient of many international and national awards and the author of more than 400 papers. Dr Sirohi is involved with research concerning optical metrology, optical instrumentation, holography, and speckle phenomena.

About the series

Optics, photonics, and optoelectronics are enabling technologies in many branches of science, engineering, medicine, and agriculture. These technologies have reshaped our outlook and our ways of interacting with each other, and have brought people closer together. They help us to understand many phenomena better and provide deeper insight into the functioning of nature. Further, these technologies themselves are evolving at a rapid rate. Their applications encompass very large spatial scales, from nanometers to the astronomical scale, and a very large temporal range, from picoseconds to billions of years. This series on advances in optics, photonics, and optoelectronics aims to cover topics that are of interest to both academia and industry. Some of the topics to be covered by the books in this series include biophotonics and medical imaging, devices, electromagnetics, fiber optics, information storage, instrumentation, light sources, CCD and CMOS imagers, metamaterials, optical metrology, optical networks, photovoltaics, free-form optics and its evaluation, singular optics, cryptography, and sensors.

About IOP ebooks

The authors are encouraged to take advantage of the features made possible by electronic publication to enhance the reader experience through the use of color, animation, and video and by incorporating supplementary files in their work.

Do you have an idea for a book you'd like to explore?

For further information and details of submitting book proposals, see iopscience.org/books or contact Ashley Gasque at Ashley.gasque@iop.org.

A full list of titles published in this series can be found here: https://iopscience.iop.org/bookListInfo/series-on-advances-in-optics-photonics-and-optoelectronics.

Full Field Optical Metrology and Applications

Fernando Mendoza-Santoyo, Manuel De la Torre-Ibarra, María del Socorro Hernández-Montes and Jorge Mauricio Flores Moreno
Centro de Investigaciones en Optica, León, Mexico

IOP Publishing, Bristol, UK

ISBN 978-0-7503-3027-5 (ebook)
ISBN 978-0-7503-3025-1 (print)
ISBN 978-0-7503-3028-2 (myPrint)
ISBN 978-0-7503-3026-8 (mobi)

DOI 10.1088/978-0-7503-3027-5

Version: 20221201

IOP ebooks

British Library Cataloguing-in-Publication Data: A catalogue record for this book is available from the British Library.

Published by IOP Publishing, wholly owned by The Institute of Physics, London

IOP Publishing, No.2 The Distillery, Glassfields, Avon Street, Bristol, BS2 0GR, UK

US Office: IOP Publishing, Inc., 190 North Independence Mall West, Suite 601, Philadelphia, PA 19106, USA

We dedicate this book to our families, their unconditional love gave us the motivation and strength to write it. We are forever grateful to them, we love you.

Contents

Preface

More than two hundred years have passed since the first publication on the interference phenomena of light, with the development of optical interferometry being closely linked to the history of wave optics based on Huygens' theory of 1690. The presentation of the first interferometric experiments by Thomas Young in 1801 could be thought of as the birth of full-field optical metrology. Optical metrology based on interferometry commonly illuminates an object with near-monochromatic or broad-spectrum light to study it in a non-invasive manner. According to the study's objective, the latter means that the object is interrogated in a non-contact or non-destructive manner when it undergoes different deformations (static or dynamic). These deformations modify the original physical conditions of this object, which could be permanent or temporary.

Interferometry has an essential role in several optical metrology methods, and it has been proven for science and industry, i.e. optical metrology allows the precise reconstruction of a wavefront distorted by an object under study. Thus it is possible to measure objects' shape, displacement, stress, strain, stiffness, rigidity, and many more physical and mechanical variables. Measuring these variables can help detect unknown external forces that modify the object.

Although optical characterization was used a long time ago, the range of possibilities was expanded in the 1960s with the introduction of two remarkable developments: the invention of the laser and the development of electronic video devices. Since then, the number of experimental set-ups has increased constantly and has evolved as a research subject that serves the scientific and technological communities in many and increasing application areas. These include biology, medicine, engineering (with industries such as aerospace, automotive, manufacturing, and materials), communications, security, photonics, art preservation, and archaeology, among many more. As optical metrology is involved in various fields and is a broad topic that involves several areas and methods, it would require vast space in a single book to analyse them in a detailed manner, in particular as more lines of research and technology are being considered as optical metrology methods/techniques.

This book is mainly devoted to interferometric and speckle-based full-field optical metrology. Thus, in making choices as to which topics to cover, the authors decided to address the basics of interference and laser speckle metrology, adding recent topics such as digital image correlation, speckle denoising, and holography with particles and photons. Once the topics were decided, we discussed how the contents should be presented, trying to bear in mind a reader looking into the field for the first time, i.e. this book is aimed at students, teachers, and researchers new to the field. We also considered those readers in optical metrology who wish to discuss and increase their understanding based on speckle interference and to identify new approaches for its use in measurement and application purposes.

Optical metrology includes several methods and techniques involving basic optics principles, such as diffraction, interferometry, coherence, Fourier optics, holography, scattering, etc. There are quite a few books that deal with all these theories at different levels. To distinguish our book from others, we took our experience and knowledge (as teachers, researchers, and technology developers) to propose several experimental set-ups based on interferometry to study a wide range of objects. Thus, we based the contents of the book on three main topics, with five chapters each, including basic topics on interference, digital holographic techniques, and recent advances in interferometric techniques:

I. *Interference*: We provide the bases of interferometry, coherence properties, speckle, phase evaluation methods, and moiré. We also introduce non-contact optical techniques based on the speckle effect, but without going into detail about the speckle phenomenon and its statistics.
II. *Digital holography*: We present topics on holography, digital speckle pattern interferometry, and digital holographic interferometry/microscopy; we also explain the optical process for contouring objects based on digital holography. The main contribution of this section is to show and explain procedures based on Fourier analysis to obtain objects' shapes and 3D information using techniques and algorithms not described elsewhere.
III. *Recent advances*: We present some techniques that have experienced increased use in several fields of knowledge, namely optical coherence tomography and digital image correlation, among others. A hot topic that we discuss is electron holography, where much of the work in retrieving the phase evolved from light-based interferometry. We also review some of the basics of quantum holography. The last two topics rely on the wave-like nature of electrons, and thus the techniques based on and developed for interferometry may be applied, enabling the study of physical phenomena within the atomic world.

In order to adhere to the state-of-the-art in the subjects dealt with, we invited several colleagues as special contributors to collaborate with us in producing this book. We, the primary authors (who became editors and co-authors of these contributions) wish to acknowledge and sincerely thank these colleagues—their outstanding contributions have made this a better book. Finally, the editors hope that all readers find this book a helpful tool in understanding and implementing a basic set-up that serves their particular needs. In particular we encourage any student involved in graduate and postgraduate programs in optical sciences to implement some (or all, why not!) of the modular set-ups based on full-field interferometric methods presented in this book.

We are deeply grateful to the following colleagues for their contributions:

Chapters 1, 3, 4 and 6: Thomas Kreis, Universität Bremen, Germany

Chapter 3: Angel Manuel Fernández Doval, Universidad de Vigo, Vigo, Spain

Chapter 5: Apolinar Muñoz, Center of Research in Optics, León Guanajuato, Mexico

Chapter 9: Myung K Kim, University of South Florida, United States

Chapter 12: Bing Pan, Institute of Solid Mechanics, Beihang University, Beijing, China

Chapter 13: Manoj Kumar, Kobe University, Japan
Yassine Tounsi, Chouaib Doukkali University, Morocco
Karmjit Kaur, Government PG College, India
Abdelkrim Nassim, Chouaib Doukkali University, Morocco
Osamu Matoba, Kobe University, Japan

Finally, this book was written during the COVID19 pandemic, and a lot happened during this time—some of us got sick, we lost friends and relatives, but we prevailed thanks to the loving care of our families. So here is to them—the first edition of our book. Thank you!

Author biographies

Fernando Mendoza Santoyo

Fernando Mendoza Santoyo is an Emeritus Professor at Centro de Investigaciones en Optica-CIO, México, where he founded the Optical Metrology Division in 1991. He is a Fellow of OPTICA and SPIE.

Manuel H De la Torre Ibarra

Manuel H De la Torre Ibarra is a Researcher with more than 16 years of experience in optical non-destructive testing applied to surface and internal inspections in engineering and biological studies.

María del Socorro Hernández-Montes

María del Socorro Hernández-Montes is a Researcher in the Division of Optical Metrology at the Center for Research in Optics (Centro de Investigaciones en Optica - CIO, México). Her Research Interests are mainly in Non-invasive and non-destructive optical methods and biomedical optics.

Jorge Mauricio Flores Moreno

Jorge Mauricio Flores Moreno is a Researcher at Optical Research Centre (CIO). He works in microscopy using techniques like holography, confocal, multiphoton, and spectroscopy.

IOP Publishing

Full Field Optical Metrology and Applications

Fernando Mendoza-Santoyo, Manuel De la Torre-Ibarra, María del Socorro Hernández-Montes and Jorge Mauricio Flores Moreno

Chapter 1

Foundations of optical interferometry

Contributing author: Thomas Kreis

The subject of optical metrology comprises several lines of study, including classical interferometry, digital holographic interferometry, white-light interferometry, speckle methods, moiré methods, and techniques such as deflectometry and photoelasticity. Optical metrology methods and techniques use light to measure physical, mechanical, and a wide variety of quantities of interest to scientists and engineers (and with time, no doubt to other professionals working in many fields and disciplines). Light is a physical phenomenon and a small part of the electromagnetic spectrum. It manifests in different ways depending on the observables in an experiment, for example, light rays and their geometric behavior are described by geometrical or Gaussian optics, with the main topics being the laws of reflection and refraction. The effects of diffraction, interference, and polarization are part of wave optics. At the same time, the behavior of light as particles, the so-called photons, e.g. those that make lasers, is understood with the help of quantum optics. In this chapter, the foundations of interferometry, as the basis for most optical metrological methods, are outlined.

1.1 Light waves

1.1.1 The wave equation: solutions

Electromagnetic (EM) radiation is conceived as a traveling wave propagating through space, carrying electric and magnetic energy vibrating at different frequencies. Light is defined as an electromagnetic wave emitting energy in the visible range of the electromagnetic spectrum (frequency distribution). Then, it must obey the Maxwell equations from which the wave equation can be derived. The wave equation for the propagation of the electric field of the EM wave in free space is defined as

$$\nabla^2 \boldsymbol{E} - \frac{1}{c^2}\frac{\partial^2 \boldsymbol{E}}{\partial t^2} = 0, \tag{1.1}$$

doi:10.1088/978-0-7503-3027-5ch1

where $\boldsymbol{E} = \boldsymbol{E}(x, y, z, t)$ is the electric field, ∇^2 the Laplace operator, (x, y, z) the Cartesian spatial coordinates, t the temporal coordinate associated with time, and c the propagation speed of the wave, which for free-space propagation equals the speed of light in a vacuum $c_0 = 299\ 792\ 458\ \mathrm{m\ s^{-1}}$. Boldface expressions denote vectors.

While longitudinal waves solve the wave equation, e.g. evanescent waves in which energy is spatially concentrated in the vicinity of the source, the most important waves in interferometry are the transverse waves that are propagated through space. They vibrate at right angles relative to the direction of propagation. They may vibrate horizontally (x-direction), vertically (y-direction), or in any direction combining these, an effect called polarization, dealt with in the next section. Fortunately, for most applications it is not necessary to use the full vectorial description of the fields so that we can assume a vibrating wave in a single plane. Such a wave is called plane-polarized. For a plane-polarized wavefield propagating in the z-direction the scalar form of the wave equation is sufficient:

$$\frac{\partial^2 E}{\partial z^2} - \frac{1}{c^2}\frac{\partial^2 E}{\partial t^2} = 0, \tag{1.2}$$

where c is the light velocity in the vacuum in this linear and second-order differential equation; its solution is a linear combination of solutions based on the superposition principle.

The most important solution of (1.2) is the harmonic wave which in real notation is $E(z, t) = E_0\cos(kz - \omega t)$. Here E_0 is the wave's amplitude, and the term $(kz - \omega t)$ gives the wave's phase, see figure 1.1(a) for spatial and figure 1.1(b) for temporal cases. The wavenumber k is associated with the wavelength λ by $k = 2\pi/\lambda$. The angular frequency ω is related to the frequency ν of the wave by $\omega = 2\pi\nu$, where ν is the number of periods per second, which means $\nu = c/\lambda$ or $\nu\lambda = c$. If the maximum amplitude is not at $x = 0$ and $t = 0$, we have to introduce the relative phase ϕ in $E(z, t) = E_0\cos(kz - \omega t + \phi)$. With the period T, the time for a full 2π-cycle, we can write

$$E(z, t) = E_0\cos\left(\frac{2\pi}{\lambda}z - \frac{2\pi}{T}t + \phi\right). \tag{1.3}$$

Calculations can be performed by using the complex exponential, which is related to the trigonometric functions by Euler's formula $e^{i\alpha} = \cos\alpha + i\sin\alpha$ where $i = \sqrt{-1}$ is the imaginary unit. Since the cosine now is $\cos\alpha = (e^{i\alpha} + e^{-i\alpha})/2$, the harmonic

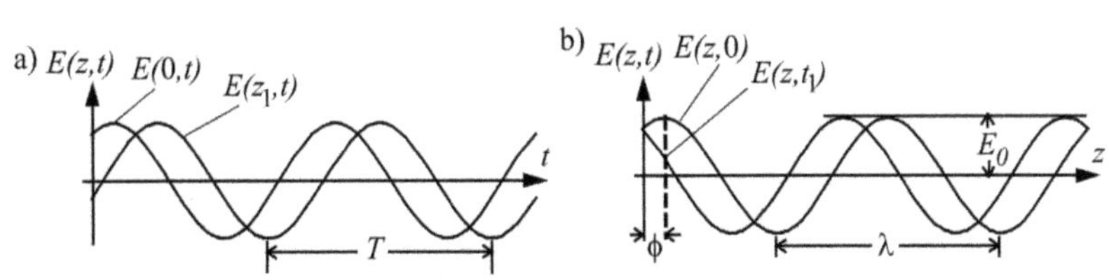

Figure 1.1. (a) Spatial and (b) temporal distribution of a scalar harmonic wave.

wave can be rewritten as $E(z, t) = (E_0 e^{i(kz-\omega t+\phi)} + E_0 e^{-i(kz-\omega t+\phi)})/2$. The second term on the right-hand side is the complex conjugate of the first term and can be omitted as long as it is understood that only the real part of $E(z, t)$ represents the physical wave. Thus the harmonic wave in complex notation is

$$E(z, t) = \frac{1}{2} E_0 e^{i(kz-\omega t+\phi)}. \tag{1.4}$$

A wavefront refers to the spatial distribution of the maxima of the wave, or other surfaces of constant phase, as these surfaces propagate. The wavefronts are normal to the direction of propagation. For instance, a plane wave is a wave that has a constant phase in all planes orthogonal to the propagation direction for a given time t. To describe the spatial distribution of the wave, we can assume $t = 0$ in an arbitrary time scale. Since $\boldsymbol{k} \cdot \boldsymbol{r}$ = constant is the equation for a plane in a three-dimensional space, with the wave vector as $\boldsymbol{k} = (k_x, k_y, k_z)$ and the spatial vector as $\boldsymbol{r} = (x, y, z)$, a plane harmonic wave at time $t = 0$ is

$$E(\boldsymbol{r}) = E_0 e^{i(\boldsymbol{k}\cdot\boldsymbol{r}+\varphi)}. \tag{1.5}$$

The expression

$$E(\boldsymbol{r}, t) = E_0 e^{i(\boldsymbol{k}\cdot\boldsymbol{r}-\omega t+\varphi)} \tag{1.6}$$

describes the temporal dependence of a plane harmonic wave propagating in the direction of the wave vector $\boldsymbol{k}$.

Another waveform often used is the spherical wave, where the phase is constant on each spherical surface around the source point. The importance of spherical waves comes from the Huygens' principle, which states that each point on a propagating wavefront can be considered radiating itself a new spherical wavelet. Each resulting wavefront is the superposition of such wavelets.

The harmonic spherical waves are described by

$$E(r, t) = \frac{1}{r} E_0 e^{i(kr-\omega t+\varphi)}. \tag{1.7}$$

One observes that the amplitude E_0/r decreases according to $1/r$. Interestingly, the spherical wave locally approximates a plane wave at a long distance from its origin.

The complex amplitudes of wavefronts scattered by a surface are generally very complicated. However, due to the superposition principle, they can be treated as the sum of plane waves or spherical waves. There are more solutions to the wave equation, e.g. the Bessel waves. However, such waves are not common in interferometry so, in the following, we restrict the discussion to the plane and spherical harmonic waves.

1.1.2 Polarization

The particular importance of polarization in interferometric metrology is that only identically polarized components (of superposed waves) can interfere since they vibrate in the same direction of propagation. As mentioned, the transverse harmonic wave solution of the wave equation propagating in the z-direction is a vector field $\boldsymbol{E}(z, t) = \frac{1}{2}\boldsymbol{E}_0 \exp[\mathrm{i}(kz - \omega t)]$, where the complex amplitude vector $\boldsymbol{E}_0$ consists of the components $\boldsymbol{E}_0 = E_{0x}\hat{x} + E_{0y}\hat{y}$ with $\hat{x}$, $\hat{y}$ as the unit vectors along the x- and y-directions. We can express the complex E_{0x} and E_{0y} by the real amplitudes a_x, a_y and phases φ_x, φ_y as

$$E_{0x} = a_x e^{\mathrm{i}\varphi_x}, \quad E_{0y} = a_y e^{\mathrm{i}\varphi_y}. \tag{1.8}$$

If the phase difference $\varphi = \varphi_x - \varphi_y$ equals an integer multiple of π, we obtain $E_{0x} = \pm(a_x/a_y)E_{0y}$ and we have linear polarization in the direction of a straight line with slope $\pm a_x/a_y$, see figure 1.2.

If $\varphi = \pm\pi/2$ and $a_x = a_y = a$ we obtain $E_{0x}^2 + E_{0y}^2 = a^2$ which is the equation of a circle with radius a. The x–y-projection of the tip of the electric field vector rotates along such a circle, and we speak about a circular polarized wave, see figure 1.3. For arbitrary a_x and a_y as well as any φ we have the general case of elliptical polarization, including the cases just mentioned of linear or circular polarization.

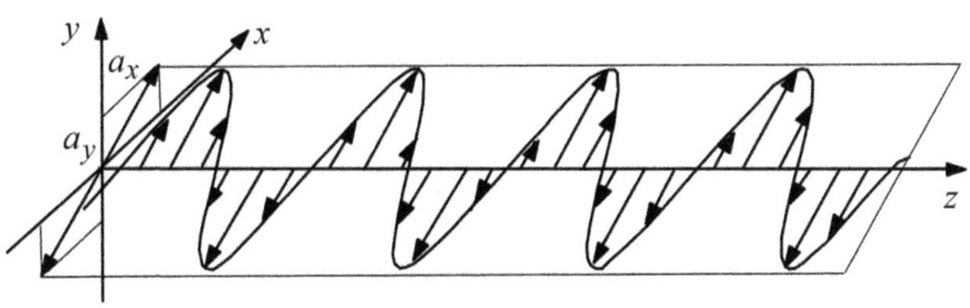

Figure 1.2. Linearly polarized wave sketch.

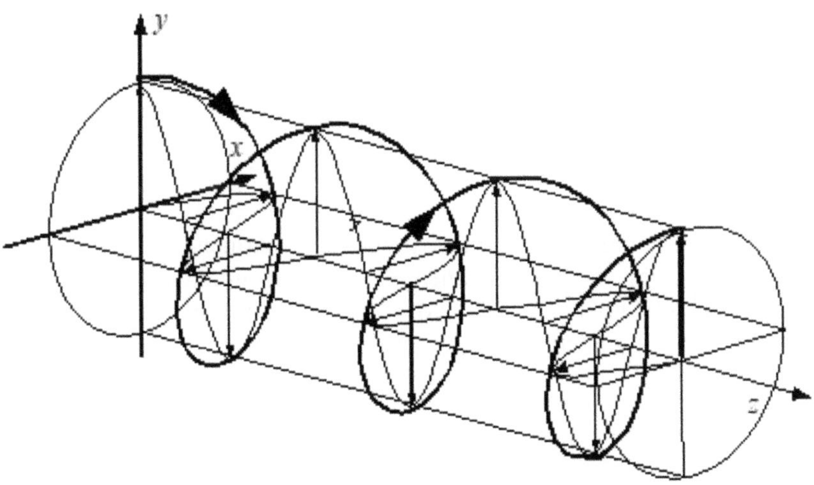

Figure 1.3. Right circularly polarized wave.

The use of linear polarizers in interferometry means that the light beams can be controlled in various ways, e.g. experimentally, to avoid the backpropagation of light into the light source or to increase the interference amplitude. Examples of such linear polarizers are those that filter any linear polarization component out of arbitrarily polarized waves or retarder plates that can rotate the polarization direction ($\lambda/2$—half-wave plate). They also change from linear to circular polarization ($\lambda/4$—quarter-wave plate) and polarizing beam splitters (PBSs), in which the deflection of waves depends on their polarization direction. The most used retardation values are $\lambda/2$, $\lambda/4$, and λ.

1.1.3 Intensity

The only parameter of light which is measurable by sensors, i.e. the eye, photodiode, CCD-target, etc, is the intensity (and, on a rough scale, the frequency as color). The energy flux defines the intensity in optics through an area per unit of time. From the Maxwell equations, $I = \varepsilon_0 c E^2$, wherein the following only the proportionality of the intensity $I \sim E^2$ is used, $\varepsilon_0 c$ being proportionality constants.

Thus, the intensity has a nonlinear dependence on the electric field strength, and since no sensor can follow the frequency of light, it must be integrated over a measuring time T_m, meaning that the momentary intensity is not measurable. So if $T_m \gg T = 2\pi/\omega$, omitting the proportionality constants we define $I = E_0 E_0^* = |E_0|^2$, where * denotes the complex conjugate. The intensity of a general stationary wavefield is

$$I(\mathbf{r}) = \langle EE^* \rangle = \lim_{T_m \to \infty} \int_{-T_m/2}^{T_m/2} E(\mathbf{r}, t')E^*(\mathbf{r}, t')\mathrm{d}t'. \tag{1.9}$$

This intensity is the limit of the short time-intensity:

$$I(\mathbf{r}, t, T_m) = \frac{1}{T_m} \int_{t-T_m/2}^{t+T_m/2} E(\mathbf{r}, t')E^*(\mathbf{r}, t')\mathrm{d}t', \tag{1.10}$$

which is a sliding average of a temporal window centered around t with a width T_m. In practical applications, the measuring time T_m always is large compared with the period of the light wave. Nevertheless, it has to be short in the time scale for many investigated processes.

1.1.4 Wavefield propagation

Full-field interferometry compares two extended wavefields. The wavefield propagation is described analytically for any numerical treatment of such effects. This is the topic of diffraction theory. This theory's general foundations and preliminaries can be found in [1]. Here a complex field $U(\xi,\eta)$ inside a finite extent aperture in the $(\xi, \eta, z = 0)$-plane is assumed. Then the scalar field $E(x, y, z)$ in the (x, y) plane at a distance z from the $(\xi, \eta, z = 0)$, the plane can be described by the Fresnel–Kirchhoff formula [1]

$$E(x, y, z) = \frac{1}{\mathrm{i}\lambda} \int_{-\infty}^{\infty} \int_{-\infty}^{\infty} U(\xi, \eta) \frac{\mathrm{e}^{ikr}}{r^2} \mathrm{d}\xi \mathrm{d}\eta. \tag{1.11}$$

If the distance z between the diffracting aperture and (x, y)-plane is large compared to the lateral distances of the points in the x-, y-, ξ-, and η-directions, then equation (1.11) is approximated by expanding the square root $r = \sqrt{z^2 + (\xi - x)^2 + (\eta - y)^2}$ into a binomial series, and omitting higher than linear terms, r is obtained as

$$r \approx z + \frac{x^2 + y^2}{2z} - \frac{x\xi + y\eta}{z} + \frac{\xi^2 + \eta^2}{2z}. \tag{1.12}$$

Substituting also $\nu = x/(z\lambda)$ and $\mu = y/(z\lambda)$ the Fresnel approximation is obtained [1]:

$$E(\nu, \mu) = \frac{\mathrm{e}^{ikz}}{\mathrm{i}\lambda z}\mathrm{e}^{\mathrm{i}\pi z\lambda(\nu^2+\mu^2)}\int_{-\infty}^{\infty}\int_{-\infty}^{\infty} U(\xi, \eta)\mathrm{e}^{\frac{\mathrm{i}\pi}{\lambda z}(\xi^2+\eta^2)}\mathrm{e}^{-2\mathrm{i}\pi(\xi\nu+\eta\mu)}\mathrm{d}\xi\mathrm{d}\eta. \tag{1.13}$$

Apart from the constant intensity factor and the phase factor preceding the integrals, this is the Fourier transform of the product of the field $U(\xi, \eta)$ and the two-dimensional chirp-function $\exp\left[\frac{\mathrm{i}\pi}{\lambda z}(\xi^2 + \eta^2)\right]$.

For even longer distances z, the term $(\xi^2 + \eta^2)/(2z)$ can be omitted and we obtain the Fraunhofer approximation:

$$E(\nu, \mu) = \frac{\mathrm{e}^{ikz}}{\mathrm{i}\lambda z}\mathrm{e}^{\mathrm{i}\pi z\lambda(\nu^2+\mu^2)}\int_{-\infty}^{\infty}\int_{-\infty}^{\infty} U(\xi, \eta)\mathrm{e}^{-2\mathrm{i}\pi(\xi\nu+\eta\mu)}\mathrm{d}\xi\mathrm{d}\eta. \tag{1.14}$$

In practice, the necessity of a large distance z can be circumvented by either illuminating the aperture by a spherical wave converging to the observation plane (x, y) or by placing a positive lens between the aperture and observer [2]. Hence, despite the phase and intensity factors preceding the integrals, which are not dependent on the diffracted field $U(\xi, \eta)$, the Fresnel–Kirchhoff formula is reduced to a pure Fourier transform.

1.2 Types of interferometers

In order to interfere, the superposed wavefields must stem from the same source and maintain a minimum coherence between them [1–3], as will be explained in chapter 2. There are two ways to produce two wavefields out of one: amplitude division and wavefront division. The Michelson interferometer mentioned previously is of the amplitude division type. The amplitude of the original wavefront is divided by the beam splitter into two separate waves, which are finally recombined on a sensor after traveling different paths. An example of wavefront division is Young's interferometer, also mentioned before. The two slits separate different parts of the wavefront and, in the end, are recombined on a screen or detector.

1.2.1 Interferometers with amplitude division

The Michelson interferometer commonly uses a coherent light source. It measures distance changes with applications ranging from displacement measurements, deformations, or vibrations of machine parts to detecting gravitational waves [4]. The Twyman–Green interferometer used to test optical components such as lenses is another example of an amplitude division device. See figure 1.4, where the light beam is expanded and travels twice through the lens to be tested. The center of curvature of the spherical mirror must coincide with the lens's focus under test; if it is necessary to test the optical quality of flat surfaces, a Fizeau interferometer is employed, see figure 1.5.

In these interferometers, the beam splitter divides the waves and also combines them. A further example is given by the Mach–Zehnder interferometer, where these two tasks are performed by different beam splitters/combiners, see figure 1.6. This type of interferometer is best suited to measure path-integrated refractive index changes in transparent media due to variations of such physical quantities as temperature, pressure, chemical concentrations, mechanical stress, etc.

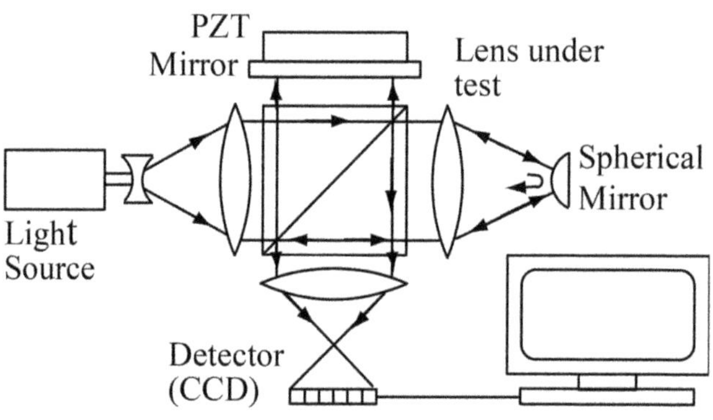

Figure 1.4. Twyman–Green interferometer.

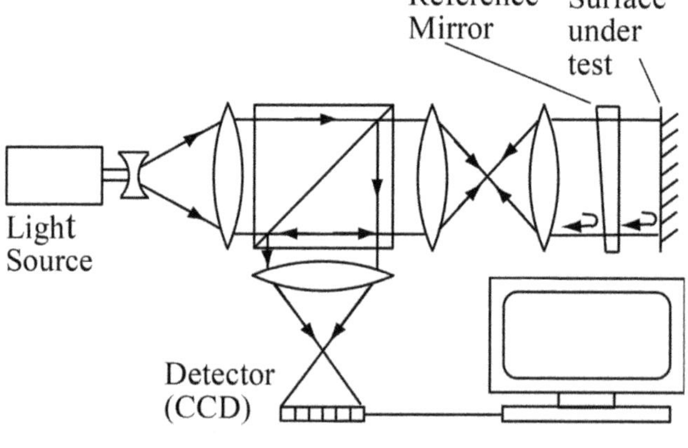

Figure 1.5. Fizeau interferometer.

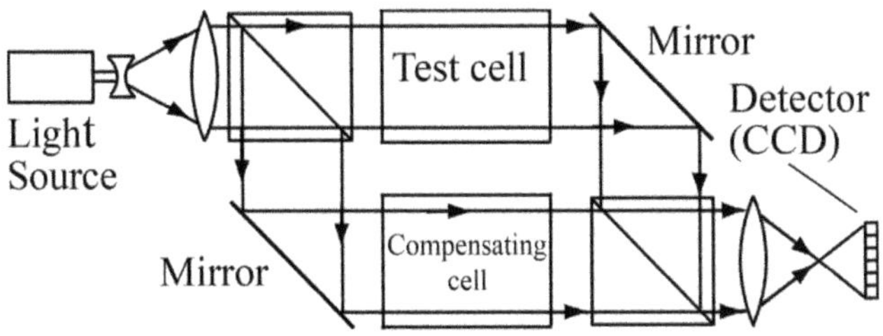

Figure 1.6. Mach–Zehnder interferometer.

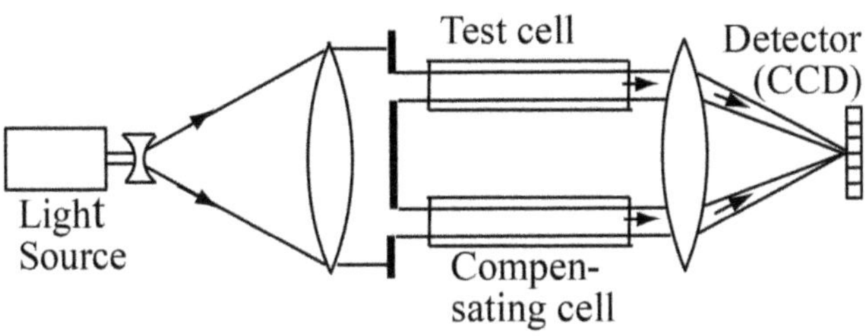

Figure 1.7. Rayleigh interferometer.

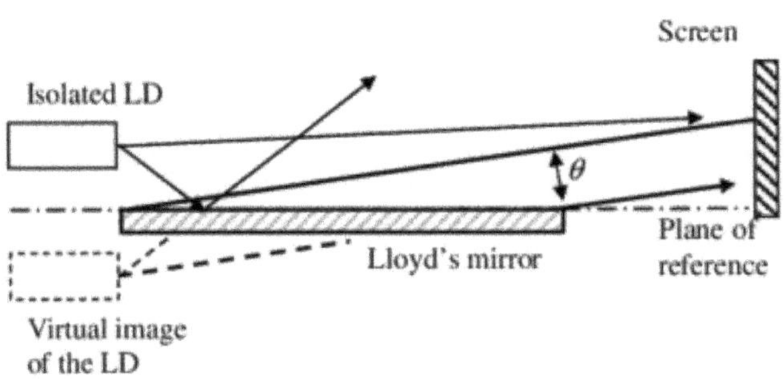

Figure 1.8. Lloyd's mirror interferometer; θ is the angle between the reflected and unreflected rays.

1.2.2 Interferometers with wavefront division

Two other examples of wavefront division interferometers, in addition to Young's interferometer, are the Rayleigh interferometer, see figure 1.7, with applications comparable to those offered by the Mach–Zehnder interferometer; and the so-called Lloyd's mirror (figure 1.8), where reflected and unreflected parts of the wave interfere, showing how during reflection at a medium with higher refractive index, $n_2 < n_1$, the wave undergoes a 180° phase change.

1.3 Interference of light waves

1.3.1 Interference of two waves with the same frequency

Let us now assume two mutually coherent waves, emitted by the same source, which differ in the directions $\boldsymbol{k}_1$ and $\boldsymbol{k}_2$, phases φ_1 and φ_2, and amplitudes E_{01} and E_{02}, but have the same frequency ω and are linearly polarized in the same direction. As seen before, in scalar notation they are represented by

$$E_1(\boldsymbol{r}, t) = E_{01}e^{i(\boldsymbol{k}_1 \cdot \boldsymbol{r} - \omega t + \varphi_1)}, \quad E_2(\boldsymbol{r}, t) = E_{02}e^{i(\boldsymbol{k}_2 \cdot \boldsymbol{r} - \omega t + \varphi_2)}. \tag{1.15}$$

To determine the interference of these two waves, we decompose the vectors $\boldsymbol{k}_1$ and $\boldsymbol{k}_2$ into components of equal and opposite directions, $\boldsymbol{k}' = (\boldsymbol{k}_1 + \boldsymbol{k}_2)/2$ and $\boldsymbol{k}'' = (\boldsymbol{k}_1 - \boldsymbol{k}_2)/2$, see figure 1.9. If θ is the angle between $\boldsymbol{k}_1$ and $\boldsymbol{k}_2$ then

$$| \boldsymbol{k}''| = \frac{2\pi}{\lambda} \sin \frac{\theta}{\lambda}. \tag{1.16}$$

In the same way, we define the mean phase $\varphi = (\varphi_1 + \varphi_2)/2$ and the half-phase difference $\varphi = (\varphi_1 - \varphi_2)/2$. Now the overlapping gives the intensity distribution:

$$\begin{aligned} I(\boldsymbol{r}, t) &= \left[E_1(\boldsymbol{r}, t) + E_2(\boldsymbol{r}, t)\right]\left[E_1(\boldsymbol{r}, t) + E_2(\boldsymbol{r}, t)\right]^* \\ &= \left[E_{01}e^{i(\boldsymbol{k}_1 \cdot \boldsymbol{r} - \omega t + \varphi_1)} + E_{02}e^{i(\boldsymbol{k}_2 \cdot \boldsymbol{r} - \omega t + \varphi_2)}\right]\left[E_{01}e^{-i(\boldsymbol{k}_1 \cdot \boldsymbol{r} - \omega t + \varphi_1)} + E_{02}e^{-i(\boldsymbol{k}_2 \cdot \boldsymbol{r} - \omega t + \varphi_2)}\right] \\ &= E_{01}^2 + E_{02}^2 + E_{01}E_{02}\left[e^{i(\boldsymbol{k}_1 \cdot \boldsymbol{r} - \boldsymbol{k}_2 \cdot \boldsymbol{r} + \varphi_1 - \varphi_2)} + e^{i(-\boldsymbol{k}_1 \cdot \boldsymbol{r} + \boldsymbol{k}_2 \cdot \boldsymbol{r} - \varphi_1 + \varphi_2)}\right] \\ &= E_{01}^2 + E_{02}^2 + E_{01}E_{02} \cos(2\boldsymbol{k}'' \cdot \boldsymbol{r} + 2\ \varphi) \\ &= I_1 + I_2 + 2\sqrt{I_1 I_2} \cos(2\boldsymbol{k}'' \cdot \boldsymbol{r} + 2\ \varphi). \end{aligned} \tag{1.17}$$

A key feature shown in this equation is that we get a temporarily constant intensity. In general, the result of superposing two waves is an intensity distribution that consists of two parts: one is the sum of the original intensities $I_1 + I_2$, and the second is the interference term defining a spatial variation around the mean intensity $(I_1 + I_2)/2$ with amplitude $2\sqrt{I_1 I_2}$ multiplied by the cosine term, also known as the fringe modulation term.

Up to now, only the superposition of parallel linear polarized waves has been considered: if the polarization direction of the two waves is perpendicular to each other, they do not interfere, and their superposition is given by the addition of the intensities $I = I_1 + I_2$, a constant and uniform intensity. For other angles between the polarization directions, the field vector has to be decomposed into parallel and

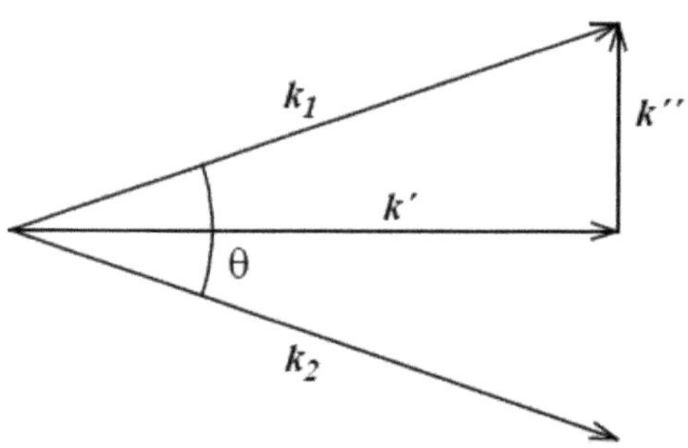

Figure 1.9. Decomposition of the wave vectors.

orthogonal polarization components, with the resulting wave superposition containing interference parts and the term giving the addition of intensities (equation (1.17)). Yet another reason for the additive intensity term to be the dominant one in the interference process is due to a low coherence between the interfering waves (due, for instance, to a light source not being temporally stable, such as gas discharge lamps). Experimentally, the superposition of waves coming from an incoherent light source renders pure addition of the intensities without interference. The additive term is often called the incoherent part, or we also call it the incoherent wave superposition.

The visibility or contrast of the interference pattern may also be defined in terms of the intensity's maxima and minima,

$$V = \frac{I_{\max} - I_{\min}}{I_{\max} + I_{\min}}. \tag{1.18}$$

If two parallel polarized waves of the same intensity interfere, we have the maximal contrast of $V = 1$, and we have minimal contrast $V = 0$ for incoherent superposition. For example, if the ratio of the intensities of interfering waves is 5:1, the contrast is 0.745.

The intensity described by (1.17) has a minimum when $\cos(2\boldsymbol{k}'' \cdot \boldsymbol{r} + 2\ \varphi) = -1$ which is equivalent to $\cos^2(\boldsymbol{k}'' \cdot \boldsymbol{r} + \ \varphi) = 0$. These are the loci where

$$\boldsymbol{k}'' \cdot \boldsymbol{r} + \ \varphi = (2n+1)\frac{\pi}{2},\ n \in \mathbb{Z}, \tag{1.19}$$

where the wavefronts are anti-phase and we speak of destructive interference. The intensity is maximal where

$$\boldsymbol{k}'' \cdot \boldsymbol{r} + \ \varphi = n\pi,\ n \in \mathbb{Z}, \tag{1.20}$$

where the wavefronts are in-phase and we have constructive interference.

The resulting time-independent pattern is called the interference pattern and the fringes are called interference fringes. For plane waves, they are oriented parallel to $\boldsymbol{k}'$ and have a distance of $\pi / | \boldsymbol{k}'' |$ in the direction $\boldsymbol{k}''$. This is shown in figure 1.10.

Using the definition of $\boldsymbol{k}''$ (1.16) we have a spacing p between the interference fringes given by

$$p = \lambda / \left(2\sin\frac{\theta}{2}\right). \tag{1.21}$$

For spherical or other wave shapes, which are not plane waves, the same expression is for local spacing, but now $\theta = \theta(P)$ is the angle between the waves at the point P of interest.

1.3.2 Interference of two waves with different frequencies

Let us now add different frequencies, $\nu_i = \omega_i/(2\pi)$, to the interference of two waves:

$$E_1(\boldsymbol{r}, t) = E_{01}\mathrm{e}^{\mathrm{i}(\boldsymbol{k}_1 \cdot \boldsymbol{r} - 2\pi\nu_1 t + \varphi_1)},\ E_2(\boldsymbol{r}, t) = E_{02}\mathrm{e}^{\mathrm{i}(\boldsymbol{k}_2 \cdot \boldsymbol{r} - 2\pi\nu_2 t + \varphi_2)}. \tag{1.22}$$

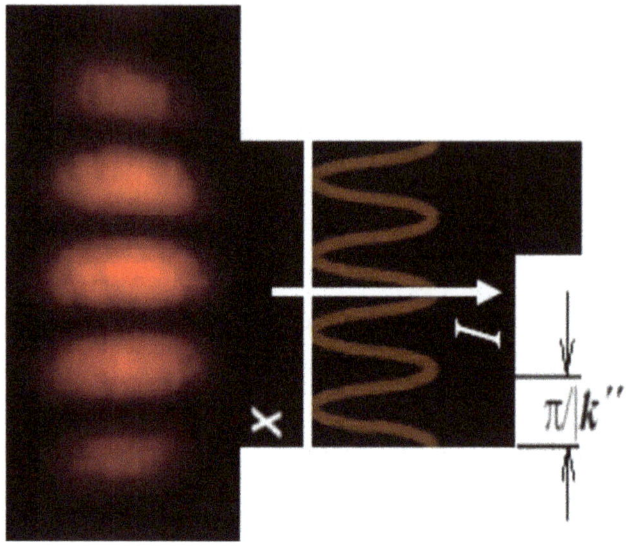

Figure 1.10. Interference fringes constant in time.

In addition the former definitions of $\boldsymbol{k}'$, $\boldsymbol{k}''$, φ, and φ now allow $\nu = (\nu_1 + \nu_2)/2$ and $\nu = (\nu_1 - \nu_2)/2$. Then the addition results

$$\begin{aligned}(E_1 + E_2)(\boldsymbol{r}, t) &= E_0\Big[\mathrm{e}^{\mathrm{i}(\boldsymbol{k}'\cdot\boldsymbol{r}+\boldsymbol{k}''\cdot\boldsymbol{r}-2\pi\nu t-2\pi\ \nu t+\varphi+\ \varphi)} \\ &\quad + \mathrm{e}^{\mathrm{i}(\boldsymbol{k}'\cdot\boldsymbol{r}-\boldsymbol{k}''\cdot\boldsymbol{r}-2\pi\nu t+2\pi\ \nu t+\varphi-\ \varphi)}\Big] \\ &= E_0\mathrm{e}^{\mathrm{i}(\boldsymbol{k}'\cdot\boldsymbol{r}-2\pi\nu t+\varphi)}\Big[\mathrm{e}^{\mathrm{i}(\boldsymbol{k}''\cdot\boldsymbol{r}-2\pi\ \nu t+\ \varphi)} + \mathrm{e}^{\mathrm{i}(-\boldsymbol{k}''\cdot\boldsymbol{r}+2\pi\ \nu t-\ \varphi)}\Big] \\ &= 2E_0\mathrm{e}^{\mathrm{i}(\boldsymbol{k}'\cdot\boldsymbol{r}-2\pi\nu t+\varphi)}\cos(\boldsymbol{k}''\cdot\boldsymbol{r} - 2\pi\ \nu t +\ \varphi),\end{aligned} \tag{1.23}$$

moreover, the intensity becomes

$$\begin{aligned}I(\boldsymbol{r}, t) &= 4E_0^2\cos^2(\boldsymbol{k}''\cdot\boldsymbol{r} - 2\pi\ \nu t +\ \varphi) \\ &= 2E_0^2[1 + \cos(2\boldsymbol{k}''\cdot\boldsymbol{r} - 4\pi\ \nu t + 2\ \varphi)].\end{aligned} \tag{1.24}$$

If the frequency difference is small enough, $\nu_1 \approx \nu_2$, a detector can register an intensity at $\boldsymbol{r}$ oscillating with the beat frequency $2\ \nu = \nu_1 - \nu_2$. The phase of this modulation is the double phase difference $2\ \varphi = \varphi_1 - \varphi_2$ of the superposed waves. Contrary to the frequencies of the generating optical waves, the beat frequency can be measured electronically and further evaluated as long as it remains in the kHz or MHz range. This is known as the heterodyne method that has many applications in optical metrology; for instance, it enables the calculation of the reflector motion via the Doppler shift or determines the phase difference φ between different points of an object where the intensity oscillates with the same constant beat frequency.

1.4 Multiple wave interference

Multiple wave interference is the effect behind Newton's rings, the colors of thin layers, Bragg reflection, or interference filters, to name just a few. A widely used device is the Fabry–Perot interferometer which is effectively employed as an etalon for selecting a single wavelength in the emission spectrum of a laser. The etalon interferometer consists of two identical parallel semitransparent mirrors separated by a distance d, see figure 1.11. Each mirror is typically a plane parallel glass plate with a silvered front surface. Let us briefly describe the interference process: a wave E_0 impinges perpendicularly onto one of the mirrors, then E_r is the reflected wave, while the transmitted wave is E_t. Note that in figure 1.11 the beams are inclined instead of normal to the mirrors; this is just to depict the interferometric effect in a more easily observable manner. Now, define the amplitude reflectivity $R = E_r/E_0$ and the amplitude transmissivity $T = E_t/E_0$. If we neglect absorption by the semi-transparent mirrors, then from conservation of energy, we obtain $R^2 + T^2 = 1$.

Passing through the first mirror gives $E_t = TE_0$ and through the second one, we have $E_1 = TE_t \exp(ikd) = T^2E_0 \exp(ikd)$. The reflected wave at the second mirror is $E_{1r} = RE_t \exp(ikd) = RTE_0 \exp(ikd)$; this wave travels back a distance d to the first mirror and then is reflected there and travels again a distance d to the second mirror. The part transmitted now is $E_2 = T^2R^2E_0\exp(ik3d)$. This inner reflection and transmission continues within the etalon cavity giving $E_n = E_0T^2R^{2(n-1)}\exp(ik(2n-1)d)$. All these waves interfere, and we obtain

$$E_{FP} = \sum_{n=1}^{\infty} E_n = T^2E_0\exp(ikd)\sum_{n=1}^{\infty} R^{2(n-1)}\exp(ik(2n-1)d)$$

$$= T^2E_0\exp(ikd)\sum_{n=0}^{\infty}[R^2\exp(ik2d)]^n = \frac{T^2E_0\exp(ikd)}{1 - R^2\exp(ik2d)}, \tag{1.25}$$

where the subscript FP is for Fabry–Perot. Using $T^2 = 1 - R^2$ as well as $\cos(k2d) = 1-2\sin^2(kd)$ and introducing the finesse factor K by

$$K = \left(\frac{2R}{1 - R^2}\right)^2, \tag{1.26}$$

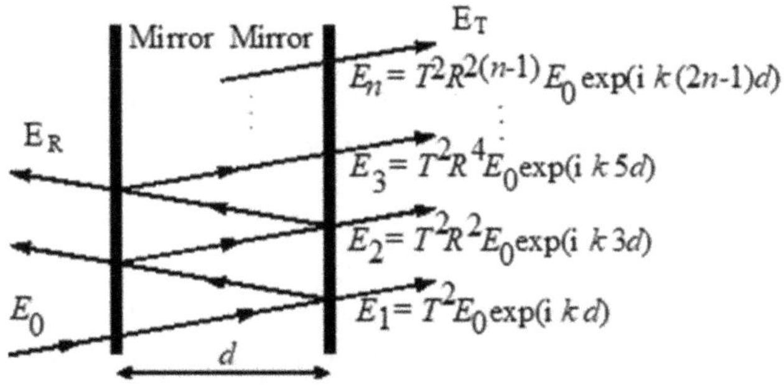

Figure 1.11. Fabry–Perot interferometer (the beams are shown inclined for clarity).

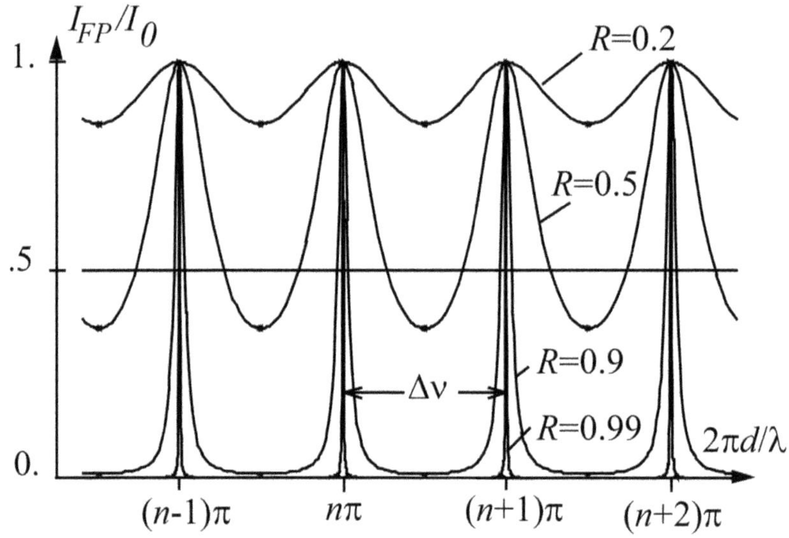

Figure 1.12. Airy function (equation (1.28)).

we calculate the intensity $I_{\mathrm{FP}} = E_{\mathrm{FP}}E_{\mathrm{FP}}^*$ as

$$I_{\mathrm{FP}} = \frac{T^4 E_0 E_0^*}{[1 - R^2\exp(\mathrm{i}k2d)][1 + R^2\exp(-\mathrm{i}k2d)]} = \frac{I_0}{1 + K\sin^2 kd}. \tag{1.27}$$

Thus, the intensity transmission T_I of a Fabry–Perot interferometer is

$$T_I = \frac{I_{\mathrm{FP}}}{I_0} = \frac{1}{1 + K\sin^2 kd}. \tag{1.28}$$

This is called the Airy function and is shown for different values of R in figure 1.12. We have transmission 1 for defined frequencies depending on the inter-mirror distance d, and transmission close to 0 for the frequencies in the range $\pm\ \nu = \pm c/(2d)$, i.e. to the right and the left of the frequencies giving maximum transmission. This means that the Fabry–Perot interferometer is a highly effective wavelength-selective filter, and thus its use in a wide variety of lasers; for instance, the ion-Ar laser is commonly employed in optical metrology since it provides multiple selective visible wavelengths.

1.5 White-light interferometry

Surface profiling can be achieved using white-light interferometry [5]. White light is the output of a broadband source (in terms of wavelengths) with very low coherence, and our Sun is a fine example. So interference is only possible if the path lengths in, say, a Michelson interferometer coincides or is very close indeed (in particular, for the Michelson optical layout, shown in figure 1.4, one of the mirrors is replaced by the surface whose height variations need to be measured). In order to get these paths to be almost identical, the path length difference in a white-light interference microscope is scanned with a piezo transducer, shown in figure 1.13, a device that allows very fine

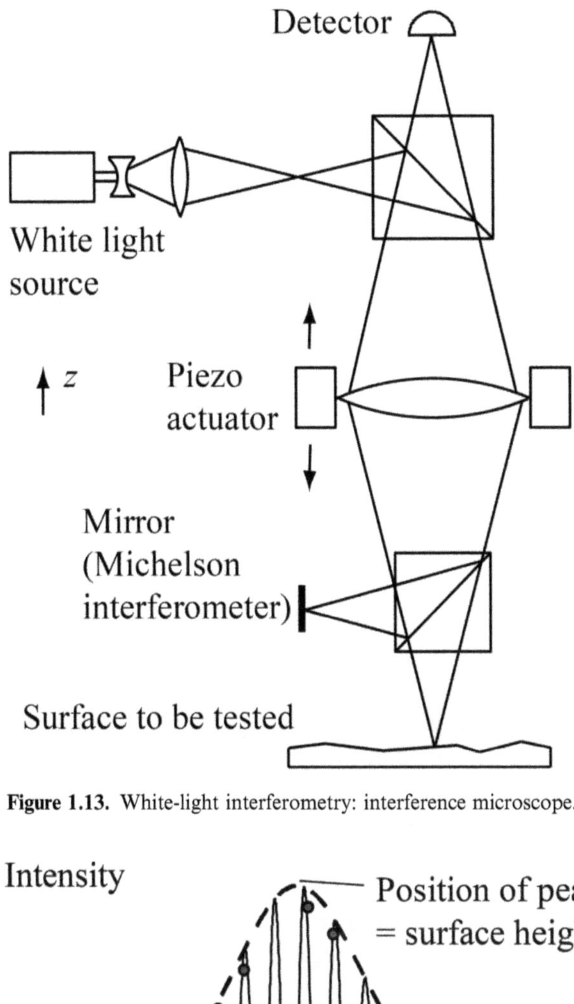

Figure 1.13. White-light interferometry: interference microscope.

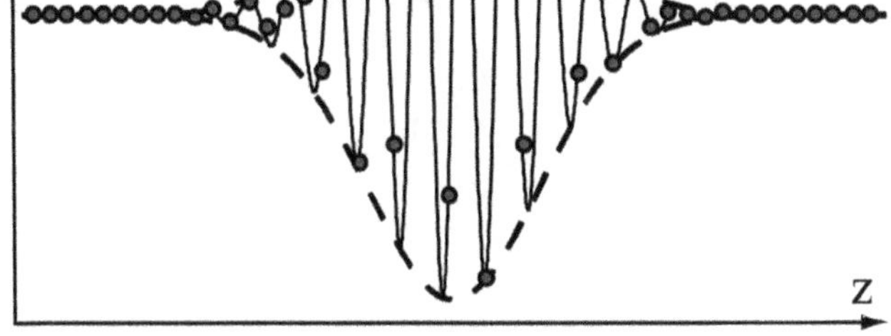

Figure 1.14. White-light interferogram.

adjustments of the optical paths. The position of the intensity peak in the resulting signal, taken from the resulting interference pattern, indicates the local surface height, figure 1.14. In the measurement of transparent films, two peaks from the front and back surface occur, so film thickness and surface height can be detected.

1.6 Holographic interferometry

The full-field interferometric methods seen so far compare two simultaneously existent wavefronts, one of which is a smooth reference wavefront, while the other one is deformed or otherwise not smooth due to the typical roughness of the object surface under test or by a varying refractive index, etc. These methods do not commonly allow comparison of the wavefront on a rough surface at different time instants, e.g. before and after the object surface has undergone a deformation. A technique was developed during the 1960s that makes it possible to compare these two object states; it is called holographic interferometry. Holography, invented in the late 1930s by Dennis Gabor, provides the needed process of recording and reconstructing full complex wavefields from such objects. This technique renders the whole field of view amplitude and phase distributions from various objects.

1.6.1 Holographic recording and reconstruction of wavefields

Holography was invented by Gabor [6] with the goal to improve the imaging capabilities of electron microscopes; the idea remained dormant until the invention of the laser, a highly coherent light source that catapulted holography to be used, at the time, in visible light metrology with wavelengths far different than those calculated by electrons. The pioneering work done by Leith and Upatnieks [7] used a laser to illuminate an object surface whose scattered-reflected light fell on a recording plane where a high-resolution photographic plate was placed, i.e. the object wavefield $E_O(x, y)$ arrived at this plane, see figure 1.15. The second wavefront needed for interference to take place was brought to the photographic plane, undisturbed from the laser: it is a mutually coherent wave $E_R(x, y)$ known as the reference wavefront. It is then is superposed with $E_O(x, y)$, and both waves must stem from the same laser. As seen previously, the superposition of both wavefronts produces the intensity distribution $I(x, y)$, which is recorded on the plate, which gives (as before, the spatial coordinates (x, y) are omitted for better readability)

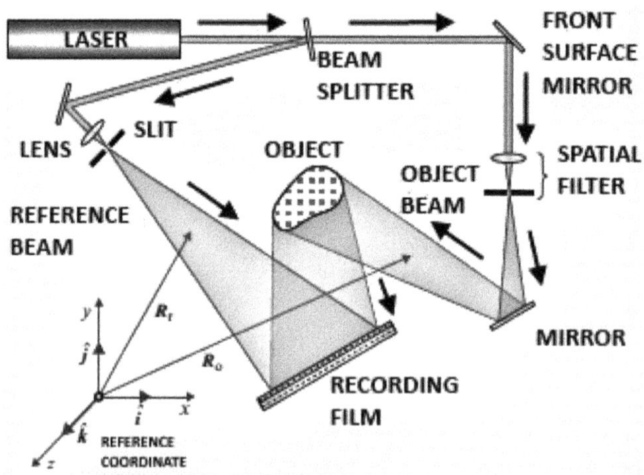

Figure 1.15. Recording of a hologram. R_o, R_f are the object and reference beam vectorial positions in a Cartesian coordinate system.

$$
\begin{aligned}
I(x, y) &= |E_O + E_R|^2 = (E_O + E_R)(E_O + E_R)^* \\
&= E_O E_O^* + E_R E_R^* + E_O^* E_R + E_O E_R^*.
\end{aligned} \tag{1.29}
$$

For reconstruction, the processed photographic plate, called the hologram, is illuminated by the reference wave $E_R(x, y)$ alone, see figure 1.16, where a holographic set-up with amplitude division is displayed. The resulting wavefield is

$$
\begin{aligned}
&E_R \cdot (|E_O|^2 + |E_R|^2 + E_O^* E_R + E_O E_R^*) = (|E_O|^2 + |E_R|^2) \\
&E_R + E_O^* E_R^2 + E_O |E_R|^2.
\end{aligned} \tag{1.30}
$$

In the last equation the first term is the zero diffraction order, which is the reconstructing reference wave E_R multiplied by the intensity term $|E_O|^2 + |E_R|^2$, the second term is the complex conjugate of the object wave and renders a so-called virtual image of the object, while the third term represents the original object wavefield E_O multiplied with the intensity $|E_R|^2$ giving, as a result, a real object image that on reconstruction can be projected or imaged on a piece of paper. If the angle between the reference and object wavefields is large enough, all three wavefields are spatially well separated [7].

1.6.2 Generation of holographic interference patterns

With holography, it is possible to compare interferometrically two object states at different time instants [8, 9]. One way is the subsequent recording of the two states on the same photographic film, one state before and one after changing the object's optical phase distribution, namely a static or dynamic deformation. After the chemical process to develop the hologram, both wavefields are reconstructed together and an interference pattern occurs, such as the one shown in figure 1.17. Such a procedure is called the double exposure method. There is an alternative approach to perform the

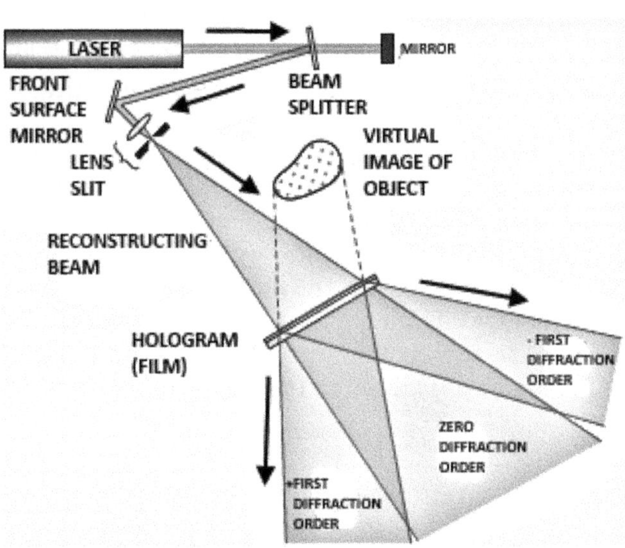

Figure 1.16. Holographic reconstruction of a wavefield.

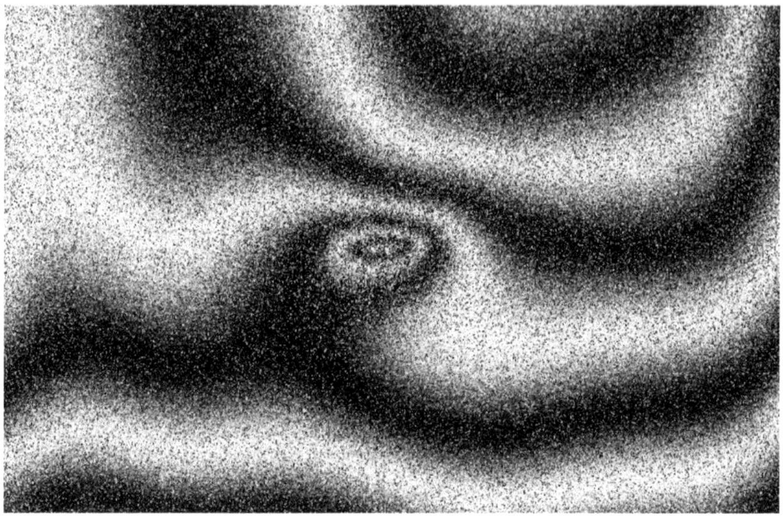

Figure 1.17. Holographic interference pattern.

interferometrical process just described: if only one initial state is recorded and reconstructed, where the hologram needs to be placed precisely at the exact location where it was recorded (alternatively, the photographic plate may be chemically developed *in situ*, to avoid errors in the relocation of the plate), then further states of the object, say upon deformation, can be seen through the hologram so these states, as well as the original object wave holographically reconstructed, interfere. This procedure is called the real-time method since the object amplitude and phase change can be observed in real-time, i.e. a moving interference pattern due to a deformation generated, for instance, by an increased mechanical load. The intensity distribution $I(x, y)$ due to the phase change $\varphi(x, y)$ in both methods obeys a cosine law as in equation (1.17). The phase change due to a deformation described by the displacement vector $\boldsymbol{d}(x, y)$ is $\varphi(x, y) = \frac{2\pi}{\lambda}\boldsymbol{d}(x, y) \cdot [\boldsymbol{b}(x, y) - \boldsymbol{s}(x, y)]$ with $\boldsymbol{b}(x, y)$ and $\boldsymbol{s}(x, y)$ unit vectors in the observation and illumination directions [2]. This subject will be treated in full detail in chapters 7, 8, and 10.

If harmonically vibrating surfaces are considered, the interference phase distributions are of the form $\varphi(x, y)\sin\omega t$, where ω is the angular frequency of the object vibration and the term $A(x, y)$ is related to the maximum amplitude of the vibration at (x, y). In principle, it should be possible to observe this with the real-time method; however, a time-averaged observation method should be used for this type of dynamic event [8, 9].

Here the hologram of the vibrating surface is recorded with an exposure time that is long compared with the period of the vibration, $T \gg 2\pi/\omega$. Then a continuum of waves is recorded and holographically reconstructed in the form

$$E_O(x, y)e^{i\ \varphi(x, y)\sin\omega t}. \tag{1.31}$$

This set of wavefronts reconstructed simultaneously will interfere to

$$E_{av}(x, y) = \lim_{T\to\infty}\frac{E_O(x, y)}{T}\int_0^T e^{i\ \varphi(x, y)\sin\omega t}dt = E_O(x, y)J_0(\ \varphi(x, y)), \tag{1.32}$$

where J_0 is the zeroth-order Bessel function. The observable intensity in the reconstructed image is then

$$I(x, y) = I_O(x, y)J_0^2(\varphi(x, y)). \tag{1.33}$$

The fringes are now contours of equal vibration amplitude of the spatial vibration modes. The maximal intensity belongs to $\varphi(x, y) = 0$, which occurs at the nodes of the vibration mode. The dark centers of the fringes refer to zeros of the Bessel function J_0. The contrast of the fringes decreases with increasing order, see figure 1.18. An example of a vibrating plate clamped at all four edges is shown in figure 1.19.

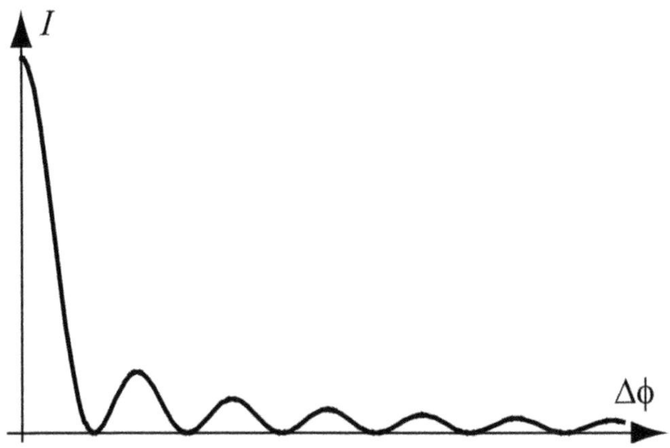

Figure 1.18. Time-averaged intensity.

Figure 1.19. Time average holographic interferogram.

References

[1] Goodman J W 1996 *Introduction to Fourier Optics* 2nd edn (New York: McGraw-Hill)
[2] Kreis T 2005 *Handbook of Holographic Interferometry* (Weinheim: Wiley)
[3] Lauterborn W, Kurz T and Wiesenfeldt M 1995 *Coherent Optics, Fundamentals and Applications* (Berlin: Springer)
[4] Thorne K S 2018 Nobel lecture: LIGO and gravitational waves III *Rev. Mod. Phys.* **90** 040503
[5] Kitagawa K 2009 Film surface and thickness profilometry *Handbook of Optical Metrology, Principles and Applications* ed T Yoshizawa (Boca Raton, FL: CRC Press), ch 29 pp 677–91
[6] Gabor D 1948 A new microscopic principle *Nature* **161** 777–8
[7] Leith E N and Upatnieks J 1964 Wavefront reconstruction with diffused illumination and three-dimensional objects *J. Opt. Soc. Am.* **54** 1295–301
[8] Powell R L and Stetson K A 1965 Interferometric vibration analysis by wavefront reconstruction *J. Opt. Soc. Am.* **55** 1593–8
[9] Stetson K A and Powell R L 1965 Interferometric hologram evaluation of real-time vibration analysis of diffuse objects *J. Opt. Soc. Am.* **55** 1694–5

Fernando Mendoza-Santoyo, Manuel De la Torre-Ibarra, María del Socorro Hernández-Montes and Jorge Mauricio Flores Moreno

Chapter 2

Coherence

Visible light is a natural phenomenon that implies random fluctuations in electronic transitions. This chapter will address essential concepts related to these fluctuations—the light coherence and light sources emitting electromagnetic waves in the visible range. In the literature devoted to optical metrology (i.e. non-destructive techniques based on interferometry, holography, speckle, and many others), coherence is a definition commonly relegated to a brief section within interference, diffraction, or any other concept in the classical theory of light phenomena as a wave. We present this chapter on the importance of the coherence property of an electromagnetic wave, mainly in the visible range, to produce interference for the study and application of light as a measuring tool.

We introduce the basic concept of coherence in light waves and basic mathematics to explain the physical concepts of light as a traveling wave. In 2005 the unified theory of coherence and polarization of light was proposed [1] and emerged as a consistent theory that considers both phenomena to be the same manifestation based on correlations between fluctuations in optical fields. This chapter will only give fundamental relations and explanations related to the coherence property of light.

2.1 Introduction: the background

Coherence may be considered a fundamental property of radiative (light) sources emitting wavefronts at different wavelengths within the electromagnetic (EM) spectrum, e.g. in the visible range. As seen in chapter 1, an EM wavefront can be described by its electric field. The coherence or optical coherence in a heuristic approximation can be thought of as the ability of a light source to emit photons or wavefronts with the same phase among them. The phase as understood here is related to the capability of a wave to repeat itself with the same spatial and time properties. A high coherence light source will emit wavefronts that are orderly and

doi:10.1088/978-0-7503-3027-5ch2

organized (harmonically, so to speak). In contrast, a light source without coherence will emit random wavefronts (sunlight could be considered an incoherent light source). Figure 2.1 shows examples of these two cases and indicates the relevance of the coherence for measuring the phase among the wavefronts; this optical phase carries physical information on different objects interacting with light.

From chapter 1, the intensity of two waves interfering is commonly written as (see the equations in section 1.3.1)

$$I = I_1 + I_2 + 2\sqrt{I_1 I_2}\cos(\delta), \tag{2.1}$$

where

$$I_1 = (E_{01})^2 = (A_1)^2, \tag{2.2}$$

$$I_2 = (E_{02})^2 = (A_2)^2, \tag{2.3}$$

and

$$\delta = k_2.\, r - k_1.\, r - \Delta\varphi. \tag{2.4}$$

It is known that the intensity I behaves proportionally to the temporal mean of the electric field amplitude. Thus, in any place of the superposition of two or more waves, the spatial length between them is a constant magnitude determined by δ. The quantities (also defined in chapter 1) k_1, k_2 correspond to the wave vector directions of the two waves, and r is the spatial position of a point in the wave. Equation (1.1) is frequently identified as the general model of interference, which shows that the irradiance of the interference varies around a background value

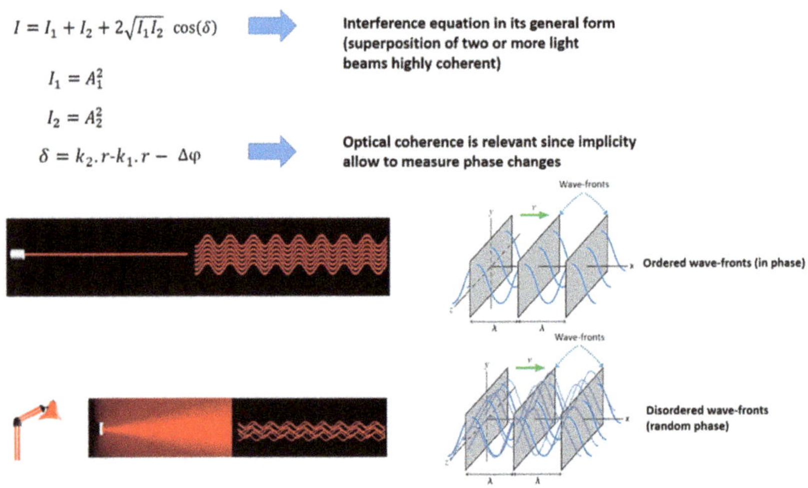

Figure 2.1. Diagrams of highly coherent (laser) and incoherent (incandescent light bulb) light sources. The laser provides a well-ordered and in-phase wavefront stream and could be considered a point light source. The incandescent light bulb emits wavefronts in all visible frequencies in a disordered fashion and is considered a broadband light source.

$I_1 + I_2$ and the interference phenomena is modulated by a cosine function that delivers a minimum intensity value $I = I_1 + I_2 - 2\sqrt{I_1 I_2}$ at all the points where $\delta = (2n + 1)\pi$, and a maximum intensity value $I = I_1 + I_2 + 2\sqrt{I_1 I_2}$ at all the points where $\delta = (2n)\pi$, with n as an integer. The resulting fringe pattern can measure the spatial distribution of the optical phase difference between both waves.

The first studies to deal with the concept of either coherence or polarization of light were based on classical studies about wavefront propagation. Here, the term classic refers to the point of view used by optical scientists before introducing the quantum theory from which arose concepts such as the quantum coherence of light, which is outside the scope of this chapter. A text compiling the historical background and efforts made by many researchers to explain the coherence (and polarization) properties of light was published in the book series *Progress in Optics* [2] and some other relevant publications [3, 4].

Let us focus on the most recent advances in coherence concepts. Polarization and coherence are concepts that somehow have been entangled since both are explained using similar statistical optics formulations. Currently, Wolf's pioneering work on the unified theory for polarization and coherence is the best known publication in the field [2]. However, for many years, theoretical physicists only addressed optical coherence theory and discarded its practical applications until the invention of the laser in the late 1950s. In fact, for many years during the last half of the twentieth century, many researchers were convinced that new, important, and breakthrough applications of optical metrology would only be possible by using highly coherent light sources. In opposition to this thought, many researchers have focused on other light sources, such as entirely incoherent or partially coherent light sources. Currently, thousands of applications in medicine, biology, and many other fields use light sources with short coherence [5–10].

Today the concept of coherence needs to be approached with classical wavefront superposition. From this approach, one of the seminal contributions explaining the meaning of partial coherence was by Zernike [11], whose proposal was to search for the best illumination sources available to obtain the sharpest possible images from phase (transparent) objects. The aim was to increase the quality of the microscopic images using hardware based on a set of diffraction apertures and other passive optical components such as pupils, corrected simple lenses, and even composed objective lenses. A second contribution during the same time was made by Peter H van Cittert, who presented his theorem to calculate the degree of spatial coherence of a light source [12]. Both researchers created the van Cittert–Zernike theorem, which is used to generate models based on the spatial coherence of a sound field that has been forward scattered from the sea surface [13]. This theorem relates the spatial coherence of an observed wavefield to the distant source intensity distribution associated with this field, as is the case of wavefronts in the visible region.

By the beginning of the 1950s, Emil Wolf [4] introduced one of his first works on the theory of interference and diffraction under the light source spatial coherence approach considering the van Cittert–Zernike theorem. The experimental procedure presented by Thompson–Wolf was used for a long time as the basis for other useful concept demonstrations [14–16]. Soon after, and with statistical optics concepts on polarization, the unified theory of coherence and polarization was demonstrated. The experiments at the time were performed without using a high temporal and spatially coherent light source. Figure 2.2 shows the main results of such experiments.

Currently, the theory of coherence (and polarization) has increased its importance in optical metrology and has become more robust in its mathematical and physical principles. In-depth and more specialized material on coherence can be found in the corresponding references as they appear in the following text.

2.1.1 Preliminary concepts on coherence and the unified theory of polarization–coherence

Some critical concepts about coherence and the unified theory of polarization and coherence will be described briefly. The intention is to show the importance of coherence and how its many properties complement other concepts derived from the wave nature of light: interference, diffraction, polarization, and classical scattering.

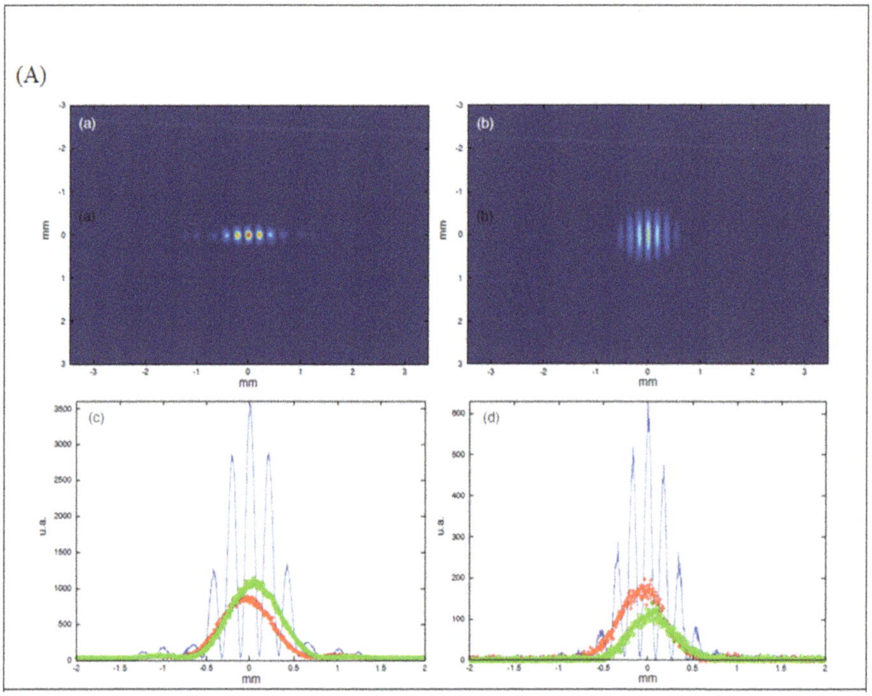

Figure 2.2. The (A) Thompson–Wolf, (B) Considine, and (C) Magyar–Mandel experiments are based on coherence concepts. (A) Interference patterns using a calibrated He–Ne laser (632 nm) corresponding to (a) two slits and (b) two pinholes; (c), (d) are the corresponding line profiles. (Reproduced with permission from [17]. Copyright 2010 Maria Luisa Calvo, David Paredes Barato.) (B) An image of a straight edge photographed with (a) a coherent light and (b) an incoherent source; (c), (d) are the traces of a coherent and incoherent image of an edge. (Reproduced with permission from [15]. Copyright 1966 Optical Society of America.) (C) Experimental fringes using two independent ruby masers, as is described in [16]; (a), (b) are the detecting fringes image and its line profile. (Reproduced with permission from [16]. Copyright 1963 Springer.)

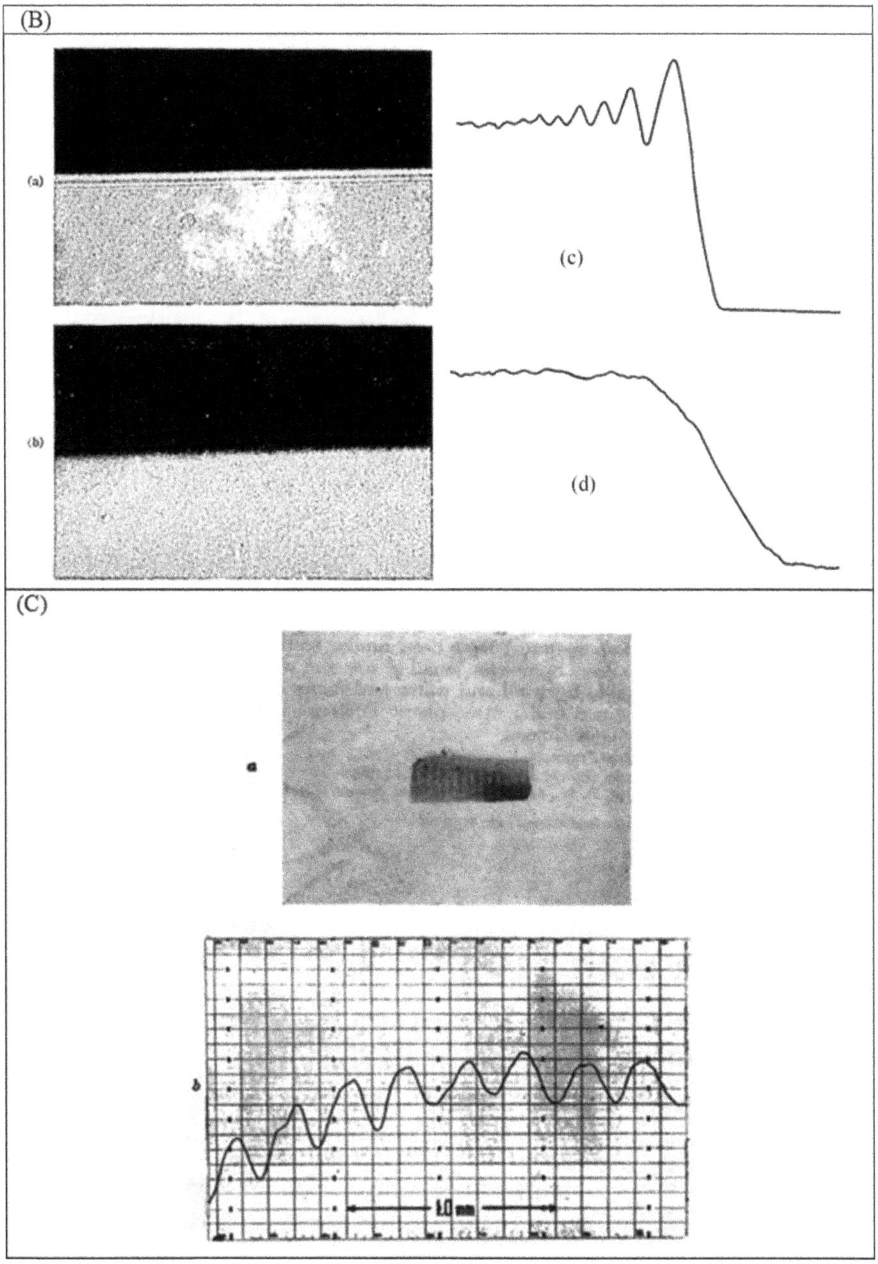

Figure 2.2. (Continued.)

The starting point is that all optical waves have some random fluctuations, and there are no entirely monochromatic or not monochromatic wave fields in the real world. The visible spectrum of electromagnetic waves, approximately 400–700 nm, results from either spontaneous emission of radiation from atoms that generate the field (such as for thermal light) or fluctuations due to controllable causes (such as the laser). Thus, the coherence phenomenon is an averaging process from many processes in a light source, which can be better understood using statistical optics. In the same way, polarization can be explained by the same statistical principles. Whereas coherence is essentially the result of correlations between some components of the fluctuating electric field at two or more points in a traveling wave, polarization manifests correlations involving components of the fluctuating electric field at a single point. Coherence is better related to the sharpness (contrast or visibility) of the interference fringes resulting from the superposition of two or more points coming from the same wavefront. The higher the contrast, the higher the coherence, while polarization may be identified using optical components such as polarizers, rotators, or phase plates.

They are mathematically modeled by correlation functions and matrices that follow basic optics propagation laws and are statistical properties of light. The coherence theory can be studied based on the scalar theory, while polarization requires a vector treatment.

Moreover, at this particular point, Wolf based his generalization theory on the concept of coherence from scalar fields to electromagnetic vector fields. Perhaps the primary motivation for Wolf to work on this unified theory is that '… optical coherence theory is probably one of the most poorly understood subjects in the whole of optics …', as he said during a 1999 conference [4]. He meant perhaps that optical coherence was devoid of any practical application. One of his main criticisms was that there existed a misguided vision about the laser since its only relevance was to be an essential source of light, forgetting any other properties, including coherence.

His approach was based on the fact that Maxwell's equations deal with deterministic situations of electromagnetic waves. In contrast, sources and fields, either in nature or laboratories, are never strictly deterministic since spontaneous emissions of photons are always present, as happens in lasers. Therefore, his primary consideration was to consider the fluctuations of random sources and fields. For Wolf, the key factor in understanding the concepts of coherence, polarization, and interference is Young's experiment (presented in chapter 1). This interferometer has many applications: it was used to establish an analogy between the interference of light waves with the phenomena of waves in water; it was used to establish the wave theory of light; and it gave some intuitive ideas on quantum mechanics (wave–particle duality or the single-photon interference, a much-discussed subject recently).

Here, due to its historical importance, let us comment on some background work. By 1670 Bartholinus observed and reported the refraction phenomena seen in the 'Iceland spar or crystal', later explained by Huygens in 1690. The same Young explained this double refraction phenomenon in 1816 using the concept of ray beams. Also, the Young experiment explained that the light vibrations are transverse, not only longitudinal or in the direction of propagation as was suggested by the Fresnel–Arago laws (1819) that can be stated as follows: two ray beams of light polarized in mutually orthogonal directions will not interfere; two parallel coherent linearly polarized waves will interfere in the same way as natural light; and the two constituent orthogonal linearly polarized states of natural light cannot interfere to form a readily observable interference pattern, even if rotated into alignment (because they are incoherent). In 1852 Stokes made investigations on the results of the Fresnel–Arago experiments. He introduced four observable parameters known as the Stokes parameters related to the degree of polarization. These parameters are a measurable ratio between the intensity of the polarized portion of the total intensity of the light beam; in other words, the degree of polarization may be different at different points in the space.

Verdet, by 1865, calculated what is known as the area of coherence related to the separation of two pinholes in a Young experiment performed using sunlight; the resulting separation was about 0.06 mm. Another step in the study of coherence was provided by Zernike in 1948, who gave a parameter to measure the coherence degree in terms of visibility. The latter is a quantity that defines the sharpness of the interference fringes. Indeed, the coherence was related to the ability of a light beam to form interference fringes. By then, concepts such as the mutual intensity of fringes, complex degree of coherence, and visibility were measured in terms of the fringe pattern produced by two or more light beams. From 1954 to 1955 Wolf introduced a general correlation function to measure the mutual degree of coherence. This function is based not only on two points of the wavefront at the same instant but also at two different time instants. Therefore, the coherence theory was carried out in the space–time domain deriving a cross-spectral density function or the spectral degree of coherence that measures the wave correlations in the space-frequency domain. This treatment meant a fundamental step forward to the unified theory of coherence and polarization. Weiner made an essential contribution to using a correlation function in 1930 since this concept allows a generalized harmonic analysis of functions based on matrices representing different polarization states of the light. Then, we can use either coherence or polarization matrices. These historical facts show that polarization and coherence are correlations between fluctuations in optical wave fields and that the unified theory can be experimentally proved with Young's interferometer. Mathematically speaking, we generalize the scalar cross-spectral density function to handle a cross-spectral density matrix that includes stochastic and statistically stationary electromagnetic waves. The elements that form such matrices can be determined from measurements employing Young's interferometer, including filters, polarizers, and rotators. As a wave propagates, it changes in its spectral density, the spectral degree of polarization, and its spectral degree of coherence. The relation between coherence and polarization from a statistical optics point of view was first predicted theoretically by Wolf in 2003 [2, 4, 18]. As an interesting example, one of the breakthrough applications of the unified theory lies in the calculation of the propagation of stochastic waves through human tissue.

2.2 Basic principles of coherence

Perhaps the most straightforward way to define coherence is the capability it provides to any two (or more) waves to interfere or produce interference. In this chapter, we are interested in studying the ability of waves to interfere within the visible light of the EM spectrum. However, the study is not limited to this range, i.e. it does apply to all waves. A more formal definition of coherence is the correlation between individual wavefronts in time or two specific traveling wave points. The correlation is a fundamental operation for linear systems, such as optical ones. It is widely used in signal and image processing. Another important key feature of correlation is that it is shift-invariant, i.e. we can perform the same operation at every point in a signal (as the traveling wave) or in an image (a filtering process, for instance). In its simplest definition, correlation measures the similarity between two signals, either the same or different. For the case of coherence, the correlation will measure the similarity between two specific points from the same wave when it is separated (such as in the case of an interferometer where a wavefront from a light source is divided into two or more beams). It also measures the similarity of the two points in a time. As can be inferred, coherence is the spatial and temporal properties of waves.

The temporal coherence is studied with either Michelson's or Twyman–Green interferometers (amplitude division, chapter 1). The spatial coherence can be analyzed using a wavefront division interferometer as a Young's interferometer. Figure 2.3 shows the main definitions related to temporal and spatial coherence. Partial coherence will be dealt with later.

In this book the words wave, wavefront, or electric field represent the same meaning as a traveling wave.

Spatial coherence correlates with electrical fields in two different positions of the same wavefront (see figure 2.4).

Mathematically, spatial coherence is defined as

$$\rho(E(x_1), E^*(x_2)) = < E(x_1, t)E^*(x_2, t) >, \tag{2.5}$$

where $E(x_1, t)$ represents the electric field at an arbitrary point x_1 in the time t, $E(x_2, t)$ is the electric field at a second arbitrary point x_2 at the same time t, * is the complex conjugate, and the symbol $\langle\ \rangle$ is the correlation operator given by ρ.

Temporal coherence is defined as the correlation of electric fields at the same position but at two different times, as shown in figure 2.4. It is then defined as

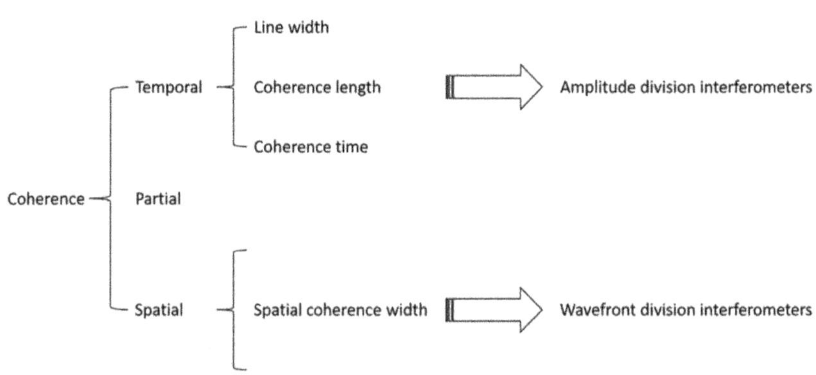

Figure 2.3. Temporal, partial, and spatial coherence and the properties (see chapter 1 for the interferometer classification).

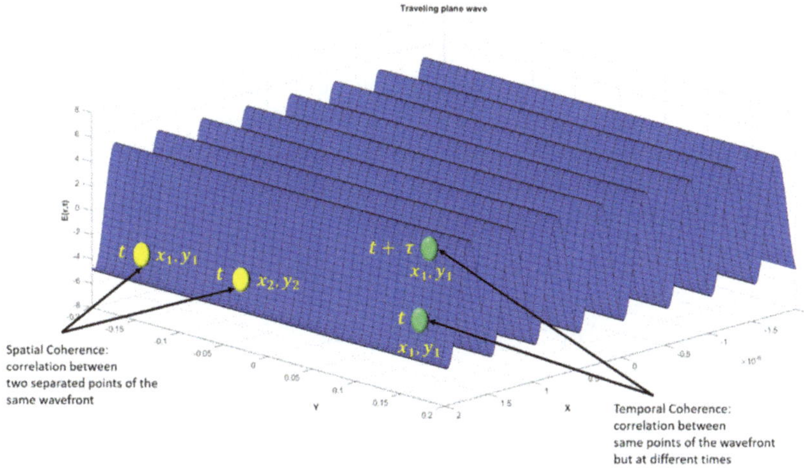

Figure 2.4. Schematic of a traveling plane wave showing the difference between spatial and temporal coherence. The spatial coherence is the correlation between two separated points on the same wavefront simultaneously; the temporal coherence is the correlation between the same points of the wavefront but at different times.

$$\rho(E(t+\tau), E^*(t)) = < E(x, t+\tau)E^*(x, t) >, \tag{2.6}$$

where $E(x, t)$ is the electric field at an arbitrary point x and time t, and $E^*(x, t+\tau)$ represents the same electric field but at time $t+\tau$.

From these definitions (spatial and temporal coherence), it is possible to calculate the correlation of electric fields at different times and positions by combining equations (2.5) and (2.6). If we do that, we speak of the mutual coherence function defined as

$$\Gamma(x_1, x_2, \tau) = < E(x_1, t+\tau)E^*(x_2, t) >. \tag{2.7}$$

This last expression contains both spatial and temporal coherence. Then the mutual coherence function depends on three variables or relations:

$$\Gamma(x_1, x_2, \tau) = \Gamma_{12}(\tau) = < E(x_1, t+\tau)E^*(x_2, t) >. \tag{2.8}$$

The letter gamma Γ represents the mutual coherence function. The sub-index implies comparing two different points in the wavefront, leaving the variable τ as representing two different times in the correlation.

In signal processing analysis, another important operation is a mathematical representation of the degree of similarity between a given time series and a lagged version of itself over successive time intervals. This is the self-correlation or autocorrelation function, the crossed correlation of a signal with itself.

If the mutual coherence function represents the correlation of two different points at two different times of an electric field, then the autocorrelation function implies the correlation of an electric field with itself—the same point at the same time. The mutual autocorrelation function can be defined as

$$\Gamma_{11}(\tau = 0) = < E(x_1, t)E^*(x_1, t) > = I_1, \tag{2.9}$$

in which I_1 represents the intensity of the electric field. It may be a complicated idea to design an experiment that correlates two electric fields separated in space. However, if we imagine a wave interference experiment where a wavefront is divided in two, according to their correlation (spatial or temporal), interference patterns will be produced. This interference phenomenon can be observed on a screen and recorded using a photodetector or a digital camera. In such cases, we will be registering the intensity of the fringe pattern:

$$I = < (E_1 + E_2)(E_1^* + E_2^*) > = < E_1E_1^* > + < E_2E_2^* > + < E_1E_2^* > + < E_1^*E_2 >,$$

$$I = < E_1E_1^* > + < E_2E_2^* > + 2\Re < E_1E_2^* >. \tag{2.10}$$

In equation (2.10) $\Re$ is the real part of the correlation of the complex fields (the imaginary part may be neglected since it is not experimentally observed), and the equation can be expressed in terms of the intensity as

$$I = I_1 + I_2 + 2\Re\{\Gamma_{12}(\tau)\}. \tag{2.11}$$

In equation (2.11) the intensity resulting from the interference between two electric fields is the addition of the individual intensities of the fields (known as the contrast bias, dc, or background term) and the real part of the mutual coherence function Γ_{12} at time τ. This is a simple way to measure the relationship between spatial and time coherence. It is possible to go a step further if we normalize equation (2.11), so we can relate the ratio of $\Gamma_{12}(\tau)$ with its autocorrelation form between two separate points at the same time ($\tau = 0$). Since the autocorrelation (equation (2.9)) is the intensity of the wave, if we perform the autocorrelation of the two different points separately and they match ($\Gamma_{11} = \Gamma_{22}$), we will obtain the ratio of the mutual coherence function and the total intensity of the interfering waves. In other words, we can express this ratio as

$$\gamma_{12}(\tau) = \frac{\Gamma_{12}(\tau)}{\sqrt{\Gamma_{11}(0)\Gamma_{22}(0)}} = \frac{\Gamma_{12}(\tau)}{\sqrt{I_1I_2}}. \tag{2.12}$$

This equation is the normalized mutual coherence (its value goes between 0 and 1) and measures the degree of coherence. Then we can relate equations (2.11) and (2.12) to obtain

$$I = I_1 + I_2 + 2\sqrt{I_1I_2}\,\Re\{\gamma_{12}(\tau)\}. \tag{2.13}$$

Recall that the real part of a complex number can be expressed as an exponential

$$Z = |\,Z\,|\,\mathrm{e}^{\mathrm{i}\varphi} = |\,Z\,|\,(\cos\varphi + \mathrm{i}\sin\varphi),$$

where (Z) is the magnitude of the complex number, φ it is a relative phase, and the real part corresponds to the cosine term. Equation (2.13) is equivalent to the interference expression:

$$I = I_1 + I_2 + 2\sqrt{I_1I_2}\,|\,\gamma_{12}(\tau)\,|\,\cos[\varphi_{12}(\tau)]. \tag{2.14}$$

Equation (2.14) represents the interference function in terms of the fringe pattern contrast, the interference magnitude, and the phase term φ, which results from the normalization of the autocorrelation function of the wavefront. Now, if $|\gamma_{12}(\tau)| = 0$ we have no fringes, and we say that the wavefront is incoherent. If $|\gamma_{12}(\tau)| = 1$ we will have well-defined fringes, the highest contrast and visibility, and a coherent wavefront. In chapter 1, we defined the visibility of the fringes [19] as

$$V = \frac{I_{\max} - I_{\min}}{I_{\max} + I_{\min}}, \tag{2.15}$$

we can calculate $I_{\max}$ and $I_{\min}$

$I_{\max} = I_1 + I_2 + 2\sqrt{I_1 I_2}\,|\gamma_{12}(\tau)|$; $\cos[\varphi_{12}(\tau) = 0] = 1$, bright fringe.

$I_{\min} = I_1 + I_2 - 2\sqrt{I_1 I_2}\,|\gamma_{12}(\tau)|$; $\cos[\varphi_{12}(\tau) = \pi] = -1$, dark fringe.

Then, replacing in equation (2.15),

$$V = \frac{2\sqrt{I_1 I_2}\,|\gamma_{12}(\tau)|}{I_1 + I_2}, \tag{2.16}$$

or

$$|\gamma_{12}(\tau)| = V\frac{I_1 + I_2}{2\sqrt{I_1 I_2}}. \tag{2.17}$$

For the case $I_1 = I_2$ then the visibility is equal to the degree of coherence $V = |\gamma_{12}(\tau)|$. This expression means that visibility measures the degree of coherence of a wavefront.

Next, the three main meanings of coherence, temporal, partial, and spatial, will be defined. These definitions imply some other essential concepts in the study of coherence, such as the line width, coherence length, coherence time, and spatial coherence width.

2.2.1 Bandwidth and lifetime

All previous quantities that define coherence are related to the basic concept of bandwidth and the lifetime of a periodic signal. These two terms can be understood when we analyze a simple and periodic (in a range) trigonometric function. Assuming a cosine function $f(t)$, with unitary amplitude and an angular frequency ω_0 defined in a range of time τ_0 (see figure 2.5). The function is defined as

$$f(t) = \begin{cases} \cos(\omega_0 t), & -\frac{\tau_0}{2} \leqslant t \leqslant \frac{\tau_0}{2} \\ 0, & \text{otherwise} \end{cases}. \tag{2.18}$$

The basic mathematical tool to analyze in the frequency space any periodic function is the Fourier transform (FT). Then, let us apply the FT to equation (2.18):

$$\mathscr{F}\{f(t)\} = F(\omega) = \int_{-\infty}^{+\infty} \cos(\omega_0 t) e^{-i\omega t} dt. \tag{2.19}$$

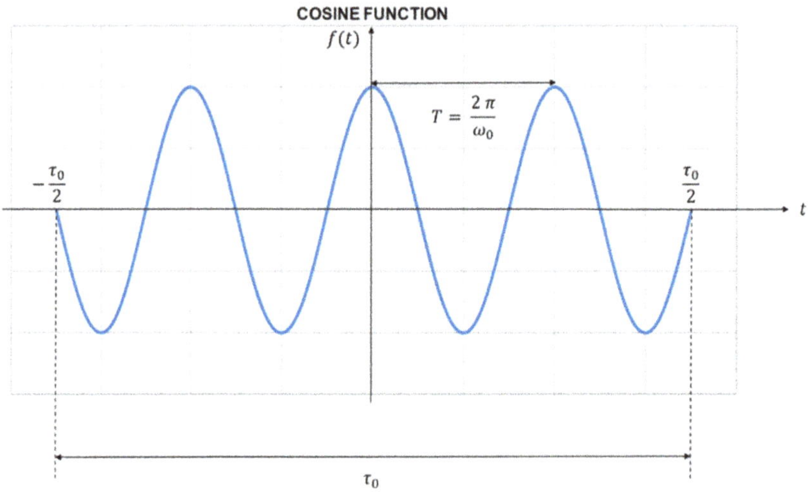

Figure 2.5. A cosine function with period T, angular frequency ω_0 and defined in a time range τ_0.

In equation (2.19) i is an imaginary number. We can write equation (2.19) limited in the range τ_0 and using Euler's identity as

$$F(\omega) = \frac{1}{2\pi} \int_{-\frac{\tau_0}{2}}^{+\frac{\tau_0}{2}} (\mathrm{e}^{\mathrm{i}\omega_0 t} + \mathrm{e}^{-\mathrm{i}\omega_0 t})\mathrm{e}^{-\mathrm{i}\omega t}\mathrm{d}t. \tag{2.20}$$

This equation can be split into two simple integrals that a simple variable change can resolve; the result is

$$F(\omega) = \frac{1}{2\pi}\left\{\left[\frac{1}{-\mathrm{i}(\omega - \omega_0)}\mathrm{e}^{-\mathrm{i}(\omega-\omega_0)t}\Big|_{-\frac{\tau_0}{2}}^{\frac{\tau_0}{2}}\right] + \left[\frac{1}{-\mathrm{i}(\omega + \omega_0)}\mathrm{e}^{-\mathrm{i}(\omega-\omega_0)t}\Big|_{-\frac{\tau_0}{2}}^{\frac{\tau_0}{2}}\right]\right\}.$$

Simplifying

$$F(\omega) = \frac{1}{(\omega - \omega_0)\pi}\left[\frac{\mathrm{e}^{\mathrm{i}(\omega-\omega_0)\frac{\tau_0}{2}} - \mathrm{e}^{-\mathrm{i}(\omega-\omega_0)\frac{\tau_0}{2}}}{2\mathrm{i}}\right] + \frac{1}{(\omega + \omega_0)\pi}\left[\frac{\mathrm{e}^{\mathrm{i}(\omega-\omega_0)\frac{\tau_0}{2}} - \mathrm{e}^{-\mathrm{i}(\omega-\omega_0)\frac{\tau_0}{2}}}{2\mathrm{i}}\right],$$

$$F(\omega) = \frac{\tau_0}{2\pi}\frac{\sin\left[(\omega - \omega_0)\frac{\tau_0}{2}\right]}{(\omega - \omega_0)\frac{\tau_0}{2}} + \frac{\tau_0}{2\pi}\frac{\sin\left[(\omega + \omega_0)\frac{\tau_0}{2}\right]}{(\omega + \omega_0)\frac{\tau_0}{2}},$$

$$F(\omega) = \frac{\tau_0}{2\pi}\mathrm{sinc}\left[(\omega - \omega_0)\frac{\tau_0}{2}\right] + \frac{\tau_0}{2\pi}\mathrm{sinc}\left[(\omega + \omega_0)\frac{\tau_0}{2}\right]. \tag{2.21}$$

If the argument of the 'sinc' function approaches zero this function is equal to 1, and the function $F(\omega)$ in the Fourier space will be defined by

$$\lim_{\omega \to \omega_0} F(\omega) = \frac{\tau_0}{2\pi}.$$

The function 'sinc' of equation (2.21) will vanish if, in general, the function sin (u) = 0, but not in $u = 0$. This means that

$\sin(u) = 0$ for $u = n\pi$, where $n = \pm 1, \pm 2, \ldots$

and

$F(\omega) = 0$, when $\omega = \omega_0 \pm \frac{2n\pi}{\tau_0}$

since

$$\sin\left\{\left[\left(\omega_0 + \frac{2n\pi}{\tau_0}\right) - \omega_0\right]\frac{\tau_0}{2\pi}\right\} = \sin(n\pi) = 0.$$

This expression implies that if the angular frequency ω increases (or decreases) in its value compared to ω_0, then $F(\omega)$ goes and crosses to zero periodically, as is seen in figure 2.6.

As seen in figure 2.6 and considering the full width at half-maximum (FWHM) criteria, the main contribution of energy is around the central lobe of the power spectrum, i.e. between $\omega - \frac{2\pi}{\tau_0}$ and $\omega + \frac{2\pi}{\tau_0}$, and therefore if τ_0 decreases, then the range of dominant frequencies will be ruled by $\frac{2\pi}{\tau_0}$. In conclusion, it is possible to obtain a measure of the bandwidth (in frequency), $\Delta\omega$, around the central angular frequency ω_0. This bandwidth will represent the harmonic components of angular frequency ω_0. Since the spatial frequency can be expressed in terms of the period of the original cosine function of frequency ω_0, we can also obtain a number that represents the 'lifetime' of the function defined by τ_0. We can express these relations as follow:

$$\omega = 2\pi\nu \text{ (the angular frequency)} \tag{2.22}$$

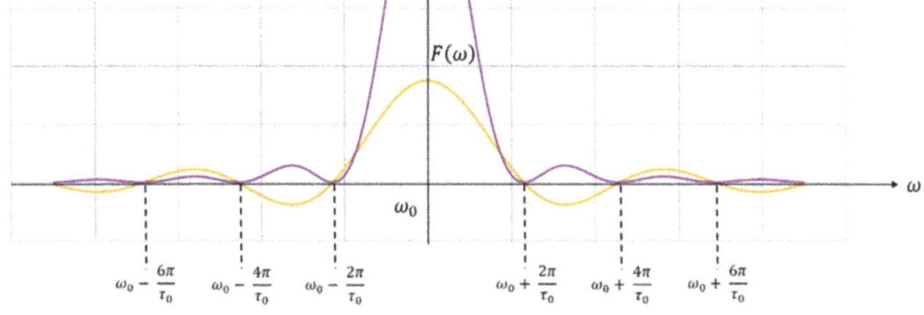

Figure 2.6. The sinc function $F(\omega)$ from equation (2.21). $|F(\omega)|^2$ is known as the power spectrum.

$$\Delta\omega = \frac{2\pi}{\tau_0} \text{ (the bandwidth)} \tag{2.23}$$

$$\Delta\nu = \frac{1}{\tau_0}\left(\text{the lifetime or spectral width}\right). \tag{2.24}$$

The angular frequency in equation (2.22) depends on the spatial frequency ν. The bandwidth and the lifetime of the signal on the spectral plane will depend on the frequency value ν. The value of $\Delta\nu$ represents the uncertainty of the spatial frequency. These parameters play an important role in defining the line width, coherence length, coherence time, and spatial coherence width.

2.3 Temporal coherence

A well-known wavefront amplitude division device is the Michelson interferometer. One of its applications is to measure the temporal coherence of a light source. The latter is possible due to its capability to detect small phase changes on the wavefronts. These changes could be an erratic process where excited atoms of the light source transit between energy levels, or in other words, due to the random emission of photons in a conventional light source. A typical Michelson configuration is shown in figure 2.7.

Temporal coherence could be defined as the correlation of electric fields at the same position but at two different times, as described by equation (2.6). Technically this property will give information about the illumination source. Since, in reality, there are no ideal or perfect monochromatic sources, any variation in their emission processes will produce phase changes, as exemplified in figure 2.8. A light source can

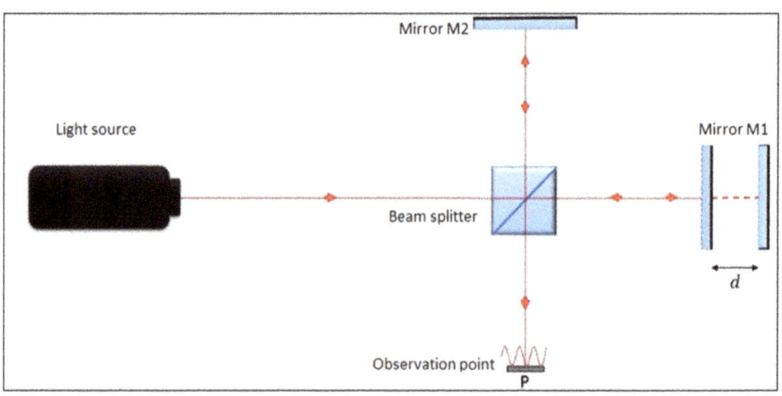

Figure 2.7. Optical layout for a classical Michelson interferometer. Depending on its coherence capabilities, the light source will form an interference pattern at an observation point (P). The mirror (M_1) can be moved at a controlled distance d to produce a path difference referenced to the position of the fixed mirror (M_2). The fringe pattern will depend on the path difference between the two arms of the interferometer and the mirror orientations (angles: tilt and rotation on each plane).

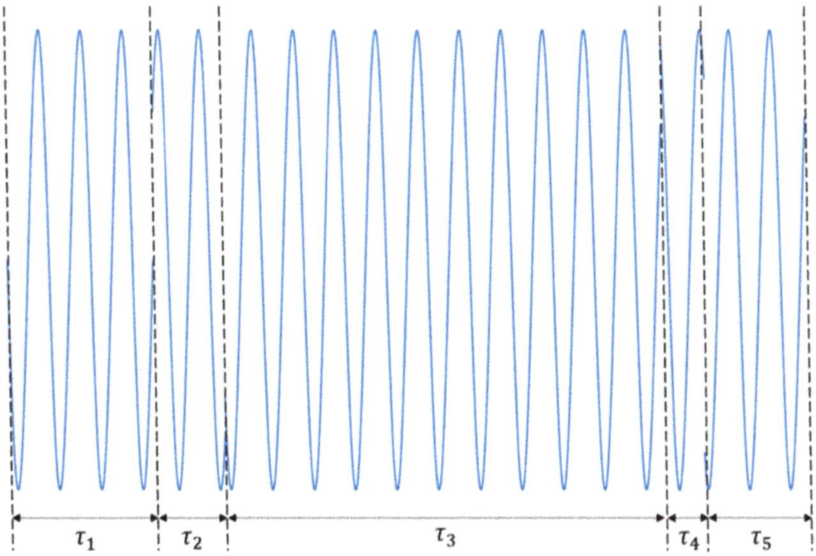

Figure 2.8. The plot shows a wave train's random fluctuations in light emission from a monochromatic source. These fluctuations provoke an unexpected change in its coherence time, producing a widening of the source's spectral line.

be characterized due to its average lifetime τ_0 which is a parameter that defines its coherence time (see equation (2.24)). Then, the larger the coherence time, the more monochromatic the light source will be since the relation between the coherence time and the spectral width, $\Delta\nu$, is inversely proportional:

$$\tau_0 = \frac{1}{\Delta\nu}. \tag{2.25}$$

The coherence length could be obtained considering the relation between the average lifetime and the spectral line (see chapter 1),

$$l_t = c\tau_0 = \frac{c}{\Delta\nu}, \tag{2.26}$$

where c is the speed of the light in a vacuum and $\Delta\nu$ can be related with $\Delta\lambda$, which represents the line width of the wave, by the expression

$$c = \lambda\nu, \tag{2.27}$$

with λ as the wavelength of the electric field. If we apply a derivative to this equation, small variations or incremental changes of spatial frequency ν with respect to the wavelength λ give

$$\Delta\nu = \left| -\frac{c}{\lambda^2}\Delta\lambda \right|. \tag{2.28}$$

Then equation (2.26) can be expressed as

$$l_t = \frac{\lambda^2}{\Delta\lambda}, \tag{2.29}$$

or

$$\Delta\lambda = \frac{\lambda^2}{l_t}. \tag{2.30}$$

The previous expression represents the line width ($\Delta\lambda$) as the average coherence time depending on the average coherence length. For example, consider that white light has a spectral emission of around 400–700 nm representing an average line width of $\Delta\lambda = 300$ nm. If we consider an average wavelength (λ_{av}) of 550 nm, then the coherent length will be

$$l_t = \frac{(550\ \text{nm})^2}{300\ \text{nm}} = 1000\ \text{nm} = 2\lambda_{av}. \tag{2.31}$$

That may be thought of as a short coherence length. As a comparison, let us look at a commercial laser such as the gas He–Ne (helium–neon), which has a nominal wavelength of 632.816 nm with a line width approximately of 0.0016 nm, then we will have

$$l_t = \frac{(632.816\ \text{nm})^2}{0.0016\ \text{nm}} = 25\ \text{cm}. \tag{2.32}$$

This laser has a much longer coherence length which is one of the reasons for its popularity and use in teaching laboratories. It is worth mentioning that this type of gas laser has an inherent Doppler effect due to the random motion of the atoms in the constituent gas mix. Then, their line width is related to the Doppler broadened line width. Furthermore, the emitting power, i.e. the optical gain at FWHM criteria and the cavity modes, plays an important role in the He–Ne laser's coherence length and line width [20]. Lasers designed and built on other technologies such as semiconductors, and solid-state and optical fibers can present high coherent lengths of hundreds of meters, particularly those emitting light in the visible range [21].

2.4 Partial coherence

If the phase difference between two mutual waves is constant, we conclude that the waves are coherent. However, in the real world, this coherence is not entirely assured. In most cases, we should talk about partial coherence and several degrees of coherence between these two waves. Next, we will complete the basic concepts of coherence discussed previously.

Assume the wavefronts emerging from a light source S, and let us focus on two waves traveling different paths towards two structures (S_1, S_2) in figure 2.9. These structures can redirect the waves by reflecting, refracting, or diffracting them, i.e. these structures change the direction of propagation of the wavefronts, which

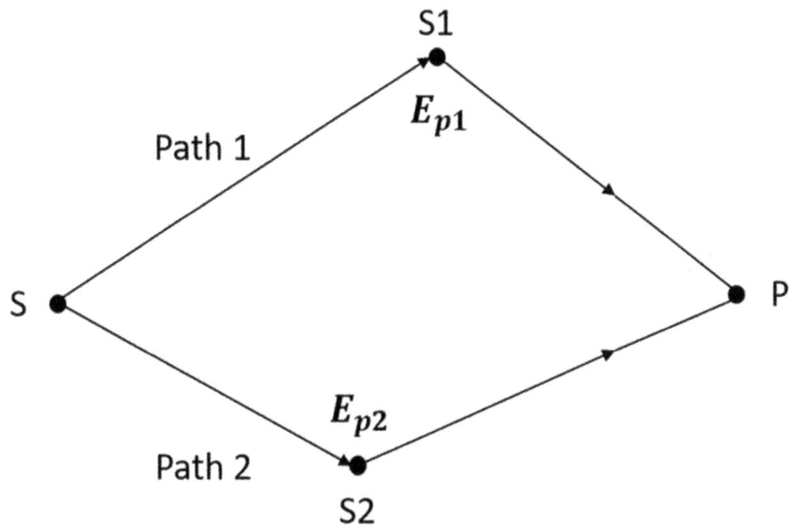

Figure 2.9. Sketch showing a light source S emitting two wavefronts toward two structures, S_1 and S_2, that redirect the waves. Both waves are superimposed in an arbitrary point P.

eventually are superimposed at an arbitrary point P. The waves have the same polarization, so they can be considered scalars.

For our analysis, allow E_S to be the field emitted by the light source S and E_{p1}, E_{p2} the waves superimposed at point P: the amplitude of the original wave is redistributed in two secondary waves traveling different paths and then interfering (equation (1.10)). Note that all the previous analysis explained in the 'basic principles of coherence' section remains valid. However, for the study of partial coherence, we will further explain the role of the mutual autocorrelation function (equation (2.9)) for the case in figure 2.9. We will find that this treatment implies the interference of the fields emerges from structures S_1 and S_2. Also, we will discuss the normalized mutual coherence function (equation (2.12)), which measures the degree of coherence (partial) of a wave related to the interference term of equation (2.13).

The mutual coherence function Γ_{12} in a time τ (equation (2.11)), determines the height of the interference term of equation (2.13), and this also depends on the existing correlation of waves emitted by the light source S at two different times. Then the degree of coherence reached $\gamma_{12}(\tau)$ expressed by equation (2.12), measures the coherence degree (in time) of the trajectories followed by the paths S_1 and S_2 regarding the average coherence time τ_0 of the same light source S. If τ_0 represents a constant coherence time, then for an arbitrary $\tau > \tau_0$ we will lose coherence. This situation is observed as random fluctuations of the phase term in equation (2.14).

Mathematically we can address this situation by using equation (2.12), with $\Gamma_{12} = \left\langle E_{p1}E_{p2}^{*} \right\rangle$ in this case and assuming that (see figure 2.9)

$$E_{p1} = E_0 \mathrm{e}^{-\mathrm{i}\omega t} \mathrm{e}^{\mathrm{i}\phi t}, \tag{2.33}$$

and

$$E_{p2}^{*} = E_{p1}^{*} = E_0 \mathrm{e}^{\mathrm{i}\omega(t+\tau)} \mathrm{e}^{-\mathrm{i}\phi(t+\tau)}, \tag{2.34}$$

at time $(t + \tau)$. Then

$$\gamma_{p1p2}(\tau) = \frac{\Gamma_{p1p2}(\tau)}{\sqrt{\Gamma_{p1p1}(0)\Gamma_{p2p2}(0)}} = \frac{\Gamma_{p1p2}(\tau)}{\sqrt{I_{p1}I_{p2}}}, \tag{2.35}$$

$$\gamma_{p1p2}(\tau) = \frac{< E_0 \mathrm{e}^{-\mathrm{i}\omega t} \mathrm{e}^{\mathrm{i}\phi t} E_0 \mathrm{e}^{\mathrm{i}\omega(t+\tau)} \mathrm{e}^{-\mathrm{i}\phi(t+\tau)} >}{E_0^2},$$

$$\gamma_{p1p2}(\tau) = \mathrm{e}^{\mathrm{i}\omega\tau} < \mathrm{e}^{-\mathrm{i}[\phi(t)-\phi(t-\tau)]} >, \tag{2.36}$$

since $I_{p1} = I_{p2} = (E_0^2)^2$. The phase term in equation (2.36) can be expressed as

$$< \mathrm{e}^{-\mathrm{i}[\phi(t)-\phi(t-\tau)]} > = \frac{1}{T} \int_0^T \mathrm{e}^{-\mathrm{i}[\phi(t)-\phi(t-\tau)]} \mathrm{d}t, \tag{2.37}$$

i.e. an averaging operation. The period T usually is larger than the time $(t + \tau)$ and the degree of coherence is calculated (or the autocorrelation case, where two wavefronts are equal only at two different times). The term $\phi(t) - \phi(t - \tau)$ represents the phase change of the wavefront along T, with random amplitude. This condition is expressed as

$$\phi(t) - \phi(t - \tau) = \begin{cases} 0, & 0 < t < (\tau_0 - \tau) \\ H_1, & (\tau_0 - \tau) < t < \tau_0 \end{cases} \tag{2.38}$$

and so on, where H_1 changes its amplitude. This situation is shown in figure 2.10.

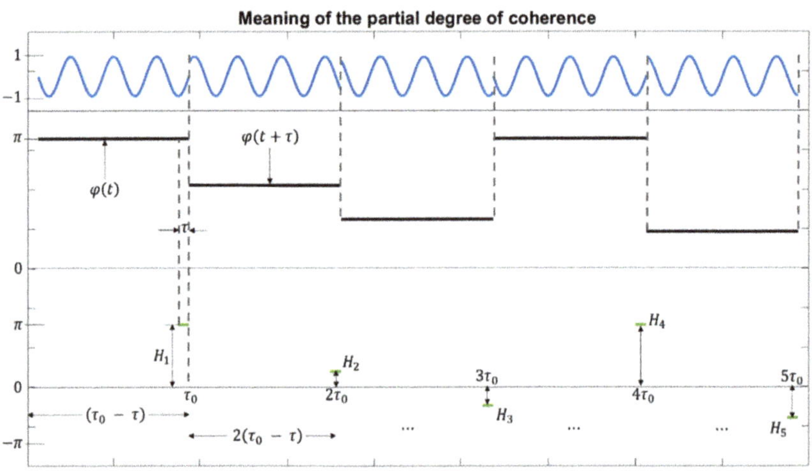

Figure 2.10. Meaning of the partial degree of coherence. See the explanation in the text that follows the figure.

For simplicity, let us consider wavefronts emerging from a laser, as shown in figure 2.10. The first train of emitted waves keeps a constant phase $\varphi(t)$ until a random variation in time t occurs. Then a random phase change is produced $\varphi(t+\tau)$ due to a temporal fluctuation of the emission at τ. Remember that there are no ideal or perfect monochromatic sources. Any variation in their emission processes will produce phase changes—we have to consider the erratic processes of excited atoms within their energy transition levels resulting in a random emission of radiation. This phase change can be represented by the height H_1 that is the measure of the phase difference $(\tau_0 - \tau)$. This process is repeated while the light source emits wavefronts. As the time difference τ between the emitted wavefronts is reduced, the degree of coherence is increased, a feature that can be measured by the height changes H_N.

If we normalize equation (2.36) for a large period T, where $T = N\tau_0$, then the degree of coherence is

$$\gamma_{p1p2}(\tau) = \mathrm{e}^{\mathrm{i}\omega\tau}\frac{1}{N\tau_0}\int_0^{\tau_0-\tau} \mathrm{e}^{-\mathrm{i}[0]}\mathrm{d}t + \int_{\tau_0-\tau}^{\tau_0} \mathrm{e}^{-\mathrm{i}[H_1]}\mathrm{d}t + \int_0^{2(\tau_0-\tau)} \mathrm{e}^{-\mathrm{i}[0]}\mathrm{d}t + \int_{2(\tau_0-\tau)}^{2\tau_0} \mathrm{e}^{-\mathrm{i}[H_2]}\mathrm{d}t$$
$$+ \int_0^{3(\tau_0-\tau)} \mathrm{e}^{-\mathrm{i}[0]}\mathrm{d}t + \int_{3(\tau_0-\tau)}^{3\tau_0} \mathrm{e}^{-\mathrm{i}[H_3]}\mathrm{d}t + \cdots. \tag{2.39}$$

In which we see that the different intervals of integration correspond to those in figure 2.10, i.e. the first interval is $\int_0^{\tau_0-\tau} \mathrm{e}^{-\mathrm{i}[0]}\mathrm{d}t + \int_{\tau_0-\tau}^{\tau_0} \mathrm{e}^{-\mathrm{i}[H_1]}\mathrm{d}t$, and so on. Then, from equation (2.39)

$$\gamma_{p1p2}(\tau) = \mathrm{e}^{\mathrm{i}\omega\tau}\frac{1}{N\tau_0}[(\tau_0 - \tau) - 0 + (\tau_0 - \tau_0 + \tau)\mathrm{e}^{-\mathrm{i}H_1} + 2(\tau_0 - \tau) - 0$$
$$+ (2\tau_0 - 2\tau_0 + 2\tau)\mathrm{e}^{-\mathrm{i}H_2}] + \cdots$$

which yields

$$\gamma_{p1p2}(\tau) = \frac{\mathrm{e}^{\mathrm{i}\omega\tau}}{N\tau_0}\left[N(\tau_0 - \tau) + \tau\sum_{p=1}^{N} p\mathrm{e}^{-\mathrm{i}H_p}\right]. \tag{2.40}$$

The H_p changes represent the amplitude of the random phase changes. For this equation, if the number N is large, then the summation factor vanishes. When we obtain the condition $\phi(t) = \phi(t - \tau)$ the terms of equation (2.39) that form the summation operation in equation (2.40) do not increase the result of the integral, so generalizing

$$\gamma_{p1p2}(\tau) = \left(1 - \frac{\tau}{\tau_0}\right)\mathrm{e}^{\mathrm{i}\omega\tau}, \tag{2.41}$$

that can be expressed as

$$\Re\left[\gamma_{p1p2}(\tau)\right] = \left(1 - \frac{\tau}{\tau_0}\right)\cos(\omega\tau). \tag{2.42}$$

This equation goes from 0 to 1 for values of τ between 0 and τ_0

Maximum in $\tau = 0$, same paths $(\text{path}_1 = \text{path}_2)$

Minimum in $\tau = \tau_0$, path difference (equivalent to the coherence length).

If we take the magnitude of equation (2.42), we will have

$$\left|\gamma_{p1p2}(\tau)\right| = 1 - \frac{\tau}{\tau_0}. \tag{2.43}$$

This means that the values of the interference term in equation (2.14) are limited by equation (2.43) and rule the contrast or visibility (equations (2.15) and (2.16)) of the interference fringes in terms of τ. If in the equation (2.16), $I_1 = I_2$ then

$$V = \left|\gamma_{p1p2}(\tau)\right|. \tag{2.44}$$

Meaning that the visibility of the fringes (equation (2.15)) is equivalent to the magnitude of the degree of the partial coherence coefficient. Then considering equations (2.14), (2.15), and (2.41), (2.42), three cases will be present.

Case A. Total incoherence, meaning that $\tau \to \tau_0$ and $\left|\gamma_{p1p2}(\tau)\right| = 0$. In terms of the visibility (equation (2.15)) and considering that at point P (figure 2.9) we have the addition of the two intensities $I_p = I_{p1} + I_{p2}$ with $I_{p1} = I_{p2} = I_0$, $V = 0$.

Case B. Total coherence, meaning that $\tau = 0$ and $\left|\gamma_{p1p2}(\tau)\right| = 1$. In terms of the visibility and considering that $I_{p1} \neq I_{p2}$, then

$I_{max} = I_{p1} + I_{p2} + 2\sqrt{I_{p1}I_{p2}}$; $\cos[\omega\tau = 0] = 1$, bright fringe,

$I_{min} = I_{p1} + I_{p2} - 2\sqrt{I_{p1}I_{p2}}$; $\cos[\omega\tau = \pi] = -1$, dark fringe.

If $I_{p1} = I_{p2} = I_0$, then the visibility $V = 1$.

Case C. Partial coherence, meaning that $0 < \tau < \tau_0$ and $1 > \left|\gamma_{p1p2}(\tau)\right| > 0$. In terms of the visibility and with $I_{p1} \neq I_{p2}$, then

$$I_p = I_{p1} + I_{p2} + 2\sqrt{I_{p1}I_{p2}}\left|\gamma_{p1p2}(\tau)\right|\cos(\omega\tau).$$

If

$$I_{p1} = I_{p2} = I_0,$$

then

$I_{max} = 2I_0\left[1 + \left|\gamma_{p1p2}(\tau)\right|\right]$; $\cos[\omega\tau = 0] = 1$, bright fringe,

$I_{min} = 2I_0\left[1 - \left|\gamma_{p1p2}(\tau)\right|\right]$; $\cos[\omega\tau = \pi] = -1$, dark fringe, and the visibility $V = \left|\gamma_{p1p2}(\tau)\right|$.

The above considerations and the mathematical formulations derived thereof mean that the visibility measures the degree of coherence of a light source or a measure of the correlation function of the wavefronts emitted by the same light source.

2.5 Spatial coherence

The *temporal coherence* gives us information about the phase correlation between different points in time and the propagation of a wavefront train (such as an electrical field) emitted by a light source. It is a measure of the source's longitudinal coherence. On the other hand, *spatial coherence* is related to the lateral or transversal wavefront coherence measurement. These terms make sense since the spatial coherence will help us to determine, at least, a coherence area. The coherence degree coefficient in a light source is related to the visibility or contrast in an interferometric fringe pattern. The analysis could be done using a Michelson interferometer as the split mechanism (amplitude division), as figures 2.7 and 2.9 show based on equation (2.14).

In order to analyze the spatial coherence, i.e. the correlation between two separated points of the same wavefront of a light source, an interferometer is used, such as Young's double-slit experiment. This instrument will indicate the degree of coherence between different regions of the wavefronts emerging from the two slits or apertures (see figure 2.11) to complete the light source's coherence study.

The basic Young's experiment is shown in figure 2.11. A light source emits wavefronts toward a couple of apertures or slits represented by S_1 and S_2. The light source S can be as coherent as a laser, gas lamp, or light bulb. Let us generalize and consider S a broadband light source such as a light bulb. If a couple of close points (separated by a distance s) from the source surface are chosen as the emitting point sources A and B, as shown in figure 2.12, then it is equivalent to consider them as a couple of slits in front of the light source and traveling until the point P_1 on the screen where they are superimposed (see figure 2.13). With this in mind, it is possible to obtain information about the coherence of the light source S. As the points on the

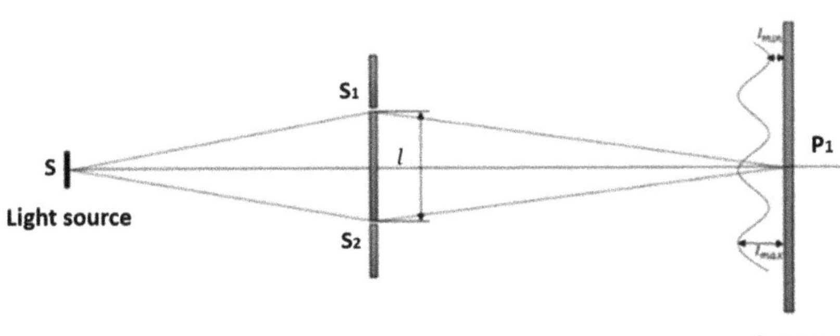

Figure 2.11. Young's double-slit experiment. S_1 and S_2 are the slits separated by a distance l, and P_1 is the observation point where the interference fringe pattern is produced.

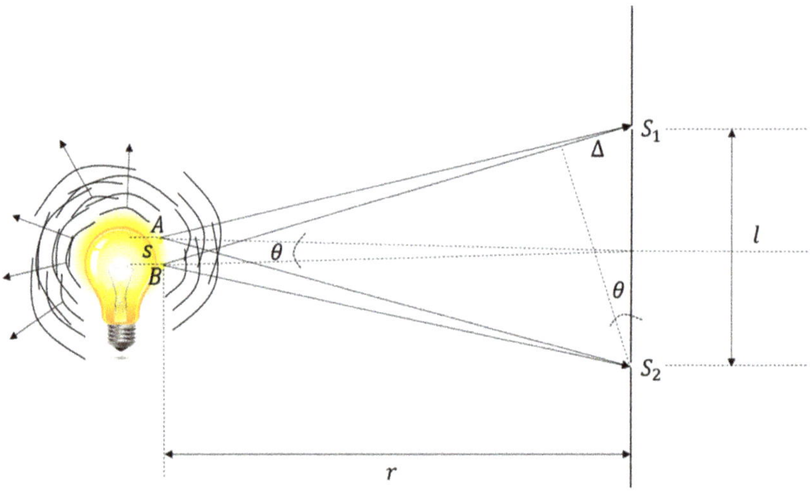

Figure 2.12. A low coherence source is considered for the analysis of spatial coherence. If a couple of close points (separated by a distance s) from the source surface are chosen as the emitting point sources A and B, it is equivalent to consider them as a couple of slits in front of the light source. The angular separation of the secondary sources A and B is given by θ and Δ represents the path difference traveled between these two sources. The distance r is the separation between the points (A, B) and the plane of the slits. Finally, the slits S_1 and S_2 are separated by a distance l.

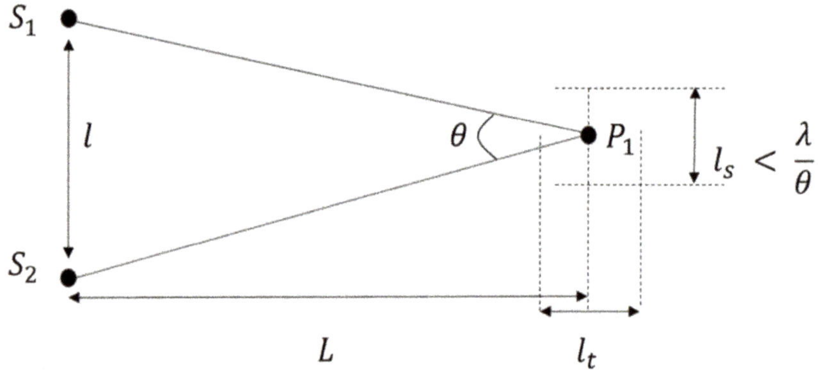

Figure 2.13. S_1 and S_2 represent secondary wave emitters (slits). Their closeness will indicate the spatial coherence of the light source S. In this figure, P_1 is the observation point in which the interference pattern occurs. l_s it is the spatial coherence width and l_t represent the coherence length from the point of view of the temporal coherence (using a Michelson interferometer).

light source are closer then we can assume that the wavefronts from A and B will be emitted simultaneously.

The waves traveling throughout the slits are superimposed, and a fringe pattern is formed at the observation point P_1 which is located on a screen placed at an arbitrary distance. If the traveled paths by the waves from S_1 to P_1 and S_2 to P_1 are different, then their path difference (Δ) is defined as

$$\Delta = \overline{S_1P_1} - \overline{S_2P_1}, \tag{2.45}$$

with $\overline{S_1P_1}$, $\overline{S_2P_1}$ as the longitude traveled by each wavefront, also related to the coherent length of the light source described in equation (2.29). We have two conditions:

$\Delta \ll l_t$, in such a case a fringe pattern on P_1 will be observed;
$\Delta \geqslant l_t$, in such case there is poor or weak definition of the fringes.

Then, it is possible to argue that the spatial coherence of the wavefronts traveling from the slits S_1, S_2 and superposed on P_1 will depend, also, on their separation l. Their closeness will tell us the spatial coherence of the light source S. Then, these slits can be considered point sources from which secondary waves emerge, see figure 2.13.

In figure 2.13 l_s and l_t are the coherence width and length, respectively. The latter is considered from the Young's interferometer's point of view (spatial, wavefront division interferometer). From this, one can see that there exists a volume (a region in the space) surrounding the point P_1 where the light source will be coherent. Then it is possible to obtain a spatial coherence measure (coherent spatial width) of the light source that will depend on the separation l between the two apertures.

First, it is necessary to analyze the fringe pattern formed on the screen when the wavefronts from S_1, S_2 are superimposed at an arbitrary observation point P (see figure 2.14), located at the distance L from the positions of the slits ($l \ll L$). The path difference is

$$\Delta = \overline{S_2P} - \overline{S_1P}. \tag{2.46}$$

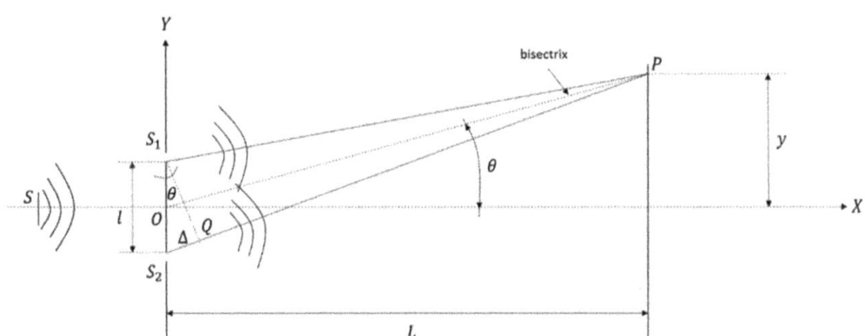

Figure 2.14. The wave from light source S travels through the slits S_1 and S_2, separated a distance l, splits there into two waves, and then both emerging waves are superimposed at point P on a screen located at a distance L from the slits' plane. θ is the angle formed between the bisectrix of the triangle $\overline{S_2PS_1S_2}$ (and the equivalent $\overline{S_1S_2QS_1}$), y is the distance of the point P from the x-axis, and Δ it is the path difference between the two waves falling on P.

From figure 2.14 a bisector is traced from point P to the origin O, and the arc $\overline{S_1Q}$ intersects the line $\overline{S_2P}$ in point Q. The angle θ is formed by S_1, S_2 and the segment $\overline{S_1Q}$. The segment OP is perpendicular to the segment $\overline{S_1Q}$. Then

$$\Delta = l\sin\theta, \tag{2.47}$$

and the optical path difference between (within a refraction index, $n = 1$) the wavefronts is

$$\delta = k(\overline{S_2P} - \overline{S_1P}), \tag{2.48}$$

with $k = \frac{2\pi}{\lambda}$ being the 'wave vector', defined in chapter 1.

Then, it is possible to write

$$S_1(r_1, t) = A_1 e^{i(k_1.r_1 - \omega t + \varphi_1)}, \tag{2.49}$$

and

$$S_2(r_2, t) = A_2 e^{i(k_2.r_2 - \omega t + \varphi_2)}, \tag{2.50}$$

where A_1, A_2 are the amplitudes of the waves and r_1, r_2 are their positions in the coordinate plane. The resulting intensity of the two superimposed waves, $(S_1 + S_2)(S_1 + S_2)^*$, can be expressed as

$$I = I_1 + I_2 + 2\sqrt{I_1 I_2}\cos\left[(k_1 r_1 - k_2 r_2) + \Delta\varphi\right], \tag{2.51}$$

where $\Delta\varphi$ is the relative phase difference of the wave, and for this particular case it may be considered zero. Then, the argument of the cosine function in equation (2.51) is related to the optical path difference δ defined by equation (1.48) (with $k_1 = k_2 = |k| = \frac{2\pi}{\lambda}$), and if we assume that $I_1 = I_2 = I_0$, then equation (2.51) is

$$I = 2I_0(1 + \cos\delta). \tag{2.52}$$

Now, since $1 + \cos\delta = 2\cos^2\left(\frac{\delta}{2}\right)$, equation (2.52) is rewritten as

$$I = 4I_0\cos^2\left(\frac{\delta}{2}\right), \tag{2.53}$$

and using equation (2.48)

$$I = 4I_0\cos^2\left(\frac{\frac{2\pi\Delta}{\lambda}}{2}\right). \tag{2.54}$$

From figure 2.14, in a set-up such as Young's experiment, $y \ll L$ and then $\sin\theta\tilde{\ }\tan\theta\tilde{\ }\frac{y}{L}$ and by using the relation (1.47), equation (1.54) is

$$I = 4I_0\cos^2\left(l\frac{\pi y}{\lambda L}\right). \tag{2.55}$$

Looking at the argument of the cosine function in equation (2.55), we found the maximum value at:

$l\frac{\pi y}{\lambda L} = n\pi$, then $y_{max} = n\frac{\lambda L}{l}$, with n an integer, meaning constructive interference;

$l\frac{\pi y}{\lambda L} = \left(n + \frac{1}{2}\right)\pi$, then $y_{min} = \left(n + \frac{1}{2}\right)\ \frac{\lambda L}{l}$, meaning destructive interference.

Then, from these conditions, the fringe separation is approximately λ and will depend on the distance L at which the fringe pattern observing screen is located. Also, the separation between fringes, Δy, is constant and is expressed as

$$\Delta y = y_{n+1} - y_n = \frac{\lambda L}{l}. \tag{2.56}$$

Now, let us obtain the spatial coherence of the light source considering the separation d between the two points A, B (see figure 2.12). If θ is the angular separation (small angle) of the secondary sources and s is their separation (see the equivalent triangle subtended by θ), then

$$\sin\theta = \frac{\Delta}{l};\ \Delta = d\theta, \tag{2.57}$$

and

$$\theta = \frac{s}{r}. \tag{2.58}$$

Combining equations (2.57) and (2.58)

$$\Delta = l\frac{s}{r}. \tag{2.59}$$

From equation (2.54), a maximum occurs if $\Delta = \frac{\lambda}{2}$, and then, from equation (2.59)

$$s = \frac{r\lambda}{2l}. \tag{2.60}$$

Then, the width l_s of the coherence length in figure 2.13 is related to the separation l of the slits (points S_1, S_2 in figure 2.11). The net separation between two consecutive maxima is then

$$s < \frac{r\lambda}{l} \quad \text{or} \quad l_s < \frac{r\lambda}{s},$$

which is equivalent (in equation (2.58), $\frac{1}{\theta} = \frac{r}{s}$) to

$$l_s < \frac{\lambda}{\theta}. \tag{2.61}$$

There is a relation between the coherent spatial width (l_s), and the coherence degree defined in equation (2.44). The van Cittert–Zernike theorem [12, 19] calculates the degree of coherence based on the geometry of figure 2.12, with as the angular separation of the two emitting points seen from the source. This theorem states that the degree of coherence between two points is the modulus of the scaled and normalized Fourier transform of the source intensity distribution (two-dimensional case) [22]:

$$| \gamma_{s1s2} | = \left| \frac{\iint_s I(\xi, \eta) e^{i\left(\frac{2\pi}{\lambda}\right)(\xi\theta_x + \eta\theta_y)} d\xi d\eta}{\iint_s I(\xi, \eta) d\xi d\eta} \right|. \tag{2.62}$$

In which θ_x and θ_y are the x and y components of the pinhole separation θ, and the integral is over the area of the emitting source, located in an arbitrary plane (ξ, η). Equation (2.62) is related to the coherent spatial width l_s and the coherent temporal length l_t, i.e. a coherence region of the light source, and its significance is that the diffraction pattern due to an aperture of the same size and shape as the source (broad or extended) will be defined by the degree of coherence coefficient (equation (2.12)) and the coherent spatial width, see figure 2.15.

From figure 2.15 the size of the primary slit s guarantees the coherence of the light source, while the slits S_1, S_2 are within the spatial coherence width (l_s) (limited by the width s). For a more specialized read on these concepts see [23]. Then, it is experimentally possible to measure the degree of coherence of a light source based on the fringe pattern formed by a Young's interferometer. Figure 2.16 shows that it is possible to determine the temporal coherence length and the spatial coherence width by using an amplitude and wavefront division interferometers, respectively, in conjunction, see figure 2.16.

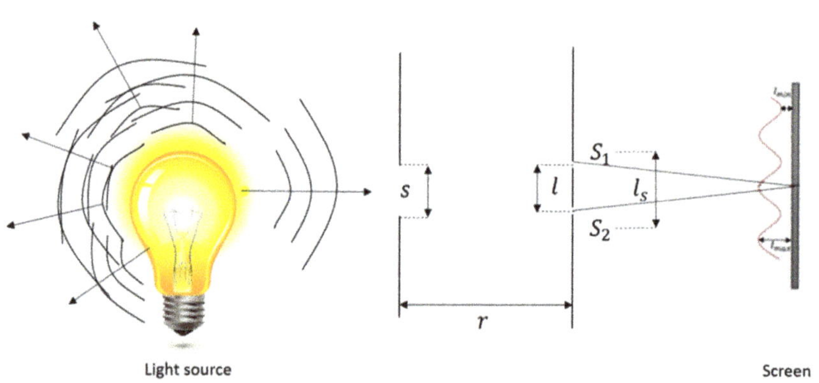

Figure 2.15. Degree of coherence and spatial coherence width for a light source (van Cittert–Zernike theorem).

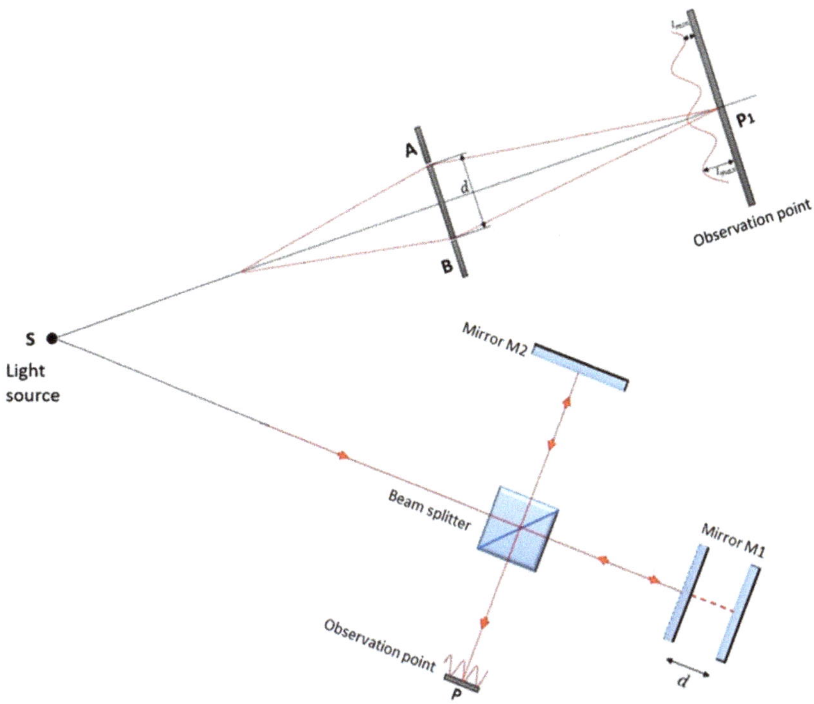

Figure 2.16. A dual optical layout using a Michelson interferometer and Young's double-slit experiment simultaneously measures the coherent degree coefficient, including the coherent temporal length or the coherent spatial width (determining the region of coherence of the light source).

References

[1] Wolf E 2003 Unified theory of coherence and polarization of random electromagnetic beams *Physics* **312** 263–7

[2] Brosseau C 2010 Polarization and coherence optics: historical perspective, status, and future directions *Progress in Optics* vol 54 (Amsterdam: Elsevier), ch 3 pp 149–208

[3] Calvo M L 1995 Coherencia óptica *Invest. Cienc.* **224** 66–73

[4] Wolf E 2001 Lecture 7.3: The development of optical coherence theory *Selected Works of Emil Wolf* (World Scientific Series in 20th Century Physics vol 29) (Singapore: World Scientific)

[5] Nakagawa K *et al* 2014 Sequentially timed all-optical mapping photography (STAMP) *Nat. Photon* **8** 695–700

[6] Jiang Z and Zhang X-C 1998 Single-shot spatiotemporal terahertz field imaging *Opt. Lett.* **23** 1114–6

[7] Suzuki T *et al* 2015 Sequentially timed all-optical mapping photography (STAMP) utilizing spectral filtering *Opt. Express* **23** 30512–22

[8] Roddier C 1979 Incoherent holography *J. Opt.* **10** 317

[9] Liu J-P, Tahara T, Hayasaki Y and Poon T-C 2018 Incoherent digital holography: a review *Appl. Sci.* **8** 143

[10] Rosen J *et al* 2019 Recent advances in self-interference incoherent digital holography *Adv. Opt. Photon.* **11** 1–66
[11] Zernike F 1942 Phase contrast, a new method for the microscopic observation of transparent objects *Physica* **9** 686–98
[12] van Cittert P H 1958 Degree of coherence *Physica* **24** 505–7
[13] Dahl P H 2004 Forward scattering from the sea surface and the van Cittert–Zernike theorem *J. Acoust. Soc. Am.* **115** 589–99
[14] Thompson B J and Wolf E 1957 Two-beam interference with partially coherent light *J. Opt. Soc. Am.* **47** 895–902
[15] Considine P S 1966 Effects of coherence on imaging systems *J. Opt. Soc. Am.* **56** 1001–9
[16] Magyar G and Mandel L 1963 Interference fringes produced by superposition of two independent maser light beams *Nature* **198** 255–6
[17] Paredes D and Calvo M L 2010 On the Thompson–Wolf experiment: a study with laser sources *J. Eur. Opt. Soc.* **5** 10051s
[18] Roychowdhury H, Ponomarenko S A and Wolf E 2005 Change in the polarization of partially coherent electromagnetic beams propagating through the turbulent atmosphere *J. Mod. Opt.* **52** 1611–8
[19] Born M, Wolf E, Bhatia A, Clemmow P, Gabor D, Stokes A and Wilcock W 1999 *Principles of Optics: Electromagnetic Theory of Propagation, Interference and Diffraction of Light* 7th edn (Cambridge: Cambridge University Press)
[20] Akhmanov S A and Nikitin S Y 1997 *Physical Optics* 1st edn (Oxford: Clarendon)
[21] Weber M J 1999 *Handbook of Laser Wavelengths* (Boca Raton, FL: CRC Press)
[22] Bass M, DeCusatis C, Enoch J, Lakshminarayanan V, Li G, MacDonald C, Mahajan V and Stryland E V 2009 *Handbook of Optics* vol 1, part 2 3rd edn (New York: McGraw-Hill Education)
[23] Wolf E 2007 *Introduction to the Theory of Coherence and Polarization of Light* (Cambridge: Cambridge University Press)

IOP Publishing

Full Field Optical Metrology and Applications

Fernando Mendoza-Santoyo, Manuel De la Torre-Ibarra, María del Socorro Hernández-Montes and Jorge Mauricio Flores Moreno

Chapter 3

Phase retrieval

Contributing authors: Angel Manuel Fernández Doval and Thomas Kreis

It is now appropriate to present how digital data are handled to obtain a diverse set of variables from the object under study. This chapter provides insight into three significant subjects dealing with data processing in interferometric systems. They are the traditional and modern methods of phase retrieval and phase demodulation.

3.1 Phase sampling evaluation

From the considerations set out in chapter 1, a full field interference pattern recorded by a CCD camera can be formulated as having an intensity distribution given by

$$I(x, y) = a(x, y) + b(x, y)\cos[\Delta\varphi(x, y)], \tag{3.1}$$

where $a(x, y)$ describes the background intensity fluctuations (as seen before, $I_1 + I_2$), e.g. an uneven illumination or reflection of the measured surface, and $b(x, y)$ is the locally varying contrast, also known as the modulation term (as before, $2\sqrt{I_1I_2}$); it can be said that $a(x, y)$ and $b(x, y)$ are the additive and the multiplicative distortions of the interference pattern. $\cos[\Delta\varphi(x, y)]$ is the oscillate term producing the interference pattern in conjunction with $b(x, y)$; its argument is the phase, which can be directly related to the physical quantity to be measured. A straightforward approach for evaluation is the detection of local intensity maxima and minima corresponding to $\Delta\varphi$ being an integer (maxima) or a half-integer (minima) times π, followed by interpolation. A more precise measurement is the direct pointwise determination of the interference phase, as shown with heterodyning in section 1.3.2. An even easier and more widely applied method is phase-shifted interferograms. Multiple interference patterns are recorded at the same object condition-state but with varying additional phase shifts. The phase shift is introduced in one arm of the interferometer by, e.g. a piezotransducer, see figure 1.4.

doi:10.1088/978-0-7503-3027-5ch3

If we introduce mutually distinct additional phase shifts ψ_n, $n = 1, \dots, N$, we obtain a set of N interferograms of the object with their corresponding equations:

$$I_n(x, y) = a(x, y) + b(x, y)\cos[\Delta\varphi(x, y) + \psi_n] \quad n = 1, \dots, N \geqslant 3. \tag{3.2}$$

3.1.1 Interference phase measurement with known phase shifts

If several intensity distributions $I_n(x, y)$ with mutual phase shifts $\Delta\psi = \psi_{n+1} - \psi_n$ are recorded at each point (x, y) a non-linear system of equations of the form in equation (3.2) has to be solved. As long as the phase shifts are known, and this is true experimentally with the use of the piezoactuator, there remain three unknowns: $a(x, y)$, $b(x, y)$, and the desired interference phase $\Delta\varphi(x, y)$. Thus at least $N = 3$ equations are necessary, meaning that at least three phase-shifted intensities have to be recorded.

One solution to the non-linear system of equation (3.2) is the Gaussian least-squares approach; this is done by introducing

$$u(x, y) = b(x, y)\cos[\Delta\varphi(x, y)] \text{ and } v(x, y) = -b(x, y)\sin[\Delta\varphi(x, y)]. \tag{3.3}$$

Then (3.2) is rewritten (with the (x, y) omitted for clarity)

$$I_n = a + b\cos[\Delta\varphi + \psi_n] = a + u\cos\psi_n + v\sin\psi_n. \tag{3.4}$$

The procedure must consider that the sum of the quadratic errors $\sum_{i=1}^{N}(a + u\cos\psi_n + v\sin\psi_n - I_n)^2$ has to be minimized. Hence, partial differentiation of this set of functions with respect to a, u, and v, has to be equaled to zero, giving the following linear system of three equations

$$\begin{pmatrix} m & \sum\cos\psi_n & \sum\sin\psi_n \\ \sum\cos\psi_n & \sum\cos^2\psi_n & \sum\sin\psi_n cos\psi_n \\ \sum\sin\psi_n & \sum\sin\psi_n\cos\psi_n & \sum\sin^2\psi_n \end{pmatrix}\begin{pmatrix} a \\ u \\ v \end{pmatrix} = \begin{pmatrix} \sum I_n \\ \sum I_n\cos\psi_n \\ \sum I_n\sin\psi_n \end{pmatrix}. \tag{3.5}$$

This system has to be solved pointwise for u and v. At the same time, the solution for a may be omitted since it represents a background intensity term with no physical utility with the object being measured. Having found $u(x, y)$ and $v(x, y)$, the interference phase is determined modulo 2π by

$$\Delta\varphi(x, y) = \arctan - \frac{v(x, y)}{u(x, y)}. \tag{3.6}$$

The pointwise solution of the system, equation (3.5), requires only a single inversion of the matrix to the left, provided the phase steps $\Delta\psi$ are the same for all pixels. Simple solutions of equation (3.5) are given for phase steps that are constant and have particular values such as 30°, 45°, 60°, or 90° because the sines and cosines of these values are easy to handle. Some formulas corresponding to equation (3.6) are given in table 3.1.

Table 3.1. Some phase shift methods for known phase shifts.

Phase shift	Start	N	Evaluation formula
30°	−30°	3	$\Delta\varphi = \arctan\frac{I_1 - I_3}{(2+\sqrt{3})(-I_1 + 2I_2 - I_3)}$
45°	−45°	3	$\Delta\varphi = \arctan\frac{\sqrt{2}(I_3 - I_1)}{(2+\sqrt{2})(I_1 - 2I_2 + I_3)}$
60°	0°	3	$\Delta\varphi = \arctan\frac{2I_1 - 3I_2 + I_3}{\sqrt{3}(I_2 - I_3)}$
90°	0°	3	$\Delta\varphi = \arctan\frac{I_1 - 2I_2 + I_3}{I_1 - I_3}$
90°	0°	4	$\Delta\varphi = \arctan\frac{I_4 - I_2}{I_1 - I_3}$
90°	0°	5	$\Delta\varphi = \arctan\frac{7(I_4 - I_2)}{4I_1 - I_2 - 6I_3 - I_4 + 4I_5}$
α	$-\alpha$	3	$\Delta\varphi = \arctan\frac{(1-\cos\alpha)(I_3 - I_1)}{\sin\alpha(I_1 - 2I_2 + I_3)}$
$2\pi k/m$	0°	m	$\Delta\varphi = \arctan\frac{-\sum_{i=1}^{m} I_i \sin\left[\frac{2\pi k}{m}(i-1)\right]}{\sum_{i=1}^{m} I_i \cos\left[\frac{2\pi k}{m}(i-1)\right]}$

3.1.2 Interference phase measurement with unknown phase shifts

Phase sampling evaluation also works with unknown phase shifts as long as these are not integer multiples of π and are constant between the recordings. Then we have one more unknown $N = 4$, so at least $m = 4$ phase-shifted intensity distributions have to be generated. A well-known solution using all four intensities at each pixel (x, y) is the Carre formula [1]

$$\Delta\varphi(x, y) = \arctan\frac{\sqrt{I_1 + I_2 - I_3 - I_4}\ \sqrt{3I_2 - 3I_3 - I_1 + I_4}}{I_2 + I_3 - I_1 - I_4}. \tag{3.7}$$

Another approach that allows the inherent suppression of errors first calculates the phase shift $\Delta\psi$ at each point by

$$\Delta\psi(x, y) = \arccos\frac{I_1(x, y) - I_2(x, y) + I_3(x, y) - I_4(x, y)}{2[I_2(x, y) - I_3(x, y)]}. \tag{3.8}$$

If we assume that $\Delta\psi$ has to be constant for all points of the interference pattern, we can take the average $\overline{\Delta\psi}$ over all pixels (x, y), so in this way the fluctuations in $\Delta\psi(x, y)$, which, for example, may be caused by speckles (an inherent noise-like interference random structure associated with rough objects being illuminated by coherent light sources; this term will become familiar in other chapters in this book), are averaged out. Outliers in $\Delta\psi(x, y)$ preferably occur where the equation (3.8) denominator is zero or near zero, and they may be detected by an outlier test and discarded before the averaging is done.

Having calculated the average phase shift $\overline{\Delta\psi}$, now only three unknowns are left and the interference phase can be determined from the first three intensities I_1, I_2, I_3 by

$$\Delta\varphi_1(x, y) = \arctan\frac{I_3 - I_2 + (I_1 - I_3)\cos\overline{\Delta\psi} + (I_2 - I_1)\cos 2\overline{\Delta\psi}}{(I_1 - I_3)\sin\overline{\Delta\psi} + (I_2 - I_1)\sin 2\overline{\Delta\psi}} + \frac{3\overline{\Delta\psi}}{2}. \quad (3.9)$$

Analogously, the interference phase may be calculated from the last three intensities I_2, I_3, I_4 by

$$\Delta\varphi_2(x, y) = \arctan\frac{I_4 - I_3 + (I_2 - I_4)\cos\overline{\Delta\psi} + (I_3 - I_2)\cos 2\overline{\Delta\psi}}{(I_2 - I_4)\sin\overline{\Delta\psi} + (I_3 - I_2)\sin 2\overline{\Delta\psi}} + \frac{\overline{\Delta\psi}}{2}. \quad (3.10)$$

The (x, y) dependence in equations (3.7)–(3.10) has been omitted for convenience. The 2π-steps of the interference phase distributions $\Delta\varphi_1(x, y)$ and $\Delta\varphi_2(x, y)$ due to the arctangent arise at different points. This additional information can be used in the subsequent demodulation by considering the continuous interference phase variation of these two-phase distributions to decide to add or subtract a further 2π. Algorithms employing unknown phase shifts can be found in [2].

3.2 Modern methods for phase retrieval

Most of today's standard methods to recover the phase from fringe patterns were developed in the second half of the twentieth century [3], mainly thanks to the fact that computers became widely available and increasingly faster and capable of storing more and more data (the big data handling scenario also applies here). Newer developments in phase retrieval methods have happened during the first two decades of this millennium, mainly focused on solving the phase extraction problem is far from ideal or poorly controlled conditions, where the initial and more traditional techniques do not provide satisfactory results. A selection of these modern methods will be described next.

3.2.1 Single fringe pattern methods

3.2.1.1 Methods for fringe patterns with a spatial carrier (open fringes)

As seen previously (see for reference chapter 1), a generic sinusoidal fringe pattern with a two-dimensional spatial carrier can be written as

$$\begin{aligned} I(x, y) &= a(x, y) + b(x, y)\cos\left[k_{cx}(x, y)x + k_{cy}(x, y)y + \varphi(x, y)\right], \\ &= a(x, y) + b(x, y)\cos\phi(x, y), \end{aligned} \quad (3.11)$$

where a is the background intensity, b is the modulation amplitude of the fringes, and ϕ is the phase to be retrieved; $\boldsymbol{k}_c = (k_{cx}, k_{cy})$ is the angular frequency of the 2D spatial carrier and φ is the modulating phase. Generally, a, b, and φ have lower frequencies than the carrier.

The standard Fourier transform method (FTM) requires a quasi-monochromatic carrier for the whole fringe pattern. If the components of the carrier frequency (k_{cx} and k_{cy}) change significantly across the fringe pattern, the spectrum at the Fourier space broadens and may become inseparable, thus limiting the applicability of the FTM. The windowed Fourier transform method (WFTM) and the wavelet transform method (WTM) relax such requirements by scanning the fringe pattern and analyzing a relatively small region of interest (ROI) in the neighborhood of each pixel. Thus the carrier should be locally quasi-monochromatic, but its frequency (k_{cx}, k_{cy}) may change smoothly from one neighborhood to the next. This allows the phase retrieval even in the regions of open fringe patterns where the fringes can be locally identified with a fringe carrier.

One relevant difference between both methods is that the size of the region used to compute the transform at a given location is the same for all the frequencies in the WFTM, resulting in fixed spatial and spectral resolutions, but decreases with frequency in the WTM, which provides a trade-off between better spatial resolution for high frequencies and better spectral resolution for low frequencies.

3.2.1.1.1 The windowed Fourier transform method

The WFTM of a two-dimensional function $f(x, y)$ at a position (u, v) and spatial frequency (k_x, k_y), and its inverse counterpart are defined as

$$Sf(u, v, k_x, k_y) = \mathscr{F}_w\{f(x, y)\} = \iint_{\mathbb{R}^2} f(x, y) g^*_{u, v, k_x, k_y}(x, y) \mathrm{d}x\mathrm{d}y \tag{3.12}$$

and

$$\begin{aligned} f(x, y) &= \mathscr{F}_w^{-1}\left\{Sf(u, v, k_x, k_y)\right\} \\ &= \frac{1}{4\pi^2} \iint_{\mathbb{R}^2} [\iint_{\mathbb{R}^2} Sf(u, v, k_x, k_y) g_{u, v, k_x, k_y}(x, y) \mathrm{d}k_x \mathrm{d}k_y] \mathrm{d}u \mathrm{d}v \end{aligned} \tag{3.13}$$

with

$$g_{u, v, k_x, k_y}(x, y) = g(x - u, y - v) \exp[\mathrm{i}(k_x x + k_y y)] \tag{3.14}$$

as the basic function of the transform. $g(x, y)$ is a real-valued window that selects the ROI; for instance, a common choice is a Gaussian window

$$g(x, y) = \frac{1}{\sqrt{\pi \sigma_x \sigma_y}} \exp\left(-\frac{x^2}{2\sigma_x^2} - \frac{y^2}{2\sigma_y^2}\right), \tag{3.15}$$

which selects an ROI with an effective size of $4\sigma_x \times 4\sigma_y$. In this particular case, the transform is known as the *Gabor transform*.

Also, the windowed Fourier transform may be implemented as a convolution [4]

$$\begin{aligned} Sf(u, v, k_x, k_y) = \mathscr{F}_w\{f(x, y)\} &= [f(u, v) ** h_{k_x, k_y}(u, v)] \\ &\cdot \exp[-\mathrm{i}(k_x u + k_y v)], \end{aligned} \tag{3.16}$$

$$f(x, y) = \mathscr{F}_w^{-1}\{Sf(u, v, k_x, k_y)\},$$

$$= \frac{1}{4\pi^2} \iint_{\mathbb{R}^2} \{Sf(x, y, k_x, k_y) \cdot \exp[i(k_x x + k_y y)]\} ** h_{k_x, k_y}(x, y) dk_x dk_y, \tag{3.17}$$

with

$$h_{k_x, k_y}(x, y) = g_{0,0, k_x, k_y}(x, y) = g(x, y) \exp\left[i(k_x x + k_y y)\right]. \tag{3.18}$$

Now, by applying the windowed Fourier transform to a fringe pattern with a spatial carrier, equation (3.11) results in a local spectral distribution (as usually expected in the FTM),

$$\begin{aligned} SI(u, v, k_x, k_y) = Sa_{u, v}(k_x, k_y) + \frac{1}{2} Sc_{u, v}(k_x - k_{cx}, k_y - k_{cy}) \\ + \frac{1}{2} Sc_{u, v}(k_x + k_{cx}, k_y + k_{cy}). \end{aligned} \tag{3.19}$$

If the window size is small enough to have an almost constant carrier frequency across the ROI, the three terms of the local spectrum are separable. Once at this point, one of the following two variants of the WFTM may be chosen:

1. The *windowed Fourier filtering* directly extends the regular FTM to the WFTM. One of the side terms of the spectrum is selected with a bandpass filter, and the inverse windowed Fourier transform returns an analytic function from which the phase can be retrieved:

$$\phi(x, y) = \arg[I_{BP}(x, y)] = \arg\left[\mathscr{F}_w^{-1}\left\{\frac{1}{2} Sc_{u, v}(k_x + k_{cx}, k_y + k_{cy})\right\}\right]. \tag{3.20}$$

   A rectangular bandpass filter may be implemented by simply changing the integration limits in equation (3.17), namely

$$I_{BP}(x, y) = \frac{1}{4\pi^2} \int_{k_{y,l}}^{k_{y,h}} \int_{k_{x,l}}^{k_{x,h}} \{SI(x, y, k_x, k_y) \cdot \exp[i(k_x x + k_y y)]\} ** h_{k_x, k_y}(x, y) dk_x dk_y. \tag{3.21}$$

2. The *windowed Fourier ridge* is the set of points $[u, v, k_{Rx}(u, v), k_{Ry}(u, v)]$ in the space-frequency domain where the modulus $|SI(u, v, k_x, k_y)|$ of the windowed Fourier transform has its local maxima. These maxima mark the highest similarity between the fringe pattern and the basic function of the transform. The values of the frequency at the ridge $\boldsymbol{k}_R(u, v) = [k_{Rx}(u, v), k_{Ry}(u, v)]$ are taken as estimates of the local spatial frequency of the fringe pattern.

The phase can be either directly retrieved from the complex value of the WFT at the ridge

$$\phi(u, v) = \arg\{SI[u, v, k_{Rx}(u, v), k_{Ry}(u, v)]\}, \tag{3.22}$$

or obtained by spatial integration of the local fringe frequencies $\boldsymbol{k}_R(u, v)$. In the first case, the transform window must be small enough to assume that the phase changes linearly; otherwise, compensation terms must be added to the right-hand side of the equation (3.22). In the second case, the resulting phase is obtained already unwrapped, but the integration process accumulates the measurement error, a possible trade-off we may not want.

3.2.1.1.2 The wavelet transform method

A two-dimensional wavelet is an oscillating function $\psi(x, y) \in L^2(\mathbb{R}^2)$, with zero mean $\iint_{\mathbb{R}^2} \psi(x, y)\mathrm{d}x\mathrm{d}y = \hat{\psi}(0,0) = 0$ (where $\hat{\psi}$ is the Fourier transform of ψ), with a limited spatial extent and centered in the neighborhood of (0,0). In practice, wavelets are often normalized to satisfy $\| \psi(x, y)\|^2 = \iint_{\mathbb{R}^2} | \psi(x, y)|^2 \,\mathrm{d}x\mathrm{d}y = 1$.

Then, a *family of wavelets* is derived from a *mother wavelet* $\psi(x, y)$ by a translation by (u, v), scaling (dilation) by a factor s, and rotation by an angle θ

$$\psi_{u,\, v,\, s,\, \theta}(x, y) = \frac{1}{s}\psi\left[\frac{\mathbf{R}_{-\theta} \cdot (x - u, y - v)^T}{s}\right], \tag{3.23}$$

with $\mathbf{R}_\theta = \begin{pmatrix} \cos\theta & -\sin\theta \\ \sin\theta & \cos\theta \end{pmatrix}$ that represents a 2D rotation matrix.

There is an ample choice of mother wavelets to select the one that best suits a particular application [5]. They may be isotropic (rotationally invariant) or directional, real- or complex-valued, and oscillate and attenuate according to different functions.

The wavelets primarily used for phase retrieval applications are analytic [6], i.e. their Fourier transforms $\hat{\psi}_M(k_x, k_y)$ must be zero in one half of the spatial frequency plane and, consequently, they are necessarily complex-valued. Among this class of wavelets, a typical choice for fringe pattern analysis is the 2D Morlet wavelet, which is directional and analytic. It essentially consists of a plane wave modulated with a Gaussian window

$$\psi_M(x, y) = \exp\left[-\frac{1}{2}\left(\frac{1}{\varepsilon}x^2 + y^2\right)\right] \exp\left[\mathrm{i}\left(k_{0x}x + k_{0y}y\right)\right], \tag{3.24}$$

$$\hat{\psi}_M\left(k_x, k_y\right) = \sqrt{\varepsilon}\, \exp\left\{-\frac{1}{2}\left[\varepsilon(k_x - k_{0x})^2 + \left(k_y - k_{0y}\right)^2\right]\right\}. \tag{3.25}$$

The magnitude of the spatial frequency vector must be $| \boldsymbol{k}_0 | = |(k_{0x}, k_{0y})| \geqslant 5.6$ to satisfy the zero-mean condition. Correction terms must be introduced in equations (3.24) and (3.25) for lower values of $| \boldsymbol{k}_0 |$. The anisotropy parameter $\varepsilon \geqslant 1$ stretches the Gaussian window in the x-direction. With $\varepsilon = 1$ the Morlet wavelet is also known as the Gabor function.

Now let us look at the continuous wavelet transform (CWT) of a two-dimensional function $f(x, y)$ with respect to the wavelet ψ, at position (u, v), with scale s and direction θ—it is the inner product of the function and each of the wavelets $\psi_{u, v, s, \theta}$ in the family of ψ

$$Wf(u, v, s, \theta) = \langle f, \psi_{u, v, s, \theta} \rangle = \frac{1}{s} \iint_{\mathbb{R}^2} f(x, y) \hat{\psi}^* \left[\frac{\mathbf{R}_{-\theta} \cdot (x - u, y - v)^T}{s} \right] \mathrm{d}x\mathrm{d}y. \tag{3.26}$$

By using the convolution theorem, it can also be written in terms of the Fourier transform [7]

$$Wf(u, v, s, \theta) = s\mathscr{F}^{-1}\left\{\hat{f}\left(k_x, k_y\right)\hat{\psi}^*\left[s\mathbf{R}_{-\theta} \cdot \left(k_x, k_y\right)^T\right)\right\}. \tag{3.27}$$

As a result of the zero-mean property of wavelets, the coefficients $Wf(u, v, s, \theta)$ resulting from the transform are non-negligible only when the wavelet $\psi_{u, v, s, \theta}$ matches the function f.

It is worth mentioning that the dilation factor s controls the frequency and the spatial extent of the wavelet simultaneously. The smaller the value of s, the higher the wavelet frequency and the smaller its spatial extent, resulting in a higher spatial resolution and a lower frequency resolution, and vice versa. This varying resolution makes the CWT well suited for analyzing fringe patterns with a wide range of local spatial carrier frequencies.

When a CWT is applied to a fringe pattern, the points in the space-scale-rotation domain $[u, v, s_R(u, v), \theta_R(u, v)]$, where the moduli of the wavelet coefficients are $|WI(u, v, s, \theta)|$, reach their local maxima—i.e. the *ridge of the wavelet transform*—corresponding to the best match between the wavelet and the fringe pattern. At each position $s_R(u, v)$ and $\theta_R(u, v)$ are the scale and direction angle, respectively, that best fit the fringe pattern. There are two main strategies to retrieve the phase:

1. *Direct phase retrieval.*

   If the wavelet is analytic, the local phase of the fringe pattern can be obtained as the argument of the complex value of the wavelet coefficients at the ridge

$$\phi(u, v) = \arg\{WI[u, v, s_R(u, v), \theta_R(u, v)]\}. \tag{3.28}$$

2. *The gradient of the phase.* The local frequency can be approximated from the value of the scaling factor at the ridge and the central frequency of the mother wavelet $\boldsymbol{k}_0 = (k_{0x}, k_{0y})$,

$$\boldsymbol{k}(u, v) = [k_x(x, y), k_y(x, y)] \approx \frac{\mathbf{R}_{-\theta} \cdot (k_{0x}, k_{0y})^T}{s_R(u, v)}. \tag{3.29}$$

   As with the WFTM, the unwrapped phase may be obtained by spatial integration of $\boldsymbol{k}(u, v)$, but the approximation and measurement errors tend to propagate and be accumulated in the process. This makes the gradient method not optimal for phase retrieval but useful when the gradient of the phase is the quantity of interest [8].

3.2.1.2 Methods for fringe patterns without a spatial carrier (closed fringes)

A generic sinusoidal fringe pattern without a clearly defined spatial carrier can be written as

$$I(x, y) = a(x, y) + b(x, y)\cos[\varphi(x, y)], \tag{3.30}$$

where a and b are the background level and modulation amplitude of the fringes, respectively, and φ is the phase to be retrieved.

Even though closed fringes do not show a globally identifiable spatial carrier, they may still behave locally as carrier fringes. This fringe pattern may use the windowed Fourier and wavelet transform methods. However, the sign of the retrieved phase becomes ambiguous in the regions where the fringes are locally closed, and no local carrier is present.

3.2.1.2.1 The spiral phase transform method

As seen in the previous subsections, the phase is often retrieved as the argument of the analytic signal corresponding to the fringe pattern. One classical method to construct the analytical representation of a 1D real-valued signal consists in taking the signal as the real part and its Hilbert transform as the imaginary part. The Hilbert transform $\mathscr{H}$ achieves the quadrature of the 1D signal, i.e. turns its cosine components into the corresponding sine components, by multiplying its spectrum by $\pm$i opposing to the sign of the frequency, thus shifting them by $\pm\pi/2$

$$\mathscr{F}[b\cos(\omega_0 t)] = \pi[\delta(\omega - \omega_0) + \delta(\omega + \omega_0)], \tag{3.31}$$

$$-\mathrm{i}\,\mathrm{sgn}\,(\omega)\mathscr{F}[b\cos(\omega_0 t)] = -\mathrm{i}\pi[\delta(\omega - \omega_0) - \delta(\omega + \omega_0)] = \mathscr{F}[b\sin(\omega_0 t)], \tag{3.32}$$

$$\mathscr{H}[b\cos(\omega_0 t)] = \mathscr{F}^{-1}\{-\mathrm{i}\,\mathrm{sgn}(\omega)\mathscr{F}[b\cos(\omega_0 t)]\} = b\sin(\omega_0 t), \tag{3.33}$$

$$b\cos(\omega_0 t) + \mathrm{i}\mathscr{H}[b\cos(\omega_0 t)] = b\exp(\mathrm{i}\omega_0 t). \tag{3.34}$$

The *spiral phase transform* is one of the possible extensions of the Hilbert transform to two dimensions [9]. The quadrature of a fringe pattern can be obtained by inverting its spectral components symmetrically to the origin. A complex sign function is defined for the angular spatial frequency plane as

$$\mathrm{sgn}\left(k_x + \mathrm{i}k_y\right) = \frac{k_x + \mathrm{i}k_y}{\left|k_x + \mathrm{i}k_y\right|} = \exp\left[\mathrm{i}\arg\left(k_x + \mathrm{i}k_y\right)\right] = \exp\left[\mathrm{i}\phi_S\left(k_x, k_y\right)\right], \tag{3.35}$$

where $\phi_S(k_x, k_y)$ is a spiral phase function around the origin. Although not defined at zero, the value $\mathrm{sgn}(0 + 0\mathrm{i}) = 0$ is conventionally assigned. This sign function has odd symmetry, unit modulus, and no preferential direction in the spatial frequency space, an advantage over other choices.

The spiral phase transform of a two-dimensional function $f(x, y)$ is defined in analogy to the Hilbert transform

$$\$[f(x, y)] = \mathcal{F}^{-1}\left\{\mathrm{sgn}\left(k_x + \mathrm{i}k_y\right)\mathcal{F}[f(x, y)]\right\}. \tag{3.36}$$

As a quadrature operator, it yields accurate results only for sinusoidal signals. Therefore, the fringe pattern's background component $a(xy)$ must be previously removed by high-pass filtering or subtracting a reference background image or a fringe pattern shifted by π, etc:

$$\tilde{I}(x, y) = I(x, y) - a(x, y) = b(x, y)\cos[\varphi(x, y)]. \tag{3.37}$$

It is easy to prove that for a monochromatic fringe pattern, i.e. $\varphi(x, y) = k_{cx}x + k_{cy}y$,

$$\$[\tilde{I}(x, y)] = \mathrm{i}\exp\left[\mathrm{i}\phi_S\left(k_{cx}, k_{cy}\right)\right]b(x, y)\sin[\varphi(x, y)], \tag{3.38}$$

and the spiral phase transform returns the quadrature of the fringe pattern multiplied by an exponential phase term according to the direction of the fringes: $\arctan(k_{cy}/k_{cx}) = \phi_S(k_{cx}, k_{cy})$.

Let us now look at the *vortex transform* of a fringe pattern, defined as

$$V[\tilde{I}(x, y)] = \exp[-\mathrm{i}\beta(x, y)]\$[\tilde{I}(x, y)], \tag{3.39}$$

with $\beta(x, y)$ the local direction angle of the fringes. The result of this transform is ideally imaginary, but noise and errors may result in a real residual part that, once removed, leaves

$$\begin{aligned}\mathrm{Im}\left\{V[\tilde{I}(x, y)]\right\} = b(x, y)\sin[\varphi(x, y)] \quad &\Rightarrow \quad \tilde{I}_a(x, y)\\ = \tilde{I}(x, y) + \mathrm{i}V[\tilde{I}(x, y)],&\end{aligned} \tag{3.40}$$

where $\tilde{I}_a(x, y) = b(x, y)\exp[\mathrm{i}\varphi(x, y)]$ is the analytic part of the background-free fringe pattern in equation (3.37), and the phase can be retrieved as

$$\varphi(x, y) = \arg[\tilde{I}_a(x, y)] = \arg\left\{\tilde{I}(x, y) + \mathrm{i}V[\tilde{I}(x, y)]\right\}. \tag{3.41}$$

Although the quadrature is exact only for spatially monochromatic fringe patterns, with a single fringe orientation a good approximation is obtained if the fringe orientation changes smoothly in space, i.e. if the curvature of the fringes is small. The method is not reliable in the stationary phase points: the centers of the fringes and the saddle points.

A key point in applying the spiral phase transform method is the estimation of the local direction angle $\beta(x, y)$. The *direction* of the fringes is the direction of the unit vector normal to the isophase lines

$$\hat{\boldsymbol{n}}_\varphi = \frac{\nabla\varphi}{|\nabla\varphi|} = \frac{(k_x, k_y)}{|(k_x, k_y)|} = (\cos\beta, \sin\beta) \quad \Rightarrow \quad \beta = \arctan\frac{k_y}{k_x}, \quad -\pi < \beta \leqslant \pi, \tag{3.42}$$

but to find $\hat{n}_\varphi$ the phase of the fringes should be known. The *orientation* of the fringes is just the orientation of the line where that vector lies, which is also the orientation of the unit vector

$$\hat{\boldsymbol{n}}_I = \frac{\nabla I}{|\nabla I|} = \frac{(\partial_x I, \partial_y I)}{|(\partial_x I, \partial_y I)|} = (\cos\theta, \sin\theta) \quad \Rightarrow \quad \theta = \arctan\frac{\partial_y I}{\partial_x I}, \\ -\frac{\pi}{2} < \theta \leqslant \frac{\pi}{2}, \tag{3.43}$$

with $\partial_x I$ and $\partial_y I$ the partial derivatives of I by respect to x and y, respectively. The orientation can, therefore, be obtained from the fringe pattern by using techniques such as, for example, the numerical estimation of the gradient in equation (3.43); the minimization of the directional derivative; the local fit of a plane to the fringe pattern; or the application of energy operators [10].

The orientation angle θ is obtained modulo π and must be unwrapped to get the direction angle β modulo 2π. This is achieved by taking into account that

$$\theta = \beta + k\pi, \quad k \in \{0,1\} \quad \Rightarrow \quad 2\theta = 2\beta + 2k\pi \quad \Rightarrow \quad \exp[\mathrm{i}(2\theta)] = \exp[\mathrm{i}(2\beta)]. \tag{3.44}$$

The angle 2θ is wrapped modulo 2π and can be unwrapped to obtain $(2\theta)_{uw} = (2\beta)_{uw}$ and eventually $\beta_{uw} = (2\theta)_{\mathrm{uw}}/2$ from which the term $\exp[-\mathrm{i}\beta(x, y)] = \exp[-\mathrm{i}\beta_{uw}(x, y)]$ can be calculated.

3.2.1.2.2 Regularized phase tracking

The regularized phase tracking is a sequential optimization method where a cost function is locally minimized to obtain the value of the phase that brings the maximum similarity with the original fringe pattern. The value of the phase that best matches a fringe pattern is not unique. Additional constraints based on *a priori* assumptions are incorporated into the cost function to ensure that the optimization problem converges to a stable solution. This is known as *regularization.* It is assumed that the fringe pattern background $a(x, y)$, modulation amplitude $b(x, y)$, and phase $\varphi(x, y)$ are spatially smooth. The smoothness of the phase is the basic regularizing constraint, but others may be added in more elaborated versions of regularized phase tracking.

The fringe pattern is locally regarded as spatially monochromatic. In a sufficiently small neighborhood L of each point (typically from 5×5 to 15×15 pixels), the spatial frequency $\boldsymbol{k}(x, y) = (k_x, k_y)$ is assumed unique, and the phase can be locally approximated by a plane

$$p(x, y, u, v) = \varphi(x, y) + k_x(x - u) + k_y(y - v). \tag{3.45}$$

A method to reduce the number of parameters to estimate and simplify the minimization problem requires that the fringe pattern should be *normalized* [11] by removing the background $a(x, y)$ and setting its amplitude to $b(x, y) = 1$, resulting in

$$I'(x, y) = \cos\varphi(x, y). \tag{3.46}$$

The most straightforward cost function for regularized phase tracking in a neighborhood L of (x, y) is [12]

$$U_{x,y}(\varphi, k_x, k_y) = \sum_{(u,v)\in L}\left(\underbrace{\{\cos[p(x, y, u, v)] - I'(u, v)\}^2}_{\text{(i)}} + \underbrace{\lambda[p(x, y, u, v) - \varphi(u, v)]^2 m(u, v)}_{\text{(ii)}}\right) \quad (3.47)$$

with

$$m(u, v) = \begin{cases} 1 \text{ if } \varphi(u, v) & \text{has already been estimated, or} \\ 0 \text{ otherwise, i.e. if yet there is no valid value for } \varphi(u, v) \end{cases}. \quad (3.48)$$

The term denoted by (i) ensures that the local plane approximation will match the fringe pattern in a least-squares sense. The term denoted by (ii) introduces the constraint that the estimated phase must be locally smooth. The value of λ, the *regularization parameter*, controls how strong this constraint is. Therefore, there is a balance between phase fidelity and phase smoothness in the result (typical values of λ range from 0.1 to 10).

The cost function is numerically minimized (e.g. with a gradient descent algorithm) with respect to the three parameters $\varphi(x, y)$, $k_x(x, y)$, and $k_y(x, y)$. The minimization is initialized with the values obtained for φ, k_x, and k_y in an already processed adjacent location (except for the first location, which is initialized to zero); this favors the algorithm's convergence and ensures that the retrieved phase will be unwrapped. In order to make the minimization fast and robust, the fringe pattern is sequentially demodulated by following the expected isophase lines, i.e. moving from one location to the next one along the local direction of the fringes.

Advanced formulations of the regularized phase tracker are obtained by adding extra constraints. For example, in [13]

$$U_{x,y}(\varphi, k_x, k_y) = \sum_{(u,v)\in L}\left(\{\cos[p(x, y, u, v)] - I'(u, v)\}^2 + \lambda[p(x, y, u, v) - \varphi(u, v)]^2 m(u, v) + \underbrace{\{k_x \sin[p(x, y, u, v) + \partial_x I'(u, v)]\}^2 + \{k_y \sin[p(x, y, u, v) + \partial_y I'(u, v)]\}^2}_{\text{(iii)}}\right). \quad (3.49)$$

The terms in (iii) introduce constraints on the consistency of the quadrature of the solution with the partial derivatives $\partial_x I'(u, v)$ and $\partial_y I'(u, v)$ of the original fringes with respect to x and y. This is intended to increase the method's reliability with low-frequency fringes, which do not allow a suitable normalization.

Other variants of the method incorporate the modulation amplitude of the fringes $b(x, y)$ and its derivatives, and even the background $a(x, y)$ at its derivatives, as minimization parameters to avoid the normalization of the fringe pattern [14]; or use quadratic modeling of the local phase to improve the behavior of the method with complicated fringe patterns [15].

3.2.2 Self-calibrating non-linear phase-shifting methods

Phase-shifting fringe demodulation methods work on a sequence of N fringe patterns,

$$I_n(x, y) = a(x, y) + b(x, y)\cos[\varphi(x, y) + \delta_n], \ n \in \{1,2, \dots, N\}, \tag{3.50}$$

where the background $a(x,y)$, the amplitude $b(x,y)$, and the phase to be retrieved $\varphi(x,y)$ can be assumed constant, and δ_n are the phase shifts that have the same value for every location on the fringe pattern.

Traditional phase-shifting algorithms rely on three or more fringe patterns with known and well-calibrated phase shifts. Modern phase-shifting methods usually relax the calibration requirements on the phase shifts, ultimately allowing the retrieval of the phase by using a set of unknown and not evenly spaced phase steps. These last are known as *self-calibrating* algorithms. Some examples of methods in this category are outlined in what follows.

3.2.2.1 Non-linear phase shifting

The use of non-evenly spaced phase steps is often called *non-linear* phase shifting. A general algorithm to retrieve the phase from a sequence of fringe patterns with non-linear and *known* phase shifts using least-squares fitting was introduced by Greivenkamp in 1984 [16]. It is worth summarizing this classical algorithm because it is the basis for the advanced iterative algorithm. It can also be used to find the phase once a set of unknown phase-shifts has been estimated using other methods.

Each fringe pattern in the sequence given by equation (3.50) is sampled as a set of K local values

$$I_{k,n} = a_k + b_k\cos(\varphi_k + \delta_n) = I_n(x_k, y_k), \quad k \in \{1,2, \dots, K\}, \tag{3.51}$$

where (x_k, y_k) refers to each location where the fringe pattern is sampled, typically (but not necessarily) from a rectangular grid. This local intensity may be rewritten as

$$I_{k,n} = a_k + \underbrace{b_k \cos \varphi_k}_{C_k} \cos \delta_n \underbrace{-b_k \sin \varphi_k}_{S_k} \sin \delta_n = a_k + C_k \cos \delta_n + S_k \sin \delta_n. \tag{3.52}$$

Greivenkamp's method obtains the unknown values of a_k, C_k, and S_k by minimizing, at each location k, the quadratic cost function

$$U_k(a_k, C_k, S_k) = \sum_{n=1}^{N}[a_k + C_k\cos\delta_n + S_k\sin\delta_n - I_{k,n}]^2, \tag{3.53}$$

with known values of δ_n. The resulting least-squares estimation is given by

$$\begin{pmatrix} a_k \\ C_k \\ S_k \end{pmatrix} = \begin{pmatrix} N & \sum_n \cos\delta_n & \sum_n \sin\delta_n \\ \sum_n \cos\delta_n & \sum_n \cos^2\delta_n & \sum_n \sin\delta_n \cos\delta_n \\ \sum_n \sin\delta_n & \sum_n \sin\delta_n \cos\delta_n & \sum_n \sin^2\delta_n \end{pmatrix}^{-1} \cdot \begin{pmatrix} \sum_n I_{k,n} \\ \sum_n I_{k,n} \cos\delta_n \\ \sum_n I_{k,n} \sin\delta_n \end{pmatrix} \tag{3.54}$$

and the phase is retrieved from the analytical part of the fringe pattern, $I_{ak} = C_k - \mathrm{i}S_k$, as

$$\varphi_k = \arg I_{ak} = \arg(C_k - \mathrm{i}S_k) = \arctan\left(\frac{-S_k}{C_k}\right). \tag{3.55}$$

3.2.2.2 The advanced iterative algorithm

The advanced iterative algorithm (AIA) extends the least-squares fitting method to the case where the phase shifts are unknown [17]. The problem of finding a set of N phase shifts δ_n compatible with the observed fringe patterns is formulated by writing their local intensities as

$$\begin{aligned} I_{k,n} &= a_n' + \underbrace{b_n' \cos\delta_n}_{C'_n} \cos\varphi_k \underbrace{-b_k' \sin\delta_n}_{S_n'} \sin\varphi_k \\ &= a_n' + C_n' \cos\varphi_k + S_n' \sin\varphi_k, \end{aligned} \tag{3.56}$$

where φ_k are known, and $a(x, y)$ and $b(x, y)$ should be spatially uniform enough to be represented by single values a_n' and b_n' for each fringe pattern in the sequence. The cost function,

$$U_n(a_n', C_n', S_n') = \sum_{k=1}^{K} [a_n' + C_n'\cos\varphi_k + S_n'\sin\varphi_k - I_{k,n}]^2, \tag{3.57}$$

is minimized for each phase step, and the corresponding least-squares estimation is given by

$$\begin{pmatrix} a'_n \\ C'_n \\ S'_n \end{pmatrix} = \begin{pmatrix} K & \sum_k \cos\varphi_k & \sum_k \sin\varphi_k \\ \sum_k \cos\varphi_k & \sum_k \cos^2\varphi_k & \sum_k \sin\varphi_k \cos\varphi_k \\ \sum_k \sin\varphi_k & \sum_k \sin\varphi_k \cos\varphi_k & \sum_k \sin^2\varphi_k \end{pmatrix}^{-1} \cdot \begin{pmatrix} \sum_k I_{k,n} \\ \sum_k I_{k,n}\cos\varphi_k \\ \sum_k I_{k,n}\sin\varphi_k \end{pmatrix}. \tag{3.58}$$

The phase steps are then estimated as

$$\delta_n = \arg(C_n' - \mathrm{i}S_n') = \arctan\left(\frac{-S_n'}{C_n'}\right). \tag{3.59}$$

The AIA starts by making a rough initial guess of the phase steps $\delta_n^{(0)}$ (the superscript, $t = 0$, indicates the number of the iteration cycle). Then, the following two steps:

1. The latest available estimates of the phase steps $\delta_n^{(t-1)}$ are used with equations (3.54) and (3.55) to obtain estimates of the phase $\varphi_k^{(t-1)}$, where t is the number of the iteration.
2. The current estimates of the phase $\varphi_k^{(t-1)}$ are used with equations (3.58) and (3.59) to obtain new estimates of the phase steps $\delta_n^{(t)}$,

are iterated until the change of the estimates of the relative phase steps is lower than a given tolerance ϵ:

$$| (\delta_n^{(t)} - \delta_1^{(t)}) - (\delta_n^{(t-1)} - \delta_1^{(t-1)}) | < \epsilon, \ \forall\, n \in \left\{1,2, \ldots , N\right\}. \tag{3.60}$$

The phase is eventually retrieved by using equations (3.54) and (3.55) with the last estimates of the phase steps δ_n.

3.2.2.3 The principal component analysis method

The principal component analysis method for phase retrieval is a non-iterative algorithm based on vector subspaces. The non-iterative character makes this algorithm more computationally efficient than the advanced iterative algorithm. Under the assumption that the unknown phase steps δ_n, taken modulo 2π, have a statistically uniform distribution in the interval $[0,2\pi)$, the background component a_k can be estimated by averaging N fringe patterns in the sampled sequence given in equation (3.51)

$$a_k \approx a_k^* = \frac{1}{N}\sum_{n=1}^{N} I_{k,\,n} \tag{3.61}$$

and a background-free fringe pattern sequence is obtained by subtracting a_k^*

$$\begin{aligned} \tilde{I}_{k,\,n} = I_{k,\,n} - a_k^* &\approx \underbrace{b_k \cos\varphi_k}_{C_k} \cos\delta_n \underbrace{-b_k \sin\varphi_k}_{S_k} \sin\delta_n \\ &= C_k \cos\delta_n + S_k \sin\delta_n. \end{aligned} \tag{3.62}$$

This expression for the fringe pattern can be written as a vector equation in $\mathbb{R}^K$

$$\boldsymbol{d}_n = \cos\delta_n\, \boldsymbol{c} + \sin\delta_n \boldsymbol{s}; \text{ with } \begin{cases} \boldsymbol{d}_n = (\tilde{I}_{1,\,n}, \ldots, \tilde{I}_{k,\,n}, \ldots, \tilde{I}_{K,\,n})^T \\ \boldsymbol{c} = (C_1, \ldots, C_k, \ldots, C_K)^T \\ \boldsymbol{s} = (S_1, \ldots, S_k, \ldots, S_K)^T \end{cases}. \tag{3.63}$$

With the further assumption that the values of the phase φ_k are well distributed in $[0, 2\pi)$, i.e. that there is more than one fringe across the fringe patterns,

$$\boldsymbol{c} \cdot \boldsymbol{s} = \sum_{k=1}^{K} C_k S_k = \sum_{k=1}^{K} b_k^2 \cos\delta_k \sin\delta_k \approx 0. \tag{3.64}$$

The vectors $\boldsymbol{c}$ and $\boldsymbol{s}$ are orthogonal and constitute a basis of a two-dimensional subspace of $\mathbb{R}^K$ where all the vectors $\boldsymbol{d}_n$ representing the fringe patterns lie. Since only two of these vectors may be linearly independent, an $N \times K$ matrix $\mathbf{D}$ with rank two is built as

$$\mathbf{D} = \begin{pmatrix} \tilde{I}_{1,1} & \cdots & \tilde{I}_{k,1} & \cdots & \tilde{I}_{K,1} \\ \vdots & \ddots & \vdots & & \vdots \\ \tilde{I}_{1,n} & \cdots & \tilde{I}_{k,n} & \cdots & \tilde{I}_{K,n} \\ \vdots & & \vdots & \ddots & \vdots \\ \tilde{I}_{1,N} & \cdots & \tilde{I}_{k,N} & \cdots & \tilde{I}_{K,N} \end{pmatrix} = \begin{pmatrix} \cos\delta_1 & \sin\delta_1 \\ \vdots & \vdots \\ \cos\delta_n & \sin\delta_n \\ \vdots & \vdots \\ \cos\delta_N & \sin\delta_N \end{pmatrix} \begin{pmatrix} C_1 & \cdots & C_k & \cdots & C_K \\ S_1 & \cdots & S_k & \cdots & S_K \end{pmatrix} \tag{3.65}$$
$$= \hat{\mathbf{U}}\hat{\mathbf{Y}}.$$

The principal component analysis is a technique to reduce the number of data sets to a smaller number of uncorrelated components and can be applied to extract $\boldsymbol{c}$ and $\boldsymbol{s}$ from $\mathbf{D}$ [18]. The $N \times N$ covariance matrix of $\mathbf{D}^T$ is diagonalized, typically by singular value decomposition

$$\mathbf{C} = \mathbf{D}\mathbf{D}^T = \mathbf{U}\mathbf{S}^2\mathbf{U}^T, \tag{3.66}$$

the resulting $N \times N$ transformation matrix $\mathbf{U}$ is orthonormal, $\mathbf{U}\mathbf{U}^T = \mathbf{I}$, and $\mathbf{S}^2$ is a diagonal matrix with the ordered eigenvalues of $\mathbf{C}$. Since $\mathbf{D}$ is of rank two, only the two higher eigenvalues will be significant and correspond to the eigenvectors in the first two columns of $\mathbf{U}$. These two columns are an estimate of $\widehat{\mathbf{U}}$ from which the phase steps δ_n may be calculated. The principal components of $\mathbf{D}$ are obtained by applying the Hotelling transformation

$$\mathbf{Y} = \mathbf{U}^T\mathbf{D} \tag{3.67}$$

, where the first two rows of $\mathbf{Y}$ are an estimate of $\widehat{\mathbf{Y}}$ and, therefore, of C_k and S_k from which the analytical part of the fringe pattern $I_{ak} = C_k - \mathrm{i}S_k$ and the phase φ_k are determined as in equation (3.55).

The order of the first two columns of $\mathbf{U}$ and, consequently, of the first two rows of $\mathbf{Y}$ depend on the relative values of the corresponding eigenvalues. The assignment of the cosine C_k and sine S_k components is arbitrary, and this may induce a global shift in the retrieved phase. On the other hand, since $C_k\cos\delta_n = (-C_k)(-\cos\delta_n)$ and $S_k\sin\delta_n = (-S_k)(-\sin\delta_n)$, the assignment of signs to the sine and cosine components according to equation (3.62) is also arbitrary. We should note that there is a sign ambiguity for each of the components, $\pm C_k$ and $\pm S_k$ that may induce an additional global shift and results in a sign ambiguity in the retrieved phase. Both ambiguities may be solved if some prior knowledge about the phase steps δ_n is available, e.g. whether they are increasing or decreasing, the limits of their relative increment, etc.

An alternative formulation referred to in the literature as the *quadrature component analysis* consists in diagonalizing the $K \times K$ covariance matrix of $\mathbf{D}$ [19]

$$\mathbf{C} = \mathbf{D}^T\mathbf{D} = \mathbf{V}\mathbf{S}^2\mathbf{V}^T. \tag{3.68}$$

The first two rows of $\mathbf{SV}^T$ are an estimate of $\widehat{\mathbf{Y}}$ from which the phase φ_k may be calculated. Alternatively, if the difference between the two larger eigenvalues in $\mathbf{S}^2$ is negligible, the phase may be retrieved from the first two rows in $\mathbf{V}^T$.

Finally, the principal/quadrature component analysis method may also be applied through the direct singular value decomposition of the data matrix $\mathbf{D}$ [20]

$$\mathbf{D} = \mathbf{USV}^T, \tag{3.69}$$

where, as in the former formulations, the phase φ_k can be retrieved from the first two rows of $\mathbf{SV}^T$ (or $\mathbf{V}^T$) and the phase steps δ_n can be estimated from the two first columns of $\mathbf{U}$.

3.3 Interference phase demodulation

Any numerical determination of interference phase distributions from interference patterns involves the arctan function, which only gives valid results between $-\pi$ and $+\pi$, see figure 3.1. Although many physical processes generate phase distributions stretching continuously over several 2π-intervals, the interferometric measurement in principle gives a sawtooth-like phase distribution, as shown in figure 3.2(a). This distribution must be expanded to a continuous one by adding or subtracting integer multiples of 2π, see figures 3.2(b) and (c), a process called phase demodulation. To achieve this, there are several different approaches, which can roughly be

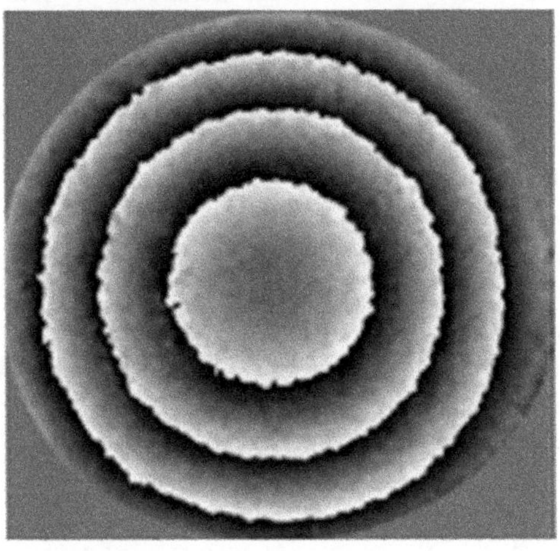

Figure 3.1. Interference pattern (wrapped phase map) modulo 2π.

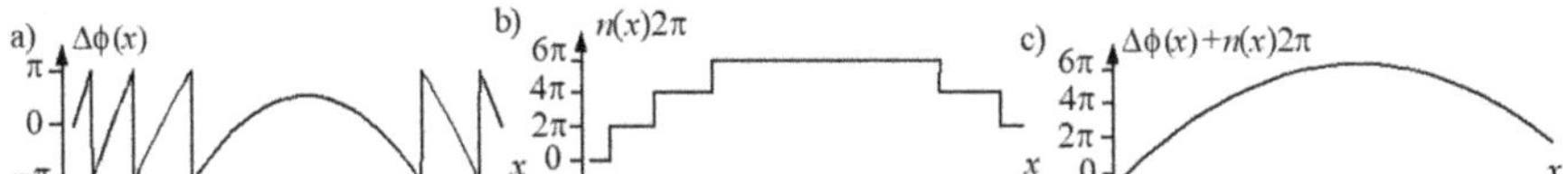

Figure 3.2. Interference phase demodulation: (a) interference phase modulo 2π, (b) step-function to be added, and (c) demodulated interference phase distribution.

categorized into path-dependent and path-independent. The interference phase difference between adjacent pixels in a fixed direction is checked in path-dependent demodulation. If this difference is greater than π or less than $-\pi$, 2π is subtracted or added from this pixel along a fixed path of pixels. Thus erroneous demodulation may spread over the rest of the path. This disadvantage is avoided by the path-independent procedures, where the direction of further demodulation at each pixel depends on the local phase differences in all directions [21].

References

[1] Carre P 1966 Installation et utilisation du comparateur photoelectrique et interferentiel du Bureau International des Poids et Mesures *Metrologia* **2** 13–23

[2] Kreis T 2005 *Handbook of Holographic Interferometry* (Weinheim: Wiley)

[3] Dorrío B V and Fernández J L 1999 Phase-evaluation methods in whole-field optical measurement techniques *Meas. Sci. Technol.* **10** R33–55

[4] Kemao Q 2007 Two-dimensional windowed Fourier transform for fringe pattern analysis: principles, applications and implementations *Opt. Lasers Eng.* **45** 304–17

[5] Watkins L R 2012 Review of fringe pattern phase recovery using the 1-D and 2-D continuous wavelet transforms *Opt. Lasers Eng.* **50** 1015–22

[6] Mallat S 2009 *A Wavelet Tour of Signal Processing* 3rd edn (Boston, MA: Academic)

[7] Antoine J-P, Murenzi R, Vandergheynst P and Twareque Ali S 2004 *Two-Dimensional Wavelets and Their Relatives* (Cambridge: Cambridge University Press)

[8] Li S, Su X and Chen W 2010 Wavelet ridge techniques in optical fringe pattern analysis *J. Opt. Soc. Am.* A **27** 1245–54

[9] Larkin K G, Bone D J and Oldfield M A 2001 Natural demodulation of two-dimensional fringe patterns. I. general background of the spiral phase quadrature transform *J. Opt. Soc. Am.* A **18** 1862–70

[10] Larkin K G 2005 Uniform estimation of orientation using local and nonlocal 2-D energy operators *Opt. Express* **13** 8097–121

[11] Quiroga J A and Servin M 2003 Isotropic *n*-dimensional fringe pattern normalization *Opt. Commun.* **224** 221–7

[12] Servin M, Marroquin J L and Cuevas F J 2001 Fringe-follower regularized phase tracker for demodulation of closed-fringe interferograms *J. Opt. Soc. Am.* A **18** 689–95

[13] Servin M, Marroquin J L and Quiroga J A 2004 Regularized quadrature and phase tracking from a single closed-fringe interferogram *J. Opt. Soc. Am.* A **21** 411–9

[14] Kai L and Kemao Q 2012 A generalized regularized phase tracker for demodulation of a single fringe pattern *Opt. Express* **20** 12579–92

[15] Tian C, Yang Y, Liu D, Luo Y and Zhuo Y 2010 Demodulation of a single complex fringe interferogram with a path-independent regularized phase-tracking technique *Appl. Opt.* **49** 170–9

[16] Greivenkamp J E 1984 Generalized data reduction for heterodyne interferometry *Opt. Eng.* **23** 350–2

[17] Wang Z and Han B 2007 Advanced iterative algorithm for randomly phase-shifted interferograms with intra- and inter-frame intensity variations *Opt. Lasers Eng.* **45** 274–80

[18] Vargas J, Quiroga J A and Belenguer T 2011 Analysis of the principal component algorithm in phase-shifting interferometry *Opt. Lett.* **36** 2215–7

[19] Vargas J and Sorzano C O S 2013 Quadrature component analysis for interferometry *Opt. Lasers Eng.* **51** 637–41

[20] Yatabe K, Ishikawa K and Oikawa Y 2016 Improving principal component analysis based phase extraction method for phase-shifting interferometry by integrating spatial information *Opt. Express* **24** 22881–91

[21] Kreis T 1993 Computer aided evaluation of fringe patterns *Opt. Lasers Eng.* **19** 221–40

Fernando Mendoza-Santoyo, Manuel De la Torre-Ibarra, María del Socorro Hernández-Montes and Jorge Mauricio Flores Moreno

Chapter 4

Laser speckle metrology

Contributing author: Thomas Kreis

When coherent light sources, e.g. a laser, illuminate optically rough surfaces the scattered light shows a granular pattern formed by bright and dark specks known as speckles, an observed granulation similar to that seen as electronic noise on old analog televisions. The speckles result from self-interference scattered light rays and are most visible when an observer's eyes move with respect to the static surface that produces the speckles. A speckle is a cumbersome web-like structure created by the scattering of coherent light by a surface that *looks rough* to the wavelength of the coherent source—it is created by the interference from microscopic surface points scattering random relative optical phases. The pattern is known as a specklegram, speckle pattern, or when viewed with a digital device, is also called a digital speckle pattern. Indeed, changes in a speckle pattern can be registered on a digital camera, a device that proves to be very useful in several experimental techniques known to date. Usually, the speckles are undesirable noise since their bright and dark structure reduces the visibility and contrast in images. However, since the speckle pattern is modified in phase and intensity when it interacts with the sample [1, 2], it contains helpful information from the microscopic surface roughness of an object. The speckle's intricate structure depends on the wavelength of the light producing it, and its size also depends on the size and shape of the aperture employed when using an imaging system to observe them. This chapter will describe topics related to the speckle phenomena and techniques that have been performed using interferometric and non-interferometric optical metrology.

4.1 Basics for laser speckle metrology

In compliance with Gaussian statistics, the speckle pattern is considered a statistical process composed of aleatory intensity and phase distributions. The speckles in space are unevenly shaped but are spatially determined and temporally constant. The complex amplitude of the speckle field $U(x, y, z)$ can be characterized at a

doi:10.1088/978-0-7503-3027-5ch4

Figure 4.1. Characteristic speckle pattern captured with a digital camera. It is a 14-bit gray-level image with an image dimension of 1392 × 1040 pixels.

certain point of observation from the summation of many fundamental scattering surface points (N), representing the light from all points of the scattering surface described by

$$U(x, y, z) = \frac{1}{\sqrt{N}} \sum\nolimits_{n=1}^{N} a_{on} e^{i\varnothing_n}, \tag{4.1}$$

where a_{on} and $\varnothing_n$ are the amplitude and phase (distributed from $-\pi$ to π) of the Nth scatter-complex wave; they are statistically independent. Figure 4.1 shows a representative experimental speckle pattern from the light scattered by an object's surface.

The intensity $I(x, y)$ is the measurable quantity. In practice, only $I \sim | U |^2$ is considered, and I contain quantities of random variations; consequently, the speckle is a statistical phenomenon. The intensity probability in the objective speckle pattern (so-called because it is observed without an imaging lens, i.e. the scattered light falls unaltered on a photographic film or the digital camera sensor) can be calculated by $(I)\mathrm{d}I = \frac{e^{-I/<I>}}{<I>}\mathrm{d}I$, where (I) is the mean speckle intensity in the dispersed field, i.e. the intensity with the highest probability is equal to 0, so we obtain dark speckles. In contrast, the lowest probability with an intensity equal to 1 gives bright specks [3, 4].

The speckle is frequently characterized by its contrast, and the formalism rests on whether the speckle is static or dynamic [5, 6]. The first condition occurs when the object is in a static state, and the pattern does not change over time, while the second happens if the object is moving and the pattern changes with time.

4.1.1 Speckle contrast characterization

The speckles' intensity fluctuation determines its contrast (C), a quantity given by the ratio of the intensity standard deviation (σ) to the speckle pattern mean intensity (both calculated over an area spatially localized in the pattern)

$$C \equiv \frac{\sigma}{\langle I \rangle} = \frac{\sqrt{\langle I^2 \rangle - \langle I \rangle^2}}{\langle I \rangle}. \tag{4.2}$$

The speckle contrast depends on the temporal and spatial coherence of the light used (coherence was dealt with in chapter 2) and the object's surface roughness. The contrast is used to assess the blurring level on the object when movement is present, the fluctuation in the intensity produced by the speckles, and the object's roughness.

In the case of Gaussian speckle patterns, the local contrast presents a distribution of values between 0 and 1. When $\sigma = \langle I \rangle$ and $C = 1$, this means that the phase is uniformly distributed from 0 to 2π, and the signal-to-noise ratio (SNR) is 1. If no speckle fluctuation occurs, the pattern has constant intensity, and the value of C is 0. When the coherence of the light source decreases, the contrast decreases as well. To minimize the error and obtain confident statistics in the contrast quantification, the speckle and pixel size, say of the digital sensor, must be the same; for instance, this is achieved by controlling the size of the imaging lens aperture.

4.1.2 Subjective and objective speckle patterns

In broad terms, speckle patterns can be classified into two types: *subjective* and *objective speckle*.

When an imaging system (it can be a single lens) is employed to convey the object's image surface, illuminated by coherent light, onto the camera sensor, the object appears speckled, a pattern known as *subjective speckle* ($\mathcal{S}_s$), see figure 4.2. The speckle size and contrast on the sensor are limited by the diffraction capabilities of the imaging lens aperture. The transverse size of the speckle ($\mathcal{S}_{st}$) for a paraxial approximation (a term used to mean light rays traveling very close to the lens optical axis) is given by

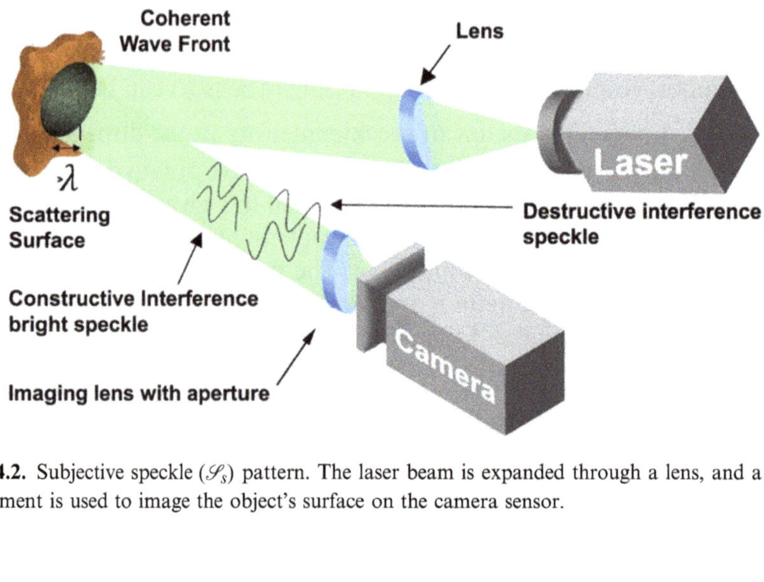

Figure 4.2. Subjective speckle ($\mathcal{S}_s$) pattern. The laser beam is expanded through a lens, and a lens–aperture arrangement is used to image the object's surface on the camera sensor.

$$\mathscr{S}_{st} \approx (1.22\lambda f\#)(\mathscr{M} + 1) \text{ or}$$
$$\mathscr{S}_{st} \approx 0.611\frac{\lambda}{\text{NA}}, \tag{4.3}$$

and the speckle length ($\mathscr{S}_{sl}$) is given by

$$\mathscr{S}_{sl} \approx (8\lambda f^2\#)(\mathscr{M} + 1)^2. \tag{4.4}$$

These equations depend on the following: the illumination wavelength (λ); the f-number ($f\#$) defined as f/D_A, where f is the lens focal length and D_A is the lens aperture diameter; (NA) is the lens numerical aperture defined as $D_A/2f$, which is related to the angle that the lens subtends from the object's surface of the observing system; and $\mathscr{M}$ is the magnification of the image system that can be given by L_i/L_o, where L_i and L_o are the distances between the lens and the sensor plane and the object and the lens, respectively.

The diameter of the speckle ($\mathscr{S}_{sd}$) corresponds to the diameter of the point spread function for a circular aperture written as

$$\mathscr{S}_{sd} = (2.44\lambda f\#)(\mathscr{M} + 1). \tag{4.5}$$

By adjusting D_A and controlling the $f\#$, the most favorable speckle size can be found, i.e. a size that can be resolved by the sensor's pixel size. In other words, by adjusting the appropriate aperture diameter D_A, the speckle size is easily adjusted to the size of the pixel of the recording camera. By fitting one speckle on one pixel, the contrast of the speckle pattern can be increased. For the case of $\mathscr{S}_s$, the speckle displacement is provoked by object movements which, due to the nature of the image forming characteristics of a lens, are opposite to each other, i.e. the object moves up, speckles move down; the effect also depends on $\mathscr{M}$.

If the speckles are captured on a photographic film or a digital sensor without using a lens, the pattern is referred to as *objective speckle* ($\mathscr{S}_o$). The backscattered light forms this objective speckle from the non-image object (scattering surface), see figure 4.3. The mean speckle size is given by the ratio of the object to camera sensor distance (z) times the illumination wavelength (λ) to the size of the light-up area on the object (A):

$$\mathscr{S}_{os} = \lambda z/A. \tag{4.6}$$

If the light-up area is a circular spot of diameter (D), the diameter for the $\mathscr{S}_o$ is given by

$$\mathscr{S}_{od} \approx 1.22\lambda z/D. \tag{4.7}$$

The speckle size is changed by either z or λ, or by D or A, or a combination thereof (as needed for the film or sensor to be able to resolve the speckle). In the case of $\mathscr{S}_o$, the displacement of the speckle is similar to the object's direction of movement [7, 8].

Hence, the range of measurable scattering surface roughness (SR) rests on the wavelength of the laser source used, and the illumination angle (α_1) defined as $\text{SR} = \lambda/\alpha_1$. The rough scattering surfaces depend on the variance in the surface height distribution, which should be of the order of the wavelength of the light used, or greater ($> \lambda$), and the intensity variations and the contrast of the speckle pattern depend on it.

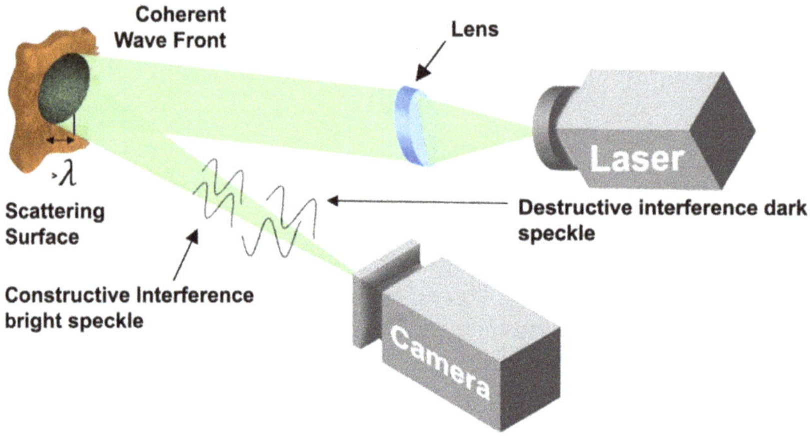

Figure 4.3. Objective speckle patterns $\mathscr{S}_0$, no lens is used to capture the scattered object light. The beam from the laser is expanded using a lens.

4.2 Laser speckle measurements

As mentioned, the speckle pattern is produced by the interference of light scattered from an object's surface: the interference or wave overlapping process (as seen in chapter 1) is the basis of several important measurement techniques and methods. Speckle measuring techniques rely on speckle patterns as they contain valuable data about the test object.

The use of laser speckles in conjunction with phase shift methods and the development of algorithms for data acquisition and handling have meant the design of a wide variety of optical systems to measure object displacement/strain, vibration, contouring, and deformation. A diverse set of physical and mechanical parameters can be derived from this information. Laser speckle techniques and methods such as speckle photography [9] or electronic speckle pattern interferometry (ESPI) have been developed over time in static and dynamic measurements. Alternative names for them are digital speckle pattern interferometry (DSPI) or electro-optic holography (EOH) [1, 10, 11]. Speckle pattern correlation has been performed using interferometric and non-interferometric optical metrology. The principle, theory, and optical configurations are presented next for each method.

4.2.1 Double-exposure speckle pattern photography

Perhaps the initial technique for metrology purposes was speckle pattern photography (SPP), which uses the subjective speckle to perform measurements along the x- and y-directions, i.e. in-plane object movements, such as translation, rotation, strain, and vibrations. The speckle pattern sensitivity to these movements is affected by the speckle size constrained by the photographic film resolution and the imaging system [12–14].

In the beginning, the object speckles were collected and conveyed by a lens–aperture combination to a high-resolution photographic film where the pattern was recorded, using the set-up seen in figure 4.4. The SPP requires taking two snap-shots

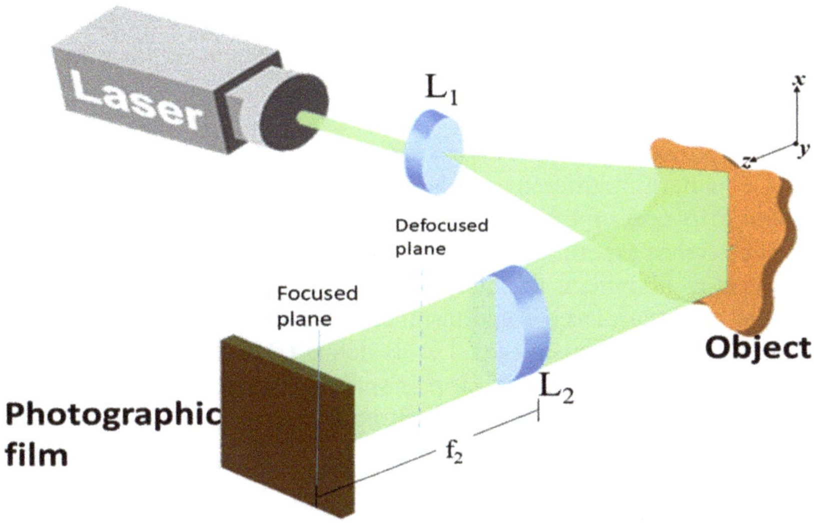

Figure 4.4. Schematic view of the recording process in speckle pattern photography. The object is illuminated with an expanded beam using the lens (L_1). The diffracted object speckles are collected and conveyed by a lens–aperture combination (L_2) with focal length (f_2) to record the pattern in a photographic film.

from the object in two different states: the first in, say, a rest state and the second in a displaced/deformed state. This method is called a double-exposure technique, and both images are captured and stored on the same film. For instance, it is possible to record changes on the object's surface due to any type of displacements and, from the acquired data, convert the information to stress suffered on the surface.

The speckle pattern produced on the film due to the double exposure presents a space between pairs of speckles (p), depending on the magnitude of the in-plane object displacement. This local displacement at a point on the object (d_o) can be determined by

$$d_o = \frac{p}{\mathcal{M}}, \tag{4.8}$$

where ($\mathcal{M}$) is the imaging lens magnification. It is seen that equation (4.8) depends on the location of the photographic film (where the object is imaged) and the object displacement magnitude when the speckle pattern is recorded on this film. Valid object displacements are found only when the speckle displacements between the two object states are at least equal to or larger than the speckle size. This condition guarantees at least one fringe produced by this speckle displacement. The interference fringes that appear are mainly sensitive to in-plane object displacements in the x- or y-direction. For instance, the very low sensitivity of SPP to out-of-plane in the z-direction to obtain strain measurements is done via a defocused speckle pattern. The SPP's benefits are: the optical set-up is easy to construct and has the advantage of having almost null susceptibility to external environmental vibrations. It belongs to the optical set of tests that are non-destructive or non-contact, and the results are easy to assess (at least qualitatively). The drawbacks are that SPP suffers mostly from low sensitivity to out-of-plane displacements and has poor fringe contrast. Two methods can be used to quantify the object displacement: point-wise analysis of Young's fringes and Fourier-transform techniques [15, 16].

4.2.1.1 Point-wise analysis of Young's fringes

The speckle pattern registered on the film can consist of speckle pairs (each corresponding to one of the two object states) acting as a slit pair, similar to Young's interference experiment. Then, to observe the fringe pattern from these slit pairs, we need to illuminate the film with a laser. As a result, the diffracted light presented on a screen placed some distance away l from the film (see chapter 1) bright and dark equidistant fringes. Let us define (s) as the fringe spacing, with s inversely proportional to each speckle pair space, p as $s = \lambda l/p$, and let θ be a small angle representing the fringe angular position, then

$$\theta = \frac{\lambda}{p} = \frac{s}{l}, \tag{4.9}$$

where λ is the laser wavelength, s is the fringe separation on the observation plane (refer to figure 4.5), and can be related to the object displacement. Thus, the magnitude of the in-plane displacement d_0 is given by

$$d_0 = \frac{l\lambda}{s\mathcal{M}}, \tag{4.10}$$

where $\mathcal{M}$ is the magnification factor determined during the exposure. The fringes' orientation is perpendicular to the object's displacement direction, and the number of fringes is directly proportional to the object's in-plane displacement.

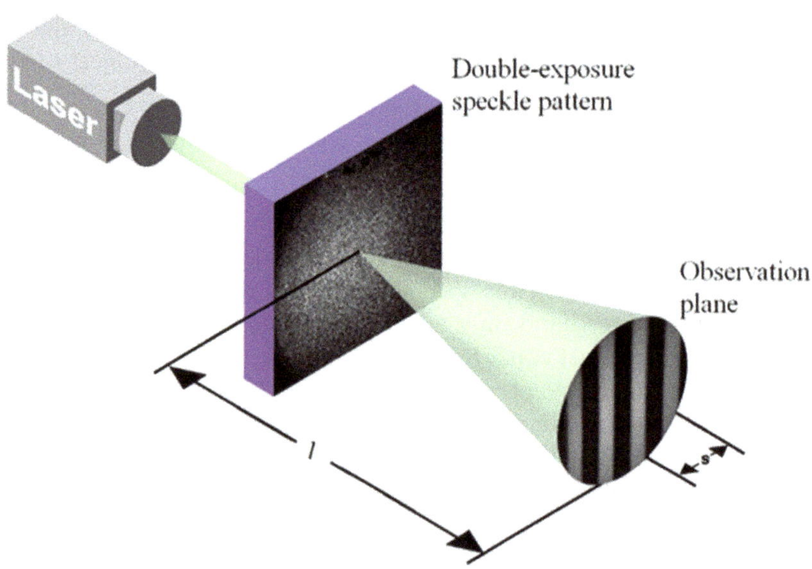

Figure 4.5. Point-wise analysis to determine the in-plane object displacement using Young's double-slit interferometer.

Let us rewrite equation (4.10) using θ and s, the fringe separation on the screen

$$d_0 = \frac{\lambda}{\mathcal{M}\sin\theta}. \tag{4.11}$$

4.2.1.2 Fourier-transform techniques

The Fourier transform of the speckle pattern can be used to determine the object displacement. One such method is considered to be full-field spatial filtering, i.e. it uses a bandpass filter, such as an aperture, at the Fourier plane of the lens (L_2), see figure 4.6. The figure also shows that the speckle pattern is illuminated with an extended plane wavefront from a laser. We can also see a Fourier-transform scheme: the light diffracted through the speckle pattern occurs at different angles (θ), and is collected by the lens (L_2); its resulting Fourier transform can be observed at the back focal plane, f, of L_2 (also called the Fourier-transform plane or the Fourier spectrum space). On this plane, one can see that the higher degree of diffraction, the higher the frequency. Following the process depicted in figure 4.6, an aperture on the Fourier plane works as the bandpass filter. This filter allows choosing the speckle displacements in either the x- or y-directions as a function of the aperture location (AL). Having AL $= f\theta$, means that by changing AL the sensitivity of the method is modified. Then, when the laser has displaced a distance d_y in the vertical direction, we obtain

$$d_y = \frac{\lambda f}{\mathrm{AL}.\mathcal{M}}, \tag{4.12}$$

a relation depending on λ, AL, the lens focal length (f), and the magnification ($\mathcal{M}$) of this lens. The same process is used to obtain d_x. The fringe pattern is reconstructed at the image plane: it represents the object displacement direction which in turn depends on the recording process, with the fringe pattern perpendicular to the spatial

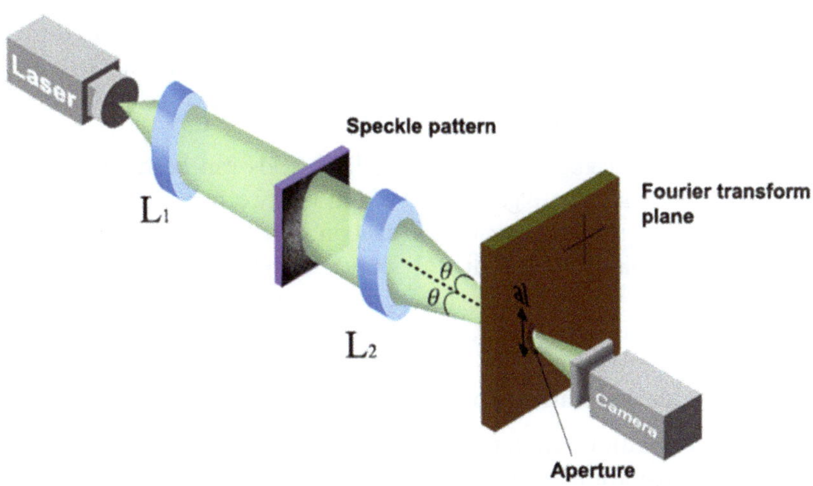

Figure 4.6. Whole-field Fourier-transform speckle pattern reconstruction used to determine the object displacement.

frequency (s_f). If the smallest displacement between neighboring fringes is considered constant, their spacing (s) is given by

$$s = \frac{1}{s_f} = \frac{\lambda}{\sin\theta}, \tag{4.13}$$

where the angle θ is shown in figure 4.6.

In this procedure, the resulting fringes for a specific frequency range are observed on the image plane located at the camera sensor. The inverse Fourier transform of the pattern recorded in the image plane should be performed to produce the image of the object with the corresponding displacement fringes.

4.2.2 Digital speckle correlation

The digital version of speckle pattern photography is called digital speckle pattern photography (DSPP). The photographic film is replaced using a digital recording device such as a high-resolution camera. DSPP is also referred to as digital laser speckle correlation (DLSC). Figure 4.2 is used for the recording of the intensity images. The size of the subjective speckles can be adjusted by choosing the appropriate dimensions of the lens–aperture combination. The speckles should be resolved under the system's parameters, such as camera sensor pixel size, the magnification $\mathscr{M}$, the lens f-number (f#), and the laser wavelength λ, see equation (4.5). When the speckle size is matched to the pixel dimensions of the camera sensor, a high contrast fringe is achieved, an important feature that helps to have smooth object deformation data. Once the sensors' pixels resolve the speckle size, a deformation is applied to the object. A series of speckle images are saved on the camera frame grabber, i.e. the acquired images are saved singly in a computer for two object states (as done for SPP). The next step in DSPP is the correlation between the acquired frames. A process of comparing and analyzing the changes among the stored images and relating them to the object displacements is performed. So the correlation between the bright and dark speckles renders data related to the response of the object surface to the applied deformation. It is worth pointing out here that DSPP can measure complex deformations, but this is only possible when the speckle motion is small to avoid de-correlation between two consecutive frames (or speckle patterns).

In-plane object displacements and their gradient fields can be resolved digitally by implementing cross-correlation algorithms between the two object-image states or fractional images by recovering the correlation function value at the highest point [17, 18].

A successful technique called digital image correlation (DIC) emerged from digital speckle pattern photography. The reader interested in this subject is referred to chapter 12.

4.2.3 Digital speckle correlation interferometry

Digital speckle correlation interferometry (DSCI) requires a smooth reference wave (U_r), falling on the recording medium (a wave that comes straight from the laser, can be a plane or divergent wave), to interference with the speckled object wavefronts (U_o), both can be expressed as

$$U_r(x, y) = u_r(x, y)e^{(i\varphi_r)} \text{ and } U_o(x, y) = u_o(x, y)e^{(i\varphi_o)}, \tag{4.14}$$

where u_r and φ_r are the amplitude and phase of the reference wave, and u_o and φ_o are the amplitude and phase of the object wave.

To adapt the micro-interference between the object and reference wave to the resolution of the camera target, colinear reference and object waves have to be employed, and an imaging system has to be used, see figure 4.7.

The object's surface is focused on the camera target—the use of the colinear reference wave results in large speckles. The camera can resolve them but degrade the resulting interference pattern noise-wise, i.e. it looks speckled. Due to focusing the object's surface onto the camera target, we record an image plane hologram, contrary to digital holography, where typically the Fresnel or Fraunhofer diffraction field of the object wavefront is registered, see chapter 6.

As we learned in chapter 1, the intensity $I(x, y)$ of the interference between these two waves can be expressed as

$$\begin{aligned} I(x, y) &= | U_r + U_o |^2 = | u_R(x, y)e^{(i\varphi_r)} + u_o(x, y)e^{(i\varphi_o)} |^2 \\ &= U_o^2 + U_r^2 + U_o^* U_r + U_o U_r^*, \end{aligned}$$

$$I(x, y) = u_o^2(x, y) + u_r^2(x, y) + 2a_o(x, y)a_r(x, y)\cos(\varphi_{o-r}(x, y)),$$

$$\begin{aligned} I(x, y) &= I_o(x, y) + I_r(x, y) + 2\sqrt{I_o(x, y)I_r(x, y)}\cos(\varphi_{o-r}(x, y)), \\ I(x, y) &= I_b(x, y) + I_m(x, y)\cos(\varphi_{o-r}(x, y)), \end{aligned} \tag{4.15}$$

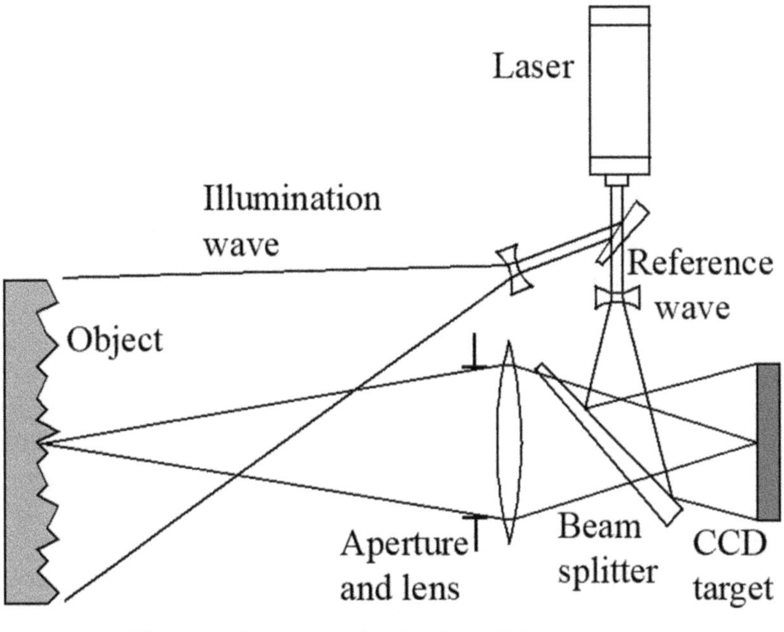

Figure 4.7. Arrangement for digital speckle interferometry.

where x and y denote the spatial coordinates of the pixels on the camera sensor, I_o and I_r represent the intensities of the object and reference beams, respectively, $\varphi_{o-r} = \varphi_o - \varphi_r$ is the random phase present in $I(x, y)$, which comes from the phase differences between the reference and the object waves; $I_b = I_o + I_r$ is the background, or dc, intensity; and $2\sqrt{I_o I_r}\cos(\varphi_{o-r})$ is the interference term, with $2\sqrt{I_o I_r}$ known as the modulation term (I_m) to the interference term. This equation can be rewritten as

$$I(x, y) = 2I_a[1 + \gamma\cos(\varphi_{o-r})], \tag{4.16}$$

where $I_a = (u_r^2 + u_o^2)/2$ is the mean intensity and $\gamma = (2u_r u_o)/(u_r^2 + u_o^2)$ is known as the interference fringe pattern's visibility (or contrast).

Additionally, to perform in-plane sensitive measurements, the smooth reference wave can be replaced by a second object illuminating wave so that two object waves fall on the object, whose intensities are I_{o1} and I_{o2}. Then,

$$I(x, y) = U_{o1}^2 + U_{o2}^2 + U_{o1}^* U_{o2} + U_{o1} U_{o2}^* = I_{o1} + I_{o2} + 2\sqrt{I_{o1} I_{o2}}\cos(\varphi_{o1-o2}). \tag{4.17}$$

DSCI optical set-ups can vary, but all render the whole field of view with highly accurate non-contact measurements of object displacements, deformations, vibrations, stress, and strain, among other physical and mechanical variables. In most cases, the quantity of interest is the optical path difference between object and reference beams, a variable known in optics as the phase difference—it is contained in the argument of the cosine function in equations (4.15)–(4.17) and appears in the form of interference fringes.

The double-exposure image acquisition in DSCI is used to quantify displacement measurements, as before: two object images in their base and modified state are captured (stored in a PC) and correlated, a process that results in a fringe pattern directly related to the object displacement. Both fringe patterns (or a set of them) are correlated by adding or subtracting the images corresponding to the base and modified states [19, 20]. The sensitivity of DSCI to determine displacements depends on the size of the subjective speckle pattern, namely the DSCI sensitivity is increased when the numerical aperture of the imaging system increases.

4.2.4 Speckle shearography

The requirement for an experimental vibration isolation environment for the measuring optical layout can be dropped to a good extent when using the speckle-shearing methods. In the following, the method is explained using an example of digital shearography. Here only the spatial variations of the displacement in a predetermined direction are measured, so the methods are relatively insensitive to rigid body motions produced by unwanted environmental noise vibrations [21, 22].

Using figure 4.8, where again two wave fields, object, and reference interfere, the main difference with DSPI is that these two waves are speckle fields slightly shifted laterally. The shifting of the speckle fields is performed by a shearing element, e.g. a glass wedge in front of one half of the imaging lens, two tilted glass plates, a Wollaston prism, or indeed a Michelson-interferometer-like arrangement with one mirror slightly tilted. The role of the reference wave is taken to be one of the two object wave fields, so one speaks about self-reference.

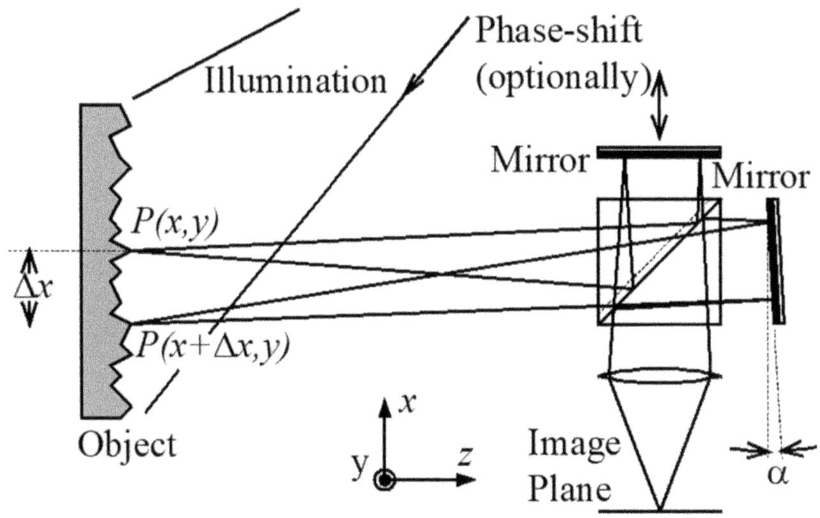

Figure 4.8. Arrangement for digital shearography.

Let the two wave fields be separated laterally by the mutual shearing Δx, and then we get the two nearly colinear wave fields

$$U_1(x, y) = U_{01}(x, y)e^{i\varphi(x, y)} \quad U_2(x, y) = U_{02}(x, y)e^{i\varphi(x+\Delta x, y)} \tag{4.18}$$

Their interference produces the speckle pattern

$$I_A(x, y) = | U_1(x, y) + U_2(x, y) |^2,$$

$$I_A(x, y) = I_1(x, y) + I_2(x, y) + 2\sqrt{I_1(x, y)I_2(x, y)} \cos\psi(x, y), \tag{4.19}$$

which is recorded, digitized, and stored. $\psi(x, y)$ is then randomly distributed phase difference $\psi(x,y) = \varphi(x,y) - \varphi(x + \Delta x, y)$. Deformation of the object leads to the wave fields

$$U_3(x, y) = U_{01}(x, y)e^{i[\varphi(x, y)+\Delta\varphi(x, y)]},$$

$$U_4(x, y) = U_{02}(x, y)e^{i[\varphi(x+\Delta x, y)+\Delta\varphi(x+\Delta x, y)]}, \tag{4.20}$$

whose superposition yields the speckle pattern

$$\begin{aligned} I_B(x, y) = I_1(x, y) + I_2(x, y) + 2\sqrt{I_1(x, y) \quad I_2(x, y)} \\ \cos[\psi(x, y) + \Delta\varphi(x, y) - \Delta\varphi(x + \Delta x, y)]. \end{aligned} \tag{4.21}$$

Point-wise subtraction gives

$$\begin{aligned} (I_A - I_B)(x, y) &= 2\sqrt{I_1 \quad I_2}\{\cos\psi(x, y) - \cos[\psi(x, y) + \Delta\varphi(x, y) - \Delta\varphi(x + \Delta x, y)]\}, \\ &= 4\sqrt{I_1 \quad I_2}\sin\left[\psi(x, y) + \frac{\Delta\varphi(x, y) - \Delta\varphi(x + \Delta x, y)}{2}\right] \\ &\quad \sin\frac{\Delta\varphi(x, y) - \Delta\varphi(x + \Delta x, y)}{2}. \end{aligned} \tag{4.22}$$

Positive values may be obtained by taking the modulus or square of this relation. Again the term in the root is the background. The first sine-term is the stochastic speckle-noise modulated by the second sine-term, which stems from the deformation. The displacement vector field $\boldsymbol{d}(x, y)$ is contained in the argument of this sine by

$$\frac{\Delta\varphi(x, y) - \Delta\varphi(x + \Delta x, y)}{2},$$

$$= \frac{\pi}{\lambda}\{\boldsymbol{d}(x, y) \cdot [\boldsymbol{b}(x, y) - \boldsymbol{s}(x, y)] - \boldsymbol{d}(x + \Delta x, y) \cdot [\boldsymbol{b}(x + \Delta x, y) - \boldsymbol{s}(x + \Delta x, y)]\},$$

$$\approx \frac{\pi}{\lambda}[\boldsymbol{d}(x, y) - \boldsymbol{d}(x + \Delta x, y)] \cdot [\boldsymbol{b}(x, y) - \boldsymbol{s}(x, y)],$$

$$= \frac{\pi}{\lambda}\left[\frac{\boldsymbol{d}(x, y) - \boldsymbol{d}(x + \Delta x, y)}{\Delta x}\right] \cdot [\boldsymbol{b}(x, y) - \boldsymbol{s}(x, y)]\Delta x,$$

$$\approx \frac{\partial(x, y)}{\partial x} \cdot \frac{\pi \Delta x}{\lambda}[\boldsymbol{b}(x, y) - \boldsymbol{s}(x, y)]. \tag{4.23}$$

The interference pattern so produced is an approximation to the derivative of the displacement field in the direction of the image shearing, here the x-direction. For rigid body motions, $\boldsymbol{d}(x, y) = \text{constant}$, implying $\partial \boldsymbol{d}(x, y)/\partial x = 0$. The latter explains the insensitivity of the shearing methods to rigid body motions. The method's sensitivity can be adjusted by controlling the amount of shearing. In addition to the shearing in the x-direction we can shear in the y-direction or other angular directions. Also, radial shearing, rotational shearing, inversion shear, or reversal shear can be performed with specific optical arrangements [23].

Figure 4.9(a) presents a DSPI-image of a plate subjected to an out-of-plane deformation by pressing from the plate's backside at a central point. Thus a deformation in the form of a Gaussian bell is produced. The fringes can be interpreted as equi-height lines of the deformed object. For comparison in

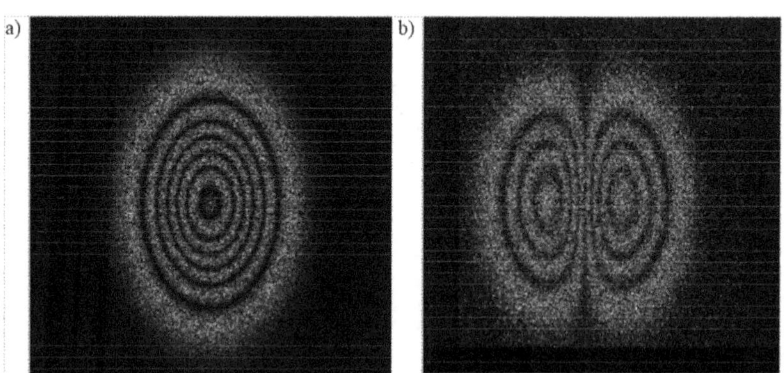

Figure 4.9. DSPI (a) and shearography (b) patterns of a plate with bell-shaped deformation.

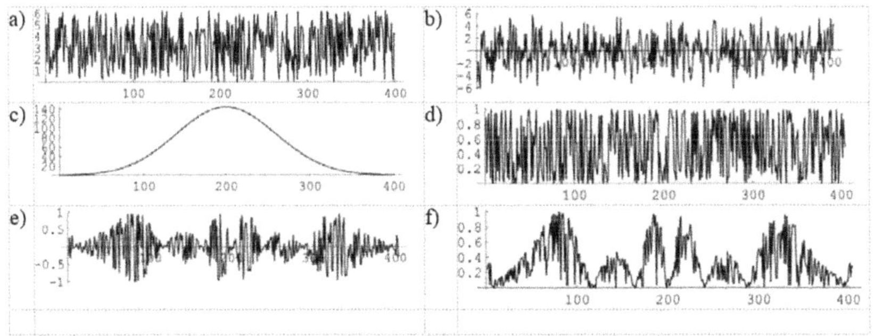

Figure 4.10. Digital speckle shearography: (a) $\varphi(x)$, (b) $\varphi(x)-\varphi(x+\Delta x)$, (c) $\Delta\varphi(x)$, (d) $I_B(x)$, (e) $I_A(x) - I_B(x)$, and (f) $|I_A(x) - I_B(x)|$.

figure 4.9(b), the same object undergoing the same deformation is investigated by digital shearography. The shear is in the horizontal direction so that the fringes can be explained as equi-contours of the derivative in the horizontal direction. It has the highest fringe order at the point of the steepest slope of the hill. At the top of the hill, we have zero slopes and thus zero fringe order in the shearography pattern.

A one-dimensional simulation of speckle shearography is shown in figure 4.10. Figure 4.10(a) displays the random phase $\varphi(x)$, while figure 4.10(b) gives the phase difference $\varphi(x) - \varphi(x + \Delta x)$. Figure 4.10(c) shows the assumed additional phase $\Delta\varphi(x)$ due to the deformation, and in figure 4.10(d), the recorded intensity $I_B(x)$ after deformation is seen. The difference $I_A(x) - I_B(x)$ of the two recorded intensities is given in figure 4.10(e), while its modulus $|I_A(x) - I_B(x)|$ is in figure 4.10(f).

References

[1] Orr J F and Shelton J C 1997 *Optical Measurement Methods in Biomechanics* 1st edn (London: Chapman and Hall)

[2] Briers D, Duncan D D, Hirst E, Kirkpatrick S J, Larsson M, Steenbergen W, Stromberg T and Thompson O B 2013 Laser speckle contrast imaging: theoretical and practical limitations *J. Biomed. Opt.* **18** 066018

[3] Bender N, Yılmaz H, Bromberg Y and Cao H 2019 Creating and controlling complex light *APL Photon.* **4** 110806

[4] Balamurugan R and Muruganand S 2013 Electronic laser speckle interferometer for displacement measurement using digital image processing technique *J. Image Graph.* **1** 59–62

[5] Draijer M, Hondebrink E, Van Leeuwen T and Steenbergen W 2009 Review of laser speckle contrast techniques for visualizing tissue perfusion *Lasers Med. Sci.* **24** 639–51

[6] Khaksari K and Kirkpatrick S J 2009 Laser speckle modeling and simulation for biophysical dynamic: influence of sample statistics using tissue perfusion *Lasers Med Sci.* **24** 639–51

[7] Gan Y and Steinchen W 2008 *Speckle Methods Springer Handbook of Experimental Solid Mechanics* (New York: Springer), pp 655–72 ch 23

[8] Meinhart C D and Wereley S T 2003 The theory of diffraction-limited resolution in microparticle image velocimetry *Meas. Sci. Technol.* **14** 1047–53

[9] Gasvik K J 2002 *Optical Metrology* 3rd edn (Chichester: Wiley)

[10] Andersson A, Runnemalm A and Sjödahl M 1999 Digital speckle-pattern interferometry: fringe retrieval for large in-plane deformations with digital speckle photography *Appl. Opt.* **38** 5408–12

[11] Jacquot P 2008 Speckle interferometry: a review of the principal methods in use for experimental mechanics applications *Strain* **44** 57–69

[12] Huntley J M 1989 Speckle photography fringe analysis: assessment of current algorithms *Appl. Opt.* **28** 4316–22

[13] Kihm K D 1997 Laser speckle photography technique applied for heat and mass transfer problems *Adv. Heat Transf.* **30** 255–311

[14] Tu M and Gielisse P J 1990 Spatial frequency in speckle metrology *Appl. Opt.* **29** 3392–8

[15] Rastogi P K 2021 *Digital Speckle Pattern Interferometry and Related Techniques* 1st edn (New York: Wiley)

[16] Tiziani H J and Pedrini G 2013 From speckle pattern photography to digital holographic interferometry *Appl. Opt.* **52** 30–44

[17] Fricke-Begemann T 2003 Three-dimensional deformation field measurement with digital speckle correlation *Appl. Opt.* **42** 6783–96

[18] Stoilov G, Kavardzhikov V and Pashkouleva D 2012 A comparative study of random patterns for digital image correlation *J. Theor. Appl. Mech. Sofia* **42** 55–66

[19] Alcalá Ochoa N, Mendoza Santoyo F, Moore A J and Pérez López C 1997 Contrast enhancement of electronic speckle pattern interferometry addition fringes *Appl. Opt.* **36** 2783–7

[20] Dvořáková P, Bajgar V and Trnka J 2007 Dynamic electronic speckle pattern interferometry in application to measure out-of-plane displacement *Eng. Mech.* **14** 37–44

[21] Hung M Y Y 1991 Electronic shearography versus ESPI for non-destructive evaluation *Proc. SPIE* **1554B** 692–700

[22] Anisimov A G and Groves R M 2021 Extreme shearography: development of a high-speed shearography instrument for quantitative surface strain measurements during an impact event *Opt. Lasers Eng.* **140** 106502

[23] Ganesan A R, Sharma D K and Kothiyal M P 1988 Universal digital speckle shearing interferometer *Appl. Opt.* **27** 4731–4

IOP Publishing

Full Field Optical Metrology and Applications

Fernando Mendoza-Santoyo, Manuel De la Torre-Ibarra, María del Socorro Hernández-Montes and Jorge Mauricio Flores Moreno

Chapter 5

Moiré interferometry

Contributing author: Apolinar Muñoz

5.1 Introduction

Moiré is a French word used to describe irregular wavy patterns that form when a silk or wool fabric exhibits patterns of light and dark bands or fringes. It can be seen in everyday things such as two overlapping window mosquito-net-screens, or a striped shirt seen on a television screen. In optics it refers to a beating pattern produced between two periodic gratings of nearly the same pitch/period, generally known as the 'moiré phenomenon', and the resulting interference patterns are called moiré fringes [1].

This chapter describes two basic methods to obtain a moiré fringe pattern. The first consists of overlapping two parallel gratings, i.e. two gratings of the same and/or different periods moved vertically (or horizontally) parallel to one another. The second method consists of overlapping two gratings at an angle, with gratings of the same and/or different periods. The moiré fringe pattern is seen as interference fringes produced by overlapping the two gratings.

Because the Moiré patterns provide a great sensitivity to detect small surface deformations, their use in optical metrology is mainly directed to shape detection, vibration analysis, strain measurement, precision measurement, alignment, and imaging. The moiré pattern is produced when two sets of line grids printed on transparent material are slightly inclined to one another. Also, this effect is produced when two gratings are overlapped in parallel form, and they become slightly shifted from one another. The line grids can also be produced by overlapping two laser fringe patterns and there are mathematical methodologies developed to generate moiré patterns graphically using computer algorithms. These line grids, or structures, are not restricted to any particular transmittance or intensity profile (i.e. sinusoidal, square, triangular, etc) and may have any spatial distribution (i.e. straight lines, curves, circles, etc) as long as they are periodic in nature. The result of this procedure is a pattern of dark and white bands, like interference fringes.

In general, the superposed grid patterns should have opaque and transparent regions. Moiré patterns are observed when images with regular grid patterns are

doi:10.1088/978-0-7503-3027-5ch5

superimposed, and the generated moiré patterns do not exist in any of the original grids. Also, the analysis of moiré patterns is used to infer the geometrical change of the original overlapped images. In a typical moiré pattern, a reference grid is kept constant, and the grid pattern on an object is allowed to vary as it is physically deformed. Thus, the moiré pattern deflection is employed to analyze the nature of the deformation occurring on the object, namely its shape or contour. One of the significant problems of moiré interferometry is the object detection derived from the moiré pattern deflection, a procedure performed by recovering the moiré pattern envelope via a low pass filter. The filtering can be performed via a Fourier method or a convolution window; both are used to retrieve the moiré pattern deflection. An interesting issue in moiré fringes is their sensitivity to changes in the grid pattern since a small change in it would result in a large change in the moiré pattern. Interestingly, the mathematical theory explaining a moiré fringe pattern is elucidated experimentally.

5.2 Moiré pattern formulation

The classical moiré pattern is produced by overlapping two straight-line gratings [2]. Thus a dark fringe is produced where the dark grid lines are out of step one-half period, and a bright fringe is produced where the dark grid lines for one of the gratings fall on top of the corresponding dark lines for the second grating. Based on these statements, a moiré fringe pattern can be generated by employing different methodologies, and in consequence, different mathematical models provide the equations to describe a moiré pattern.

The first method to generate a moiré fringe pattern is by overlapping two gratings, $I_1(x, y)$ and $I_2(x, y)$, vertical (or horizontal) parallel to one another [3]. Consider for this case that the period of the gratings is the same and is represented by T. Thus, the moiré pattern is produced by shifting the grating $I_2(x, y)$ a small distance Δx with respect to the grating $I_1(x, y)$. A cosine function depicts the line profile of the gratings, and their intensity profile is represented by the expressions $I_1(x, y) = 1 + \cos(2\pi fx)$ and $I_2(x, y) = 1 + \cos(2\pi fx + \Delta x)$, with f the spatial frequency of the line grid. The intensity profile of these gratings is shown in figure 5.1, where the varying values for $I_1(x, y)$ and $I_2(x, y)$ can be represented in grayscale, with 0 for black and 2 for white.

The period is the same for these gratings, and $f = 1/T$ represents the spatial frequency. The moiré pattern $M(x, y)$ can be obtained and represented by the sum or multiplication of the gratings. The sum of the gratings is performed using the following expression

$$M(x, y) = I_1(x, y) + I_2(x, y) = 2 + \cos(2\pi fx) + \cos(2\pi fx + \Delta x). \quad (5.1)$$

The intensity profile of the moiré pattern produced by equation (5.1) is known as additive moiré, shown in figure 5.2(a). Additionally, equation (5.1) can be rewritten by employing the trigonometric identity $\cos(2\pi fx) + \cos(2\pi fx + \Delta x) = 2\cos([2\pi fx + 2\pi fx + \Delta x]/2)\cos([2\pi fx - 2\pi fx - \Delta x]/2)$, resulting in

$$M(x, y) = 2 + 2\cos(2\pi fx + \Delta x/2)\cos(\Delta x/2). \quad (5.2)$$

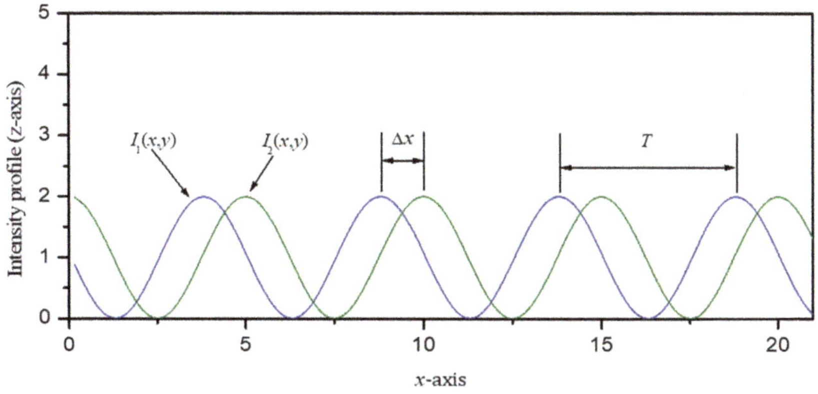

Figure 5.1. Intensity profile of the gratings $I_1(x, y)$ and $I_2(x, y)$.

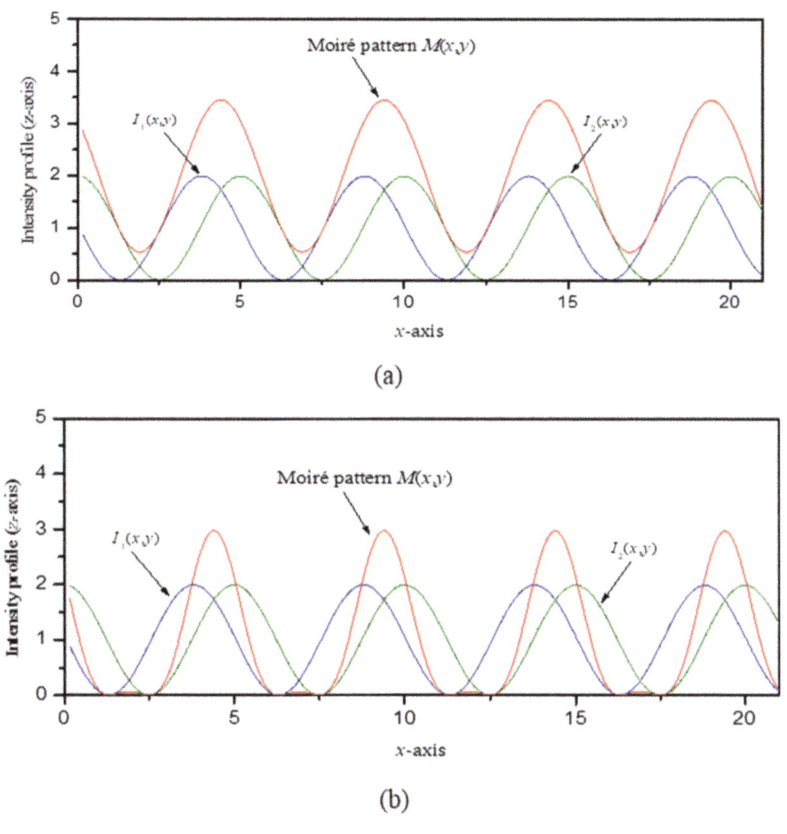

Figure 5.2. (a) Additive moiré pattern produced by equation (5.1) and (b) multiplicative moiré pattern produced by equation (5.3).

The moiré pattern produced by the multiplication of the gratings is represented through the next expression:

$$M(x, y) = I_1(x, y) \times I_2(x, y) = 1 + \cos(2\pi fx + \Delta x)[1 + \cos(2\pi fx)] + \cos(2\pi fx). \tag{5.3}$$

The intensity profile of the moiré pattern produced by equation (5.3) is known as the multiplicative moiré pattern, shown in figure 5.2(b).

The second method to produce a moiré pattern is by overlapping two gratings, $I_1(x, y)$ and $I_2(x, y)$, in parallel form, but this time we allow for the gratings' periods T_1 and T_2 to be different [4]. The line profile of these gratings is represented by the expressions $I_1(x, y) = 1 + \cos(2\pi f_1 x)$ and $I_2(x, y) = 1 + \cos(2\pi f_2 x)$, where f_1 and f_2 are the corresponding line grid spatial frequencies. The intensity profile of these gratings is shown in figure 5.3(a). As before, we can represent the magnitude of the variation of the functions $I_1(x, y)$ and $I_2(x, y)$ in grayscale, where 0 is black and 2 white. Thus, the additive moiré pattern $M(x, y)$ is determined utilizing the expression

$$M(x, y) = I_1(x, y) + I_2(x, y) = 2 + \cos(2\pi f_1 x) + \cos(2\pi f_2 x). \tag{5.4}$$

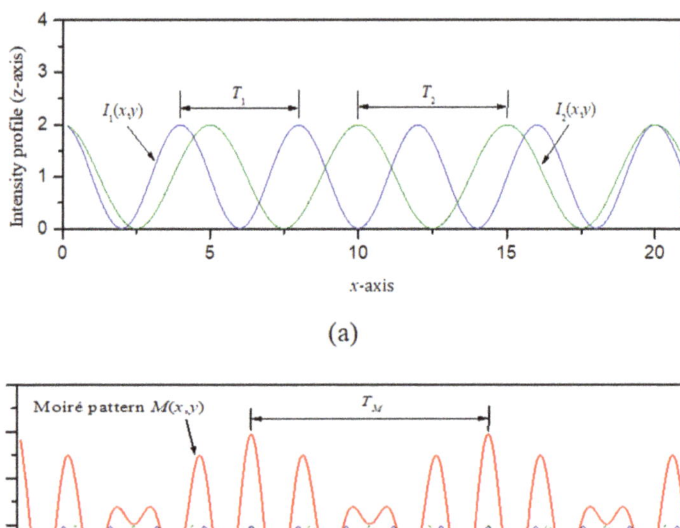

(a)

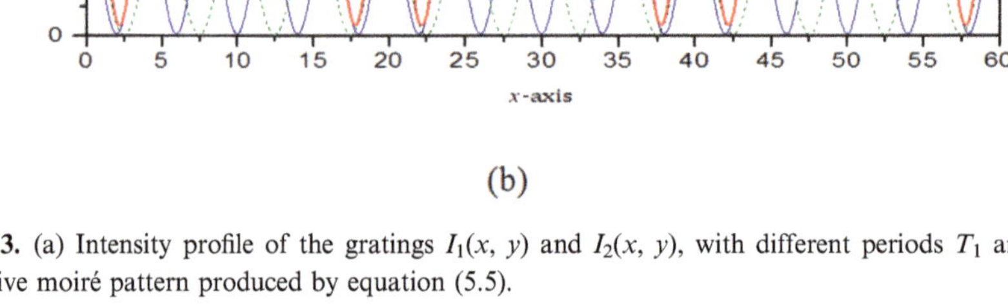

(b)

Figure 5.3. (a) Intensity profile of the gratings $I_1(x, y)$ and $I_2(x, y)$, with different periods T_1 and T_2 and (b) additive moiré pattern produced by equation (5.5).

Equation (5.4) is rewritten by employing the trigonometric identity $\cos(2\pi f_1 x) + \cos(2\pi f_2 x) = 2\cos([2\pi f_1 x + 2\pi f_2 x]/2)\cos([2\pi f_1 x - 2\pi f_2 x]/2)$, and the following equation is obtained:

$$M(x, y) = 2 + 2\cos\left(\pi x\left[f_1 + f_2\right]\right)\cos\left(\pi x\left[f_1 - f_2\right]\right). \tag{5.5}$$

The intensity profile of the additive moiré pattern $M(x, y)$ computed via equation (5.5) is shown in figure 5.3(b). In this case, the moiré period T_M represents the distance between two overlapped maxima.

The multiplicative moiré pattern is represented through the expression

$$M(x, y) = I_1(x, y) \times I_2(x, y) = 1 + \cos\left(2\pi f_1 x\right)\left[1 + \cos\left(2\pi f_2 x\right)\right] + \cos\left(2\pi f_2 x\right). \tag{5.6}$$

The two-dimensional image of the additive moiré pattern produced by equation (5.5) is shown in figure 5.4(a). For example, two-line grid transparencies with different periods are shown in figure 5.4(b). These gratings are overlapped to obtain the additive moiré pattern shown in figure 5.4(c).

Figure 5.4. (a) Additive moiré pattern produced by equation (5.5); (b) the gratings printed on transparent material, typically acetate film; and (c) moiré pattern produced by overlapping the transparencies with different periods.

Another method to obtain a moiré pattern is overlapping two gratings at an angle [5]. For this purpose, the period of the gratings T is the same, so the spatial frequency is represented by $f = 1/T$. The intensity profile of these gratings is represented by the expressions $I_1(x, y) = 1 + \cos(2\pi fx)$ and $I_2(x, y) = 1 + \cos(2\pi fx + 2\alpha x)$, where 2α is the angle of the lines between the two gratings. The images of the gratings $I_1(x, y)$ and $I_2(x, y)$ are shown in figures 5.5(a) and (b), respectively. The additive moiré pattern $M(x,y)$ is determined by

$$M(x, y) = I_1(x, y) + I_2(x, y) = 2 + \cos(2\pi fx) + \cos(2\pi fx + 2\alpha x), \quad (5.7)$$

This equation is rewritten by employing the trigonometric identity $\cos(2\pi fx) + \cos(2\pi fx + 2\alpha x) = 2\cos([2\pi fx + 2\pi fx + 2\alpha x]/2)\cos([2\pi fx - 2\pi fx - 2\alpha x]/2)$, thus

$$M(x, y) = 2+2\cos(2\pi fx + \alpha x)\cos(\alpha x). \quad (5.8)$$

The multiplicative moiré pattern is represented utilizing the expression

$$M(x, y) = I_1(x, y) \times I_2(x, y) = 1 + \cos(2\pi fx + 2\alpha x)[1 + \cos(2\pi fx)] + \cos(2\pi fx). \quad (5.9)$$

The image of the multiplicative moiré pattern produced by equation (5.9) is shown in figure 5.5(c).

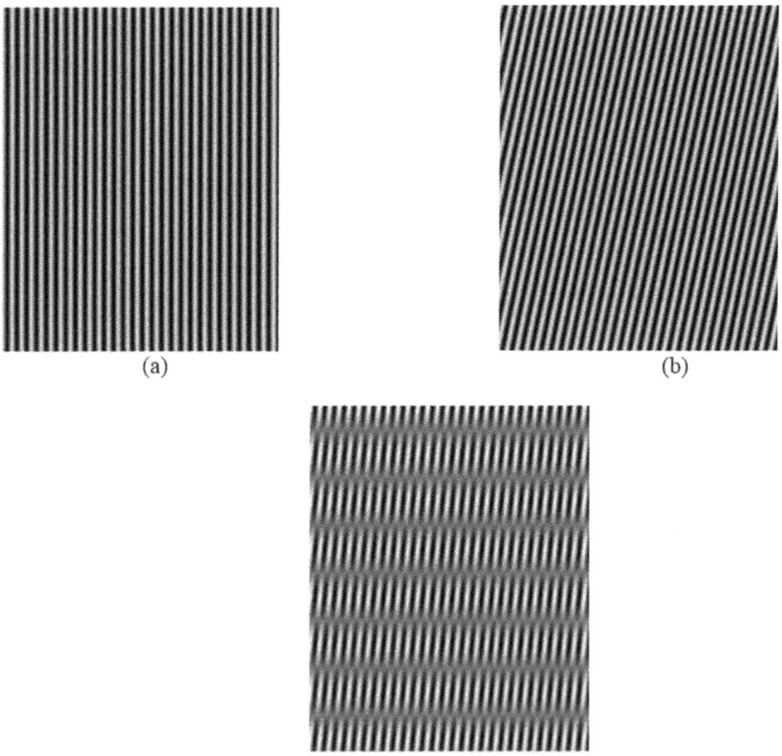

Figure 5.5. Intensity profiles of the gratings (a) $I_1(x, y)$ and (b) $I_2(x, y)$; and (c) the multiplicative moiré pattern produced by equation (5.9).

Also, this moiré pattern is generated via transparencies. The pattern is obtained when the gratings $I_1(x, y)$ and $I_2(x, y)$ are printed on transparent material. An image of these gratings is shown in figure 5.6(a). Then, these gratings are overlapped to obtain the moiré pattern shown in figure 5.6(b). Thus the moiré pattern produced by equation (5.9) is corroborated.

The last method we will present to generate a moiré pattern is by overlapping two gratings at an angle, with their period being different [6]. The profile of these gratings is represented by the expressions $I_1(x, y) = 1 + \cos(2\pi f_1 x)$ and $I_2(x, y) = 1 + \cos(2\pi f_2 x + 2\alpha x)$, where 2α is the angle of the lines between the two gratings, and f_1 and f_2 represent their spatial frequency. Thus, the additive moiré pattern $M(x,y)$ is represented by

$$M(x, y) = I_1(x, y) + I_2(x, y) = 2 + \cos\left(2\pi f_1 x\right) + \cos\left(2\pi f_2 x + 2\alpha x\right). \quad (5.10)$$

This equation is rewritten by employing the trigonometric identity $\cos(2\pi f_1 x) + \cos(2\pi f_2 x + 2\alpha x) = 2\cos([2\pi f_1 x + 2\pi f_2 x + 2\alpha x]/2)\cos([2\pi f_1 x - 2\pi f_2 x - 2\alpha x]/2)$, as

$$M(x, y) = 2{+}2\cos\left(\pi x\left[f_1 + f_2\right] + \alpha x\right]\right)\cos\left(\pi x\left[f_1 - f_2\right] - \alpha x\right]\right). \quad (5.11)$$

(a)

(b)

Figure 5.6. (a) Gratings printed on transparencies and (b) a moiré pattern is produced by overlapping the printed gratings.

(a) (b)

Figure 5.7. (a) Multiplicative moiré pattern produced by equation (5.12) and (b) the corresponding moiré pattern produced by overlapping the printed gratings.

The multiplicative moiré pattern is represented by

$$M(x, y) = I_1(x, y) \times I_2(x, y) = 1 + \cos(2\pi f_2 x + 2\alpha x)\left[1 + \cos(2\pi f_1 x)\right] + \cos(2\pi f_1 x). \tag{5.12}$$

The image of the moiré pattern produced by equation (5.12) is shown in figure 5.7(a). The two physically printed gratings are overlapped to obtain the moiré pattern shown in figure 5.7(b). Thus, the moiré pattern produced by equation (5.12) is corroborated.

5.3 Moiré pattern alignment

The moiré pattern alignment is achieved by determining the direction of the moiré fringe pattern and the moiré period [7]. This procedure employs gratings with different periods and a relative angular displacement 2α of the gratings. Let us look at the procedure based on these two variables. The moiré period (T_M) is determined via the geometry shown in figure 5.8. From this geometry, the period T_1 is defined by the expression $T_1 = b \cdot \sin(2\alpha)$, the period T_2 is defined by the expression $T_2 = c \cdot \sin(2\alpha)$ and the period T_M is defined by the expression $T_M = c \cdot \sin(\beta)$.

Also, the expression $a \cdot \sin(\beta) = b \cdot \sin(2\alpha)$ is obtained from the sine law. Then, $\sin(\beta) = b \cdot \sin(2\alpha)/a$ is replaced in the expression $T_M = c \cdot \sin(\beta)$ to obtain $T_M = c \cdot b \cdot \sin(2\alpha)/a$. Also, the expression $a^2 = b^2 + c^2 - 2b \cdot c \cdot \cos(2\alpha)$ is obtained from the cosine law. This is replaced in the period T_M to obtain $T_M = c \cdot b \cdot \sin(2\alpha)/[b^2 + c^2 - 2b \cdot c \cdot \cos(2\alpha)]^{1/2}$. Let us now use the periods T_1 and T_2 to obtain the equation

$$T_M = \frac{T_1 T_2}{\sqrt{T_1^2 + T_2^2 - 2T_1 T_2 \cos(2\alpha)}}. \tag{5.13}$$

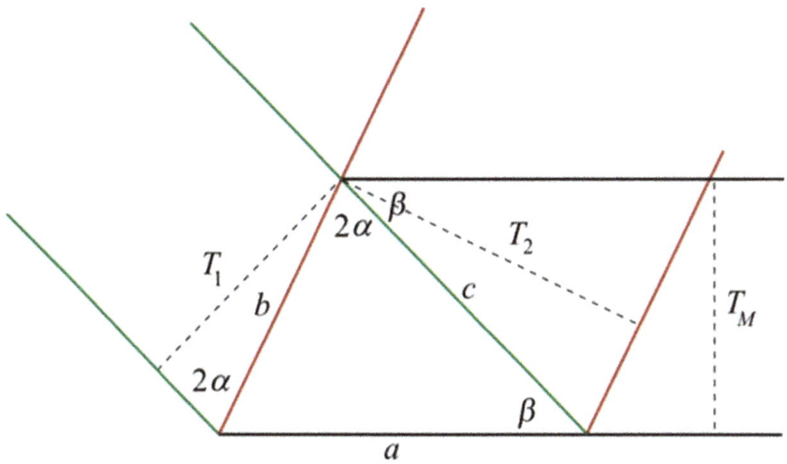

Figure 5.8. Moiré pattern produced by overlapping two gratings of different periods with a relative angle of 2α between them.

This equation can be rewritten by employing the identity $\cos^2(2\alpha) + \sin^2(2\alpha) = 1$ to obtain [8]

$$
\begin{aligned}
T_M &= \frac{T_1T_2}{\sqrt{T_1^2[\cos^2 2\alpha + \sin^2 2\alpha] + T_2^2 - 2T_1T_2\cos(2\alpha)}}\\
T_M &= \frac{T_1T_2}{\sqrt{T_1^2\sin^2 2\alpha + T_1^2\cos^2 2\alpha - 2T_1T_2\cos(2\alpha) + T_2^2}}.\\
T_M &= \frac{T_1T_2}{\sqrt{T_1^2\sin^2 2\alpha + (T_1\cos 2\alpha - T_2)^2}}
\end{aligned}
\tag{5.14}
$$

The moiré period for gratings with the same period is deduced by replacing the period $T = T_1 = T_2$ in equation (5.14), hence $T_M(2[1 - \cos(2\alpha)])^{1/2} = T$. Then, the identities $\cos^2(\alpha) + \sin^2(\alpha) = 1$ and $\cos(2\alpha) = \cos^2(\alpha) - \sin^2(\alpha)$ are replaced in this expression to obtain $T_M(2[\cos^2(\alpha) + \sin^2(\alpha) - \cos^2(\alpha) + \sin^2(\alpha)])^{1/2} = T$, which can be rewritten as $T_M[4\sin^2(\alpha)]^{1/2} = T$. Therefore, the moiré period for two gratings with the same period T is

$$T_M = \frac{T}{2\sin\alpha}. \tag{5.15}$$

In particular, the moiré period for the gratings overlapped in parallel form and whose periods T_1 and T_2 are different is deduced from figure 5.3(b). Here, the expression $1 + T_M/T_1 = T_M/T_2$ is obtained. Rewriting it, we obtain the moiré period as $T_M = -T_1T_2/(T_1 - T_2)$. This expression is also defined as [9]

$$T_M = \frac{T_1T_2}{|T_2 - T_1|}. \tag{5.16}$$

The direction of the moiré pattern is deduced based on the angle between the grating $I_1(x, y)$ and the moiré fringes. This direction is represented by the angle β shown in figure 5.8. To determine the angle β, the expressions $T_1 = \tan(\beta)(a^2 - T_1^2)^{1/2}$ and $a^2 = b^2 + c^2 - 2b \cdot c \cdot \cos(2\alpha)$ are obtained from the geometry. From these expressions, the moiré pattern direction is defined by the expression $\tan(\beta) = T_1/(b^2 + c^2 - 2b \cdot c \cdot \cos(2\alpha) - T_1^2)^{1/2}$. In this expression, $b = T_1/\sin(2\alpha)$ and $c = T_2/\sin(2\alpha)$ are replaced to obtain

$$\tan(\beta) = \frac{T \sin(2\alpha)_1}{\sqrt{T_1^2 + T_2^2 - 2T_1T_2\cos(2\alpha) - \sin^2(2\alpha)T_1^2}}, \tag{5.17}$$

which may be rewritten as

$$\tan(\beta) = \frac{T_1 \sin(2\alpha)}{\sqrt{T_1^2 + T_2^2 - 2T_1T_2\cos(2\alpha) - [1 - \cos^2(2\alpha)]T_1^2}}$$

$$\tan(\beta) = \frac{T_1 \sin(2\alpha)}{\sqrt{T_2^2 - 2T_1T_2\cos(2\alpha) + \cos^2(2\alpha)T_1^2}} = \frac{T_1 \sin(2\alpha)}{T_1\cos(2\alpha) - T_2}. \tag{5.18}$$

The moiré sensitivity is found using figure 5.3(b). For instance, if the moiré period is computed via equation (5.16) by employing $T_1 = 4$ and $T_2 = 5$ it results in $T_M = 20$. If T_2 is changed to $T_2 = 6$, the moiré period is $T_M = 12$. As we can see, a change of one unit produces a change of eight units in the moiré pattern. Therefore, a small change in any one parameter of equation (5.16) produces a significant change in the moiré pattern. In other words, the magnification effect of the moiré fringe relates to a small displacement in a large moiré fringe shift.

5.4 Shape detection via the moiré pattern

The moiré pattern is represented via equations (5.1) and (5.3), which is the case when two gratings are projected and overlapped in a parallel form on an object surface $\varphi(x, y)$. We take the period of the gratings to be the same, and the profile of these gratings is then represented by the expressions $I_1(x, y) = 1 + \cos(2\pi fx - \varphi(x,y))$ and $I_2(x, y) = 1 + \cos(2\pi fx + \Delta x)$, where the term $\varphi(x, y)$ relates to the object shape (contour). Thus, the additive moiré pattern is deduced via equation (5.1) through the expression

$$\begin{aligned} M(x, y) = I_1(x, y) + I_2(x, y) = 2 + \cos(2\pi fx - \phi(x, y)) + \\ \cos(2\pi fx + \Delta x). \end{aligned} \tag{5.19}$$

The multiplicative moiré pattern is represented via equation (5.3) by

$$\begin{aligned} M(x, y) = 1 + \cos(2\pi fx + \Delta x) + \cos(2\pi fx - \phi(x, y)) \\ + \cos(2\pi fx - \phi(x, y))\cos(2\pi fx + \Delta x). \end{aligned} \tag{5.20}$$

The term $\varphi(x, y)$ produces a moiré pattern deflection, or change in its line-like form, and is directly related to the object shape: the envelope of the moiré fringe pattern depicts the object shape $\varphi(x, y)$ [10, 11]. Let us look at this in detail: a moiré pattern is generated by employing the particular object shape shown in figure 5.9(a), so we need to replace the object shape $\varphi(x, y)$ in equation (5.20). The result is the moiré pattern shown in figure 5.9(b), whose envelope depicts the object shape $\varphi(x, y)$.

In order to test the utility of the moiré pattern envelope, a moiré pattern is generated with equation (5.20) by employing the object shape shown in figure 5.10(a). The object shape $\varphi(x, y)$ is converted to values in the interval between 0 and $\pi/2$. This procedure is carried out by the expression $\varphi(x, y) = \pi[\varphi(x,y) - \min]/[2(\max - \min)]$, where the variables max and min are the maximum and minimum values of the object shape $\varphi(x, y)$, respectively. Thus, the normalization to values $\pi/2$ avoids values larger than 2.55, which destroys the fringe pattern when $\varphi(x, y)$ is substituted in equation (5.20). In this way, a destroyed fringe pattern is avoided. Additionally, the term Δx is defined as $\pi/2$ so the identity $\cos(2\pi fx + \pi/2) = \sin(2\pi fx)$ is replaced in equation (5.20). Thus, the multiplicative moiré pattern is determined

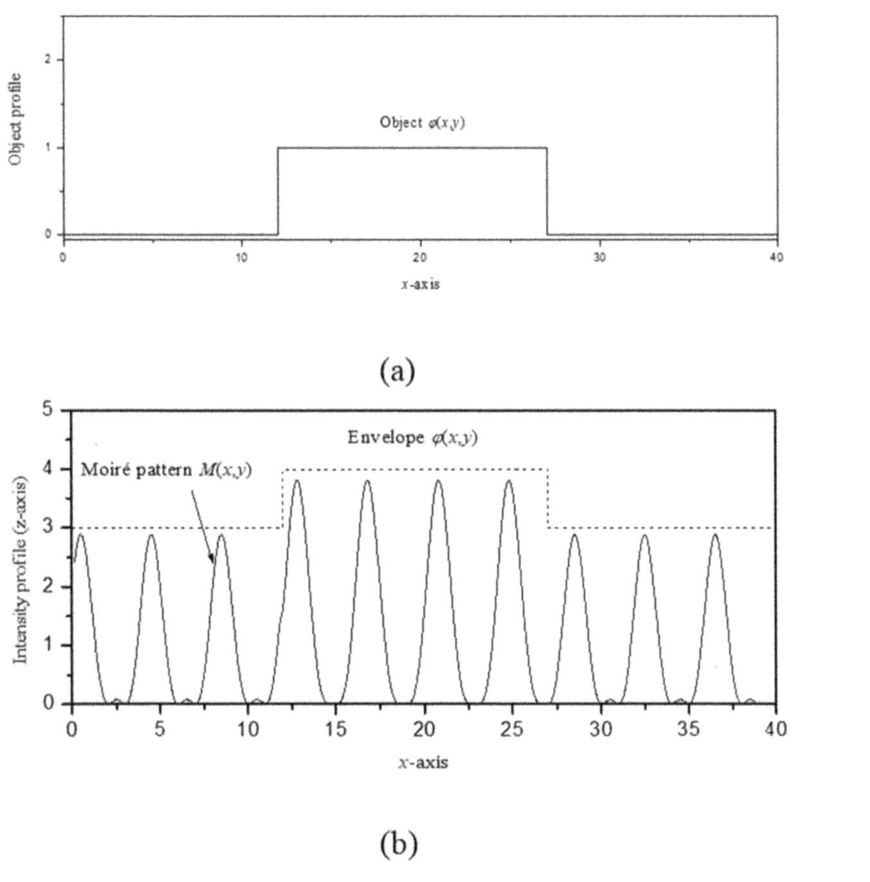

Figure 5.9. (a) Object shape employed to generate a moiré pattern and (b) moiré pattern envelope that provides the object shape $\varphi(x, y)$.

by the expression $M(x, y) = 1 + \sin(2\pi fx) + \cos(2\pi fx - \varphi(x, y)) + \cos(2\pi fx - \varphi(x, y))\sin(2\pi fx)$. The moiré fringe pattern shown in figure 5.10(b) is obtained by computing this expression.

As we can see, the moiré pattern envelope depicts the object shape shown in figure 5.10(a). Thus, the moiré pattern envelope has been validated. Additionally, the moiré pattern envelope is retrieved by applying a low pass filter on the moiré fringe pattern to obtain the object shape $\varphi(x, y)$. In order to elucidate this procedure, a moiré fringe pattern is generated via equation (5.20) by employing the object shape shown in figure 5.11(a). In this procedure, the object shape $\varphi(x, y)$ is converted to values between 0 and $\pi/2$. Again, the term Δx is defined as $\pi/2$ and the $\cos(2\pi fx + \pi/2) = \sin(2\pi fx)$ is replaced in equation (5.20). Thus, the moiré pattern is computed by the expression $M(x, y) = 1 + \sin(2\pi fx) + \cos(2\pi fx - \varphi(x, y)) + \cos(2\pi fx - \varphi(x, y))\sin(2\pi fx)$, resulting in the moiré fringe pattern shown in figure 5.11(b). As we can see, the moiré pattern envelope corresponds to the object shape shown in figure 5.11(a). The envelope of this moiré pattern is retrieved using a low pass filter, which can be performed through the Fourier method or a convolution window. A convolution window was used to retrieve the moiré pattern envelope in the case shown. The result

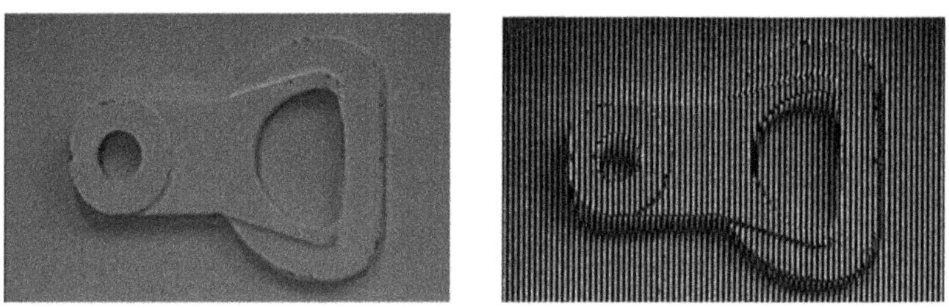

(a) (b)

Figure 5.10. (a) Metallic piece employed as object shape to produce a moiré pattern and (b) moiré pattern produced by equation (5.20).

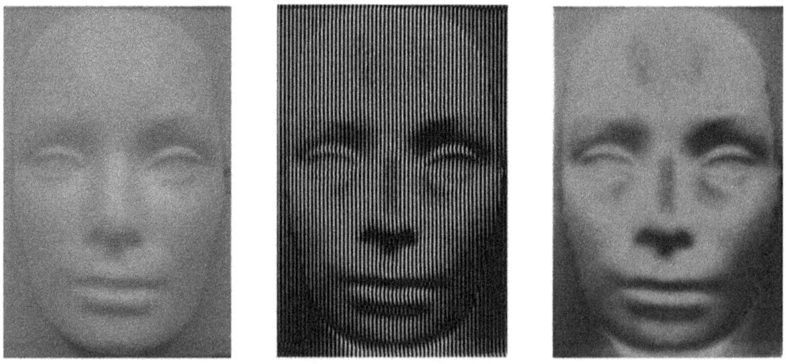

(a) (b) (c)

Figure 5.11. (a) Dummy face employed as object shape $\varphi(x, y)$; (b) moiré pattern produced by equation (5.20); and (c) envelope retrieved from the moiré pattern via a low pass filter.

Figure 5.12. Moiré pattern produced by overlapping two physically printed line gratings.

of this procedure is shown in figure 5.11(c). As we can see, the envelope corresponds to the object's shape $\varphi(x, y)$. Then, we can safely claim that the moiré pattern envelope has been successfully retrieved.

The moiré pattern envelope represents a version of the object shape $\varphi(x, y)$. This statement is corroborated by generating a moiré pattern via the transparencies $I_1(x, y)$ and $I_2(x, y)$, overlapping them to obtain the *real* moiré pattern shown in figure 5.12. Thus, the moiré pattern produced by equation (5.20) is corroborated.

The moiré pattern envelope depicts the object shape $\varphi(x, y)$, retrieved from the moiré pattern via a low pass filter. When the moiré fringe pattern contains deflections larger than 2π, a phase unwrapping procedure should be applied after the low pass filter to retrieve the object shape. It is worth pointing out that the actual great potential of the moiré fringe pattern is to retrieve the shape of tiny objects—the shape of a small object produces a large moiré pattern deflection. Furthermore, the moiré fringe pattern can be realized via laser fringe projection to retrieve the true 3D object shape, i.e. a laser beam used with a fast scanner can produce the two-line gratings (two fringe patterns) on the object surface, so there is no need to print.

References

[1] Khan M, Kazuhiko M, Teramoto K and Hasan M 2012 Precise measurement of moving object by moiré-based image processing technique *Open J. Fluid Dyn.* **2** 202–7

[2] Creath K and Wyant J C 2007 Moiré and fringe projection techniques *Optical Shop Testing* ed D Malacara 3rd edn (New York: Wiley), ch 16 pp 559–652

[3] Shao J, Liu H, Ding Y, Wang L and Lu B 2008 Alignment measurement method for imprint lithography using moiré fringe pattern *Opt. Eng.* **47** 113604
[4] Huang B, Wang C, Fang H, Zhou S and Suga T 2019 Moiré-based alignment using centro symmetric grating marks for high-precision wafer bonding *Micromachines* **10** 339
[5] Abolhassani M 2011 Formulation of moiré fringes based on spatial averaging *Optik* **122** 510–3
[6] Abolhassani M and Mirzaei M 2007 Unification of formulation of moiré fringe spacing in parametric equation and Fourier analysis methods *Appl. Opt.* **46** 7924–6
[7] Shao J, Ding Y, Tian H, Li X, Li X and Liu H 2012 Digital moiré fringe measurement method for alignment in imprint lithography *Opt. Laser Technol.* **44** 446–51
[8] Saveljev V, Kim S K and Kim J 2018 Moire effect in displays: a tutorial *Opt. Eng.* **57** 030803
[9] Zhu J, Hua S, Yu J, Tang Y, Xu F, Hea Y, Zhou S and Li L 2013 Influence of tilt moiré fringe on alignment accuracy in proximity lithography *Opt. Lasers Eng.* **51** 371–81
[10] Muñoz-Rodríguez J A and Rodríguez-Vera R 2004 Image encryption based on moiré pattern performed by computational algorithms *Opt. Commun.* **236** 295–301
[11] Ragulskis M, Aleksa A and Saunoriene L 2007 Improved algorithm for image encryption based on stochastic geometric moiré and its application *Opt. Commun.* **273** 370–8

IOP Publishing

Full Field Optical Metrology and Applications

Fernando Mendoza-Santoyo, Manuel De la Torre-Ibarra, María del Socorro Hernández-Montes and Jorge Mauricio Flores Moreno

Chapter 6

Holography

Contributing author: Thomas Kreis

Dennis Gabor reported the invention of holography in 1949. The lack of coherent electron sources at the time meant that hologram reconstruction was performed using quasi-coherent light sources. The invention of the laser in the late 1950s gave a quantum leap to holography. A laser is highly coherent and led to the invention of holographic interferometry (HI) during the first part of the 1960s. In the beginning, the hologram was recorded on a silver halide photographic plate that had to undergo the chemical process necessary to develop and fix the film. These silver halide holographic plates were replaced by video cameras and eventually digital cameras (CCD and CMOS), which circumvent the wet chemical processing involved in conventional holography. With these technological advances in electronics and devices to store data, digital holography (DH) emerged during the 1980s. More technical advances in these types of cameras improved the speed and performance of hologram recording and reconstruction. With time, it will be possible to reconstruct them in real-time. Also, the large volume of data embedded in digital holograms is benefitting from the availability of more extensive data memory banks, which eases the numerical reconstruction and retrieval of the object wavefront's intensity and phase [1–3].

6.1 Digital holographic interferometry

The digital holographic interferometry (DHI) technique derives from holographic interferometry (HI) principles. DHI provides an interferometric comparison of object events separated in time and space. For example, recording two different object states measures their amplitude and phase variations between the two states. The hologram is created from overlapping the reference and the scattered object waves. For DHI we need two: one hologram corresponds to the object in a known base state, while a second is acquired when the object has undergone a displacement or deformation. Both holograms are recorded on the CCD sensor and stored in a computer for further manipulation to retrieve the amplitude and phase from the said object changes. The object amplitude and phase are encoded in the hologram. Using numerical algorithms based on classical diffraction theory (a process called hologram numerical reconstruction) retrieves both the object's amplitude and phase.

6.1.1 Background

In the earliest set-ups, the intensity distribution $I(x_H, y_H)$ in HI is recorded on a photographic plate, as shown in figure 6.1.

For reconstruction, the processed photographic plate, called the hologram, is illuminated by the reference wave $U_r(x_H, x_H)$ alone, see figure 6.2, where a holographic set-up with amplitude division is displayed. The resulting wave field is

$$U_r(x_H, x_H) \cdot I(x, y). \tag{6.1}$$

Currently, holograms are recorded on digital sensors such as CCD cameras, referred to as digital holograms.

6.1.1.1 Numerical reconstruction of wave fields from a digital hologram

Digital holograms can be recorded if the CCD pixel pitch $\Delta\xi$ and $\Delta\eta$ can resolve the spatial frequency of the interferogram pattern forming the hologram [4–6]. This becomes an object fulfilled resolution if the angle θ between the reference and object waves, for all points of the object wave field and all points of the reference wave, at all points of the CCD, is restricted to an upper limit given by (please see equation (1.21))

$$\theta < \frac{\lambda}{2\Delta\xi}, \ \theta < \frac{\lambda}{2\Delta\eta}. \tag{6.2}$$

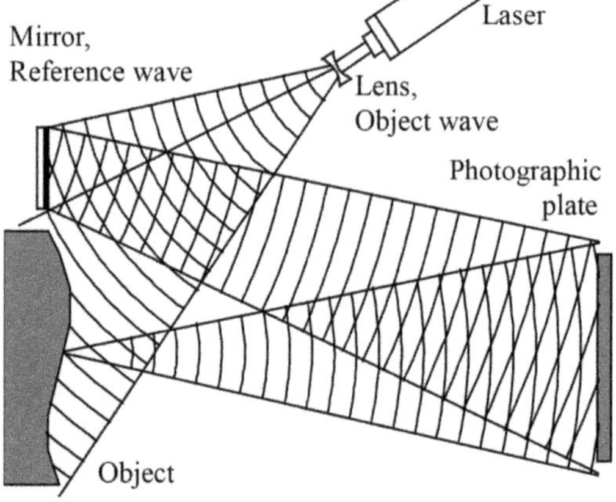

Figure 6.1. Typical optical set-up for the recording of a hologram.

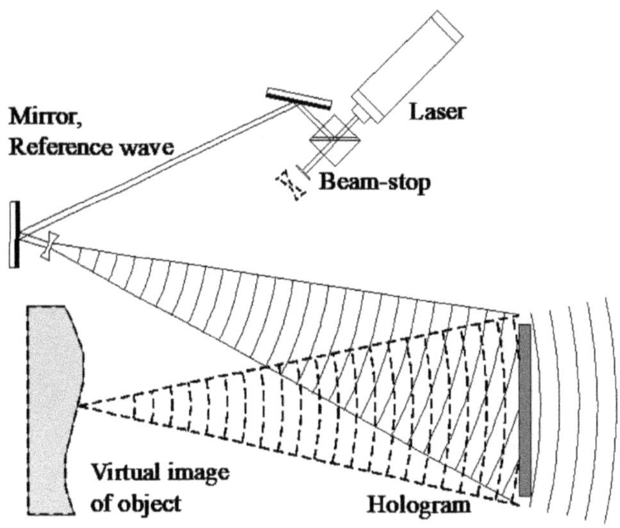

Figure 6.2. Holographic reconstruction of a wave field.

Thus the object must be placed either at a large distance from the CCD or its lateral dimensions, as seen from the CCD, are reduced by an object imaging lens [7].

Optical reconstruction of the wave fields stored holographically on photographic film is performed by illumination of the hologram with the reference wave. In digital holography, this step is replaced by a multiplication of the recorded and stored intensity distribution $h(\xi, \eta)$ with the numerical model of the reference wave field $r(\xi, \eta)$ and the calculation of the resulting diffraction field $b'(x', y')$ in the image plane. This field is given by the diffraction integral [8]

$$b'(x', y') = \frac{1}{\mathrm{i}\lambda} \iint h(\xi, \eta) r(\xi, \eta) \frac{\exp(\mathrm{i}k\rho)}{\rho} \mathrm{d}\xi \mathrm{d}\eta, \tag{6.3}$$

with $\rho = \sqrt{d'^2 + (\xi - x')^2 + (\eta - y')^2}$. The coordinates x, y, ξ, η, x', y', and d' are as defined in figure 6.3.

If a plane reference wave is normally incident, the factor $r(\xi, \eta)$ is a real constant and can be omitted [6]. Here without loss of generality, a rough plane surface in the (x, y)-plane that reflects the wave field $b(x, y)$ is assumed, keeping in mind that the procedures also apply to surfaces of any shape.

If the distance d between object and hologram plane, and equivalently d' between the hologram and image plane, is large compared to $(\xi - x')$ and $(\eta - y')$, then ρ in the denominator of equation (6.3) can be replaced by d'. Consequently, for ρ in the exponential argument, one can use the first two terms of the binomial expansion $\sqrt{1 + a} = 1 + \frac{1}{2}a - \frac{1}{8}a^2 + - \cdots$ to obtain

$$\rho \approx d' \left[1 + \frac{1}{2} \left(\frac{\xi - x'}{d'} \right)^2 + \frac{1}{2} \left(\frac{\eta - y'}{d'} \right)^2 \right]. \tag{6.4}$$

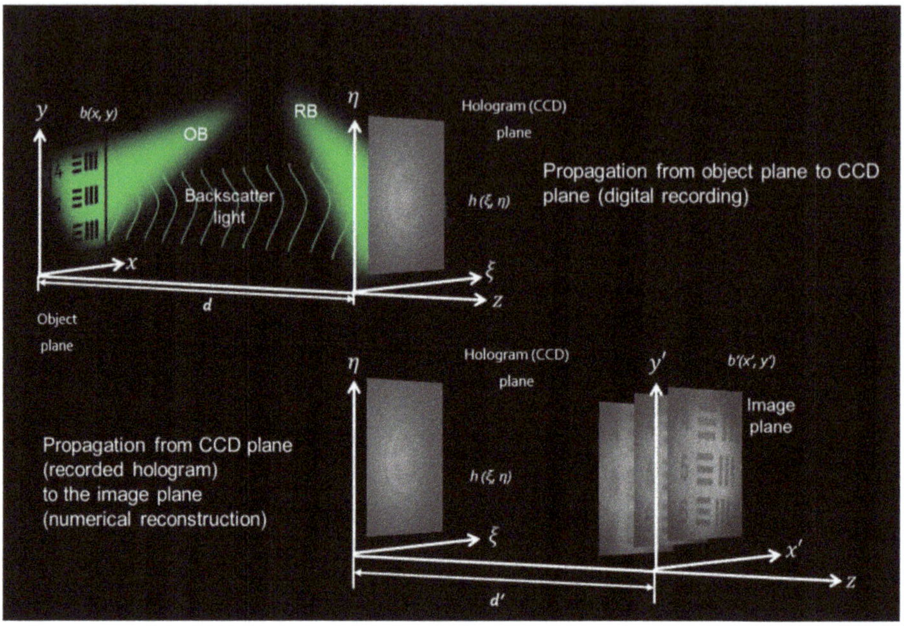

Figure 6.3. Geometry for digital holography reconstruction.

With this approximation, the diffraction integral becomes

$$b'(\nu, \mu) = \frac{\exp(\mathrm{i}kd')\exp[\mathrm{i}\pi d'\lambda(\nu^2 + \mu^2)]}{\mathrm{i}\lambda d'}$$

$$\times \iint h(\xi, \eta) r(\xi, \eta) \exp\left[\frac{\mathrm{i}\pi}{d'\lambda}(\xi^2 + \eta^2)\right] \exp[-2\mathrm{i}\pi(\xi\nu + \eta\mu)]\mathrm{d}\xi\mathrm{d}\eta, \tag{6.5}$$

where $\nu = x'/(d'\lambda)$ and $\mu = y'/(d'\lambda)$. This awkward-looking integral, nevertheless, is relatively easy to handle: the recorded hologram $h(\xi, \eta)$ is multiplied pointwise with the reference wave $h(\xi, \eta)$. This multiplication is unnecessary if a plane wave is impinging normally since it does not vary spatially.

The subsequent multiplication is with the chirp function $\exp[\mathrm{i}\pi(\xi^2 + \eta^2)/(d'\lambda)]$. Then the Fourier transform of this product is taken. The result of this Fourier transform is multiplied by the spatially constant intensity factor $1/(\mathrm{i}\lambda d')$ and a phase factor that does not depend on the evaluated hologram, so the factors outside the integral can typically be neglected. The stored digital hologram consists of $N \times M$ discrete values, each representing the intensity in a pixel of size $\Delta\xi \times \Delta\eta$. The given number and size of the elements of the CCD array define the discretization of equation (6.5):

$$b'(n, m) = \sum_{k=0}^{N-1}\sum_{l=0}^{M-1} h(k\Delta\xi, l\Delta\eta) r(k\Delta\xi, l\Delta\eta) \exp\left[\frac{\mathrm{i}\pi}{d'\lambda}(k^2\Delta\xi^2 + l^2\Delta\eta^2)\right]$$
$$\exp\left[-2\mathrm{i}\pi\left(\frac{kn}{N} + \frac{lm}{M}\right)\right], \tag{6.6}$$

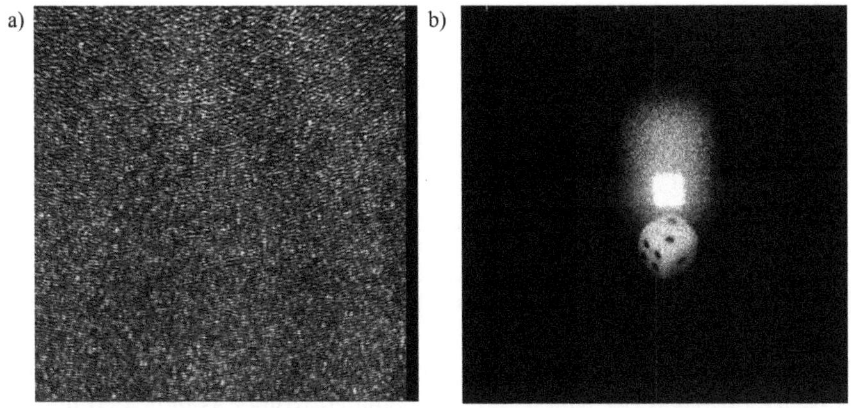

Figure 6.4. (a) Digital hologram and (b) its reconstructed intensity distribution.

where, for better readability, the constant factor $\exp(ikd')/(i\lambda d')$ and the pure phase factor $\exp\{i\pi d'\lambda[n^2/(N^2\Delta\xi^2) + m^2/(M^2\Delta\eta^2)]\}$ have been omitted. The field reconstructed by this numerical Fresnel transform in equation (6.6) is complex. Intensity and phase distributions can be calculated by

$$I(n, m) = b'(n, m) \cdot b'^{*}(n, m) \quad \varphi(n, m) = \arctan \frac{\text{Im}(b'(n, m))}{\text{Re}(b'(n, m))}. \tag{6.7}$$

The pixel spacing, or equivalently pixel size, in the reconstructed field is

$$\Delta x' = \frac{d'\lambda}{N\Delta\xi}, \quad \Delta y' = \frac{d'\lambda}{M\Delta\eta}. \tag{6.8}$$

Figure 6.4 displays a digital hologram and its intensity distribution, calculated with equation (6.7) from the wave field reconstructed by the Fresnel transform, equation (6.6). The real image of the object is seen, a dice, the bright dc-term, and the unsharp virtual image of the dice.

The diffraction integral (6.3) is a superposition integral

$$b'(x', y') = \iint h(\xi, \eta) r(\xi, \eta) g(x', y', \xi, \eta) d\xi d\eta, \tag{6.9}$$

with the impulse response $g(x', y', \xi, \eta)$ given by

$$g(x', y', \xi, \eta) = \frac{1}{i\lambda} \frac{\exp\left[ik\sqrt{d'^2 + (\xi - x')^2 + (\eta - y')^2}\right]}{\sqrt{d'^2 + (\xi - x')^2 + (\eta - y')^2}}. \tag{6.10}$$

The linear system characterized by $g(x', y', \xi, \eta) = g(x' - \xi, y' - \eta)$ is space-invariant, so the superposition integral, equation (6.9), is a convolution. For the determination of $b'(x', y')$ now we can invoke the convolution theorem, which states that the convolution of the product $h \cdot r$ with the impulse response g is the inverse Fourier transform of the product of the individual Fourier transforms of $h \cdot r$ and g:

$$b' = \mathcal{F}^{-1}\{\mathcal{F}\{h \cdot r\} \cdot \mathcal{F}\{g\}\}. \tag{6.11}$$

The Fourier transform numerical calculation is performed effectively by a fast Fourier transform (FFT) algorithm. The discrete numerical realization of the impulse response for free-space propagation with coordinate shifts by $N/2$ and $M/2$, using symmetry, is

$$g(k, l) = \frac{1}{\mathrm{i}\lambda} \frac{\exp\left[\frac{2\mathrm{i}\pi}{\lambda}\sqrt{d'^2 + (k - N/2)^2\Delta\xi^2 + (l - M/2)^2\Delta\eta^2}\right]}{\sqrt{d'^2 + (k - N/2)^2\Delta\xi^2 + (l - M/2)^2\Delta\eta^2}}. \tag{6.12}$$

The Fourier transform of the impulse response $g(k, l)$ is the transfer function $G(n, m)$ of free-space propagation, which can be calculated from g as indicated in equation (6.12) or may be defined directly by

$$G(n, m) = \exp\left[\frac{2\mathrm{i}\pi d'}{\lambda}\sqrt{1 - \frac{\lambda^2\left(n + \frac{N^2\Delta\xi^2}{2d'\lambda}\right)}{N^2\Delta\xi^2} - \frac{\lambda^2\left(m + \frac{M^2\Delta\eta^2}{2d'\lambda}\right)}{M^2\Delta\eta^2}}\right]. \tag{6.13}$$

The latter allows skipping one Fourier transform

$$b' = \mathcal{F}^{-1}\{\mathcal{F}\{h \cdot r\} \cdot G\}. \tag{6.14}$$

With equations (6.12) or (6.13), the diffraction integral is determined precisely. Nevertheless, the square roots in equations (6.12) and (6.13) can be replaced by the first terms of their binomial series to give their Fresnel approximations [9].

It is worth pointing out here that there is a conceptual difference between Fresnel transform reconstruction and the convolution. If we interpret the (ξ, η) plane of the digital hologram as the spatial domain, the procedure of equation (6.6) results in the spatial frequency domain due to the single Fourier transform. Thus the pixels values in the (x', y')-plane, the image plane, are spatial frequencies; their spacing varies with distance d' and wavelength λ. On the other hand, the algorithms based on convolution first transform into the frequency domain, where the multiplication with the transfer function is performed. However, then the product is Fourier transformed back into the spatial domain. As a consequence, the pixel size in the reconstructed image now is $\Delta x' = \Delta\xi$ and $\Delta y' = \Delta\eta$. Contrary to the Fresnel case, see equation (6.8), the size of the reconstructed image does not depend on the reconstruction distance d'. Thus, the field reconstructed by the convolution approach represents only a part of the scene that the Fresnel transform would reconstruct. However, it is possible to reconstruct shifted parts and join them together like a mosaic or puzzle or increase the pixel number and vary the pixel size to change the dimensions of the reconstructed parts [9].

6.1.1.2 Suppression of the dc-term

When reconstructing from a Fresnel hologram, the reconstructed field suffers from a bright rectangular or square dc-term. Its size is $N^2\,\Delta\xi^2/(d\lambda)$ pixels in the ξ-direction and $M^2\,\Delta\eta^2/(d\lambda)$ pixels in the η-direction. From a computational standpoint, it is the Fourier transform of the chirp function. Optically we see the reconstructed CCD array illuminated by the reference wave or, physically speaking, the zero diffraction order. Checking on equation (6.6), the Fresnel transform is the inverse Fourier transform of the product of the hologram times the reference wave $h \cdot r$ and the chirp function $w(\xi, \eta) = \exp\,[\mathrm{i}\pi(\xi^2 + \eta^2)/(d'\lambda)]$, which can also be written as the convolution of the individual inverse Fourier transforms

$$b' = \mathcal{F}^{-1}\{h \cdot r \cdot w\} = \mathcal{F}^{-1}\{h \cdot r\}\mathcal{F}^{-1}\{w\}. \tag{6.15}$$

Now the dc-term in b' is the dc-term of $h \cdot r$ convolved with the Fourier transform of the chirp function w. The dc-term in $\mathcal{F}^{-1}\{h \cdot r\}$ is just the sum over all values of $h \cdot r$, so if we can guarantee that this sum will be zero, the dc-term of b' will vanish. Therefore after the digital hologram h is recorded, we can form a modified hologram h' which has a zero dc-term by first calculating the average intensity h_{av}

$$h_{\mathrm{av}} = \frac{1}{NM}\sum_{k=0}^{N-1}\sum_{l=0}^{M-1} h(k\Delta\xi, l\Delta\eta), \tag{6.16}$$

and then subtracting this value from each stored intensity value, obtaining

$$h'(k\Delta\xi, l\Delta\eta) = h(k\Delta\xi, l\Delta\eta) - h_{\mathrm{av}} \quad k = 0, \ldots, N-1 \quad l = 0, \ldots, M-1 \tag{6.17}$$

While the hologram h has only positive intensities, h' exhibits negative intensities as well which, although optically impossible, is feasible numerically. The relations between the intensities of the individual pixels remain the same; only the whole hologram is downshifted in intensity [10]. In figure 6.5 we see two reconstructions,

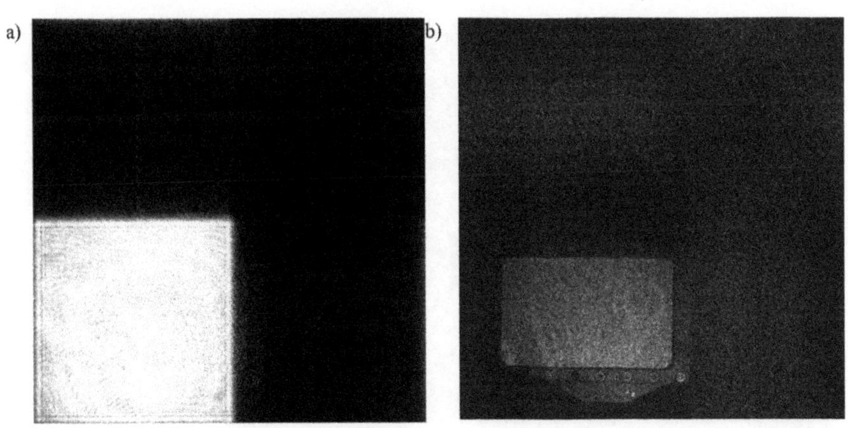

Figure 6.5. Reconstructed real image (a) without and (b) with suppression of the dc-term.

one from h without suppression of the dc-term, figure 6.5(a), the other from $h - h_{av}$ with suppression, figure 6.5(b). The object is an aluminum plate that is clamped at the edges. Used of a concave lens gives a reduced virtual image of the object. The dc-term covers nearly the whole object in figure 6.5(a), while the full object surface is seen in figure 6.5(b).

6.1.1.3 Numerical interference phase determination

In holographic interferometry, the two wave fields to be interferometrically compared are reconstructed simultaneously and an interferogram results. A direct translation of this approach to digital holography would consist of an addition of the two recorded digital holograms $h_1(k, l)$ and $h_2(k, l)$ producing a digital double exposure hologram $(h_1 + h_2)(k, l)$. Its numerical reconstruction gives a field whose intensity is the interferogram. Figure 6.6(a) shows the reconstructed intensity distribution of such a digital sum-hologram. Here the aluminum plate of figure 6.5 is pushed from behind the plate, at its center. The dc-term is eliminated as described before. Such an intensity interferogram requires an evaluation to give the interference phase distribution.

In digital holography, we have numerical access to the phase distributions of the wave fields without the need for complicated techniques such as phase shifting or Fourier transform evaluation. If the two digital holograms are individually reconstructed and their phase distributions $\varphi_1(n, m)$ and $\varphi_2(n, m)$ are calculated according to equation (6.7), then the interference phase can be determined with modulo 2π:

$$\Delta\varphi(n, m) = \begin{cases} \varphi_2(n, m) - \varphi_1(n, m) + 2\pi & \text{if} \quad \varphi_2(n, m) - \varphi_{12}(n, m) < -\pi \\ \varphi_2(n, m) - \varphi_1(n, m) - 2\pi & \text{if} \quad \varphi_2(n, m) - \varphi_{12}(n, m) \geqslant -\pi. \\ \varphi_2(n, m) - \varphi_1(n, m) & \text{else} \end{cases} \quad (6.18)$$

A more direct calculation of $\Delta\varphi(n, m)$ without the need for a case of discrimination is

a)

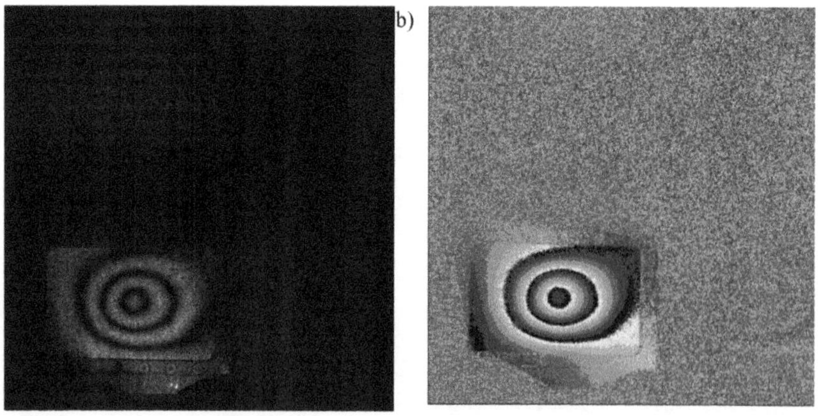

b)

Figure 6.6. Intensity reconstructed from (a) sum-hologram and (b) phase difference modulo 2π.

$$\Delta\varphi(n, m) = \arctan\frac{\mathrm{Re}\left[b_1'(n, m)\right]\mathrm{Im}\left[b_2'(n, m)\right] - \mathrm{Re}\left[b_2'(n, m)\right]\mathrm{Im}\left[b_1'(n, m)\right]}{\mathrm{Re}\left[b_1'(n, m)\right]\mathrm{Im}\left[b_2'(n, m)\right] + \mathrm{Re}\left[b_2'(n, m)\right]\mathrm{Im}\left[b_1'(n, m)\right]},$$

$$\Delta\varphi(n, m) = \arctan\frac{\mathrm{Im}\left[b_2'(n, m)\cdot b_1'^{*}(n, m)\right]}{\mathrm{Re}\left[b_2'(n, m)\cdot b_1'^{*}(n, m)\right]}. \tag{6.19}$$

Without the detour via the intensity, which destroys the sign information due to $\cos(-\varphi) = \cos(+\varphi)$, here the calculated interference phase distribution is sign-correct. Increasing as well as decreasing phases in any direction are correctly detected in figure 6.6(b) show the interference phase distribution modulo 2π determined in this way. The phase factors preceding the double sum of (6.5) can be neglected in the calculation since they cancel out during the phase subtraction, see equation (6.18) or (6.19).

The availability of the individual reconstructed phase distributions in digital holographic interferometry offers flexible metrology applications. Thus one may record a series of digital holograms with gradually increased load amplitude. In the evaluation stage, the convenient states can be compared interferometrically [11]. Motion components that lead to high fringe densities can be numerically compensated. One may add or subtract a certain phase to/from the calculated phase difference. Although we will not obtain additional information by doing this, sometimes fine details in the display of the interference phase are better recognized [11].

6.1.2 General working principle of DHI

The DHI set-up can be configured in either mode, transmission to look at transparent objects (also called phase objects) or scattering from diffuse or non-specular objects.

The transmission set-up is shown in figure 6.7. The optical layout resembles a Mach–Zehnder interferometer, where a laser can be CW or pulsed, divided by the beam splitter (BS1) into the reference and object beams. The reference wave is guided through an optical fiber (OF) to the camera sensor; the object wave is expanded with a microscope objective and reflected at a mirror (M1) that directs this beam through the sample. With another beam splitter (BS2), the reference and object waves are overlapped on the CCD sensor, thus creating a hologram.

Figure 6.8 shows a typical set-up for diffuse objects. The hologram acquisition procedure is the same as for the phase objects. The fiber optics in both arrangements can be replaced with a mirror to guide the reference beam. However, the use of optical fibers allows more straightforward arrangements for DHI and avoids the use of lenses to manipulate the size of the reference beam. An aperture–lens combination can be placed just before the second beam splitter (BS2) and conveys the object image to the camera sensor for both layouts. The aperture–lens combination aids in increasing the depth of focus of the imaging system. In addition, it controls the

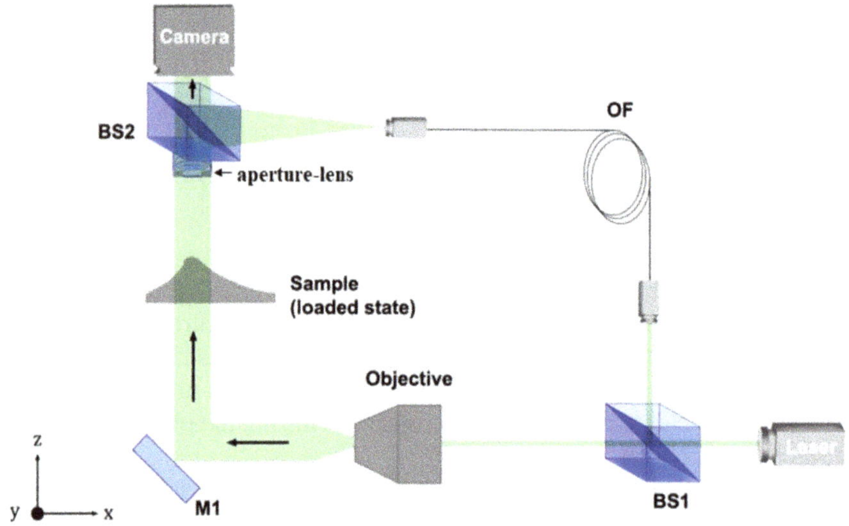

Figure 6.7. Digital holographic interferometry in transmission mode for transparent object measurements.

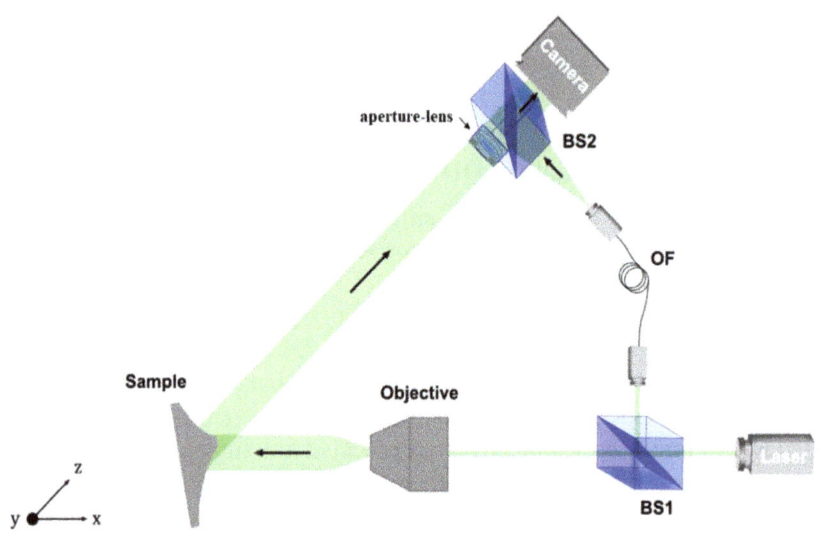

Figure 6.8. Digital holographic interferometry in scattering mode to study diffuse or opaque objects.

amount of light reaching the sensor and acts as a bandpass filter for the light scattered by the object. The aperture itself is also used during the holographic reconstruction process in the Fourier space when a filter is applied.

The optical phase retrieval in DHI is based on the interference of a smooth reference wave and an object wave expressed as $U_r(x_H, y_H)$ and $U_o(x_H, y_H)$, respectively. The hologram intensity recorded on the camera sensor is given by

$$I(x_H, y_H) = |U_r(x_H, y_H) + U_o(x_H, y_H)|^2$$
$$= |U_r(x_H, y_H)|^2 + |U_o(x_H, y_H)|^2 + U_r(x_H, y_H)U_o^*(x_H, y_H) \quad (6.20)$$
$$+ U_r^*(x_H, y_H)U_o(x_H, y_H),$$

where x_H and y_H are the x and y coordinates at the image hologram plane H (the camera sensor), and the symbol * denotes the complex-conjugate amplitude. The first two terms are related to the reference and object beam intensities. The last two terms contain information corresponding to the amplitude and the phase of the object wave, data that can be retrieved using the fast Fourier transform method, phase-stepping algorithms, or other types of algorithms, see chapter 3. The object and the reference waves in terms of their amplitude and phase are described by

$$U_o(x_H, y_H) = u_o(x_H, y_H)e^{i\varphi(x_H, y_H)}$$

$$U_r(x_H, y_H) = u_r(x_H, y_H)e^{-2\pi i(f_x x_H + f_y y_H)}, \quad (6.21)$$

where u_o and u_r are the object and reference amplitudes, respectively. $f_x x_H$ and $f_y y_H$ represent the spatial carrier frequencies along with the x_H and y_H coordinates [12, 13], introduced here in the smooth reference beam because they aid in simplifying the spectrum at the Fourier space (as seen in the next section). Experimentally, these carrier frequencies are introduced as a tiny tilt of the reference beam with respect to the object beam as they fall on the camera sensor.

6.1.3 Fourier transform method

The Fourier transform method (FTM) is used in DHI to retrieve the optical phase produced by the change in object states. The method requires the introduction of a spatial carrier frequency, which can be introduced experimentally in the reference beam by tilting it slightly according to the optical axis, namely the axis from the camera sensor center to the object geometrical center (or the center of the object area under study). To further ease the introduction of the spatial carriers, experimentally, the location of the reference beam is selected such that its distance to the CCD sensor is the same as the distance of the aperture center to the CCD sensor. To correctly record the holograms, the size of the aperture and its distance from the CCD sensor are determined by the object image spatial frequencies set by this combination that has to fulfill the sampling theorem as set by the *Nyquist criterion* [14–16]. The maximum spatial frequency (f_m) is given by the CCD sensor resolution and its pixel size ($\Delta\xi$):

$$f_m \leqslant 1/2\Delta\xi, \quad (6.22)$$

f_m is resolved by the maximum angle (α_m) resulting from the superposition of the reference and object beams, given as

$$\alpha_m = \lambda f_m \approx \lambda/2\Delta\xi, \quad (6.23)$$

where α_m is stated by the pixel size and the λ used.

Equation (6.20) represents the hologram intensity recorded by the CCD camera; upon substitution of the relationships in equation (6.21), we obtain

$$I(x_H, y_H) = a(x_H, y_H) + c(x_H, y_H)e^{2\pi i(f_x x_H + f_y y_H)} + c^*(x_H, y_H)e^{-2\pi i(f_x x_H + f_y y_H)}, \quad (6.24)$$

with

$$a(x_H, y_H) = u_o^2(x_H, y_H) + u_r^2(x_H, y_H) \quad (6.25)$$

$$c(x_H, y_H) = u_o(x_H, y_H)u_r(x_H, y_H)e^{i\varphi(x_H, y_H)}. \quad (6.26)$$

The last two terms of equation (6.24) indicate that the spatial carrier frequencies are modulated by the optical phase term $\varphi(x_H, y_H)$ coming from the object. In order to retrieve this optical phase, equation (6.24) is Fourier transformed in two dimensions,

$$\begin{aligned}\mathscr{F}\{I(x_H, y_H)\} = A\left(f_{x_H}, f_{yH}\right) + C\left(f_{x_H} - f_x x_H, f_{yH} - f_y y_H\right) \\ + C^*\left(f_{x_H} + f_x x_H, f_{yH} + f_y y_H\right).\end{aligned} \quad (6.27)$$

It can be seen that the C and C^* terms are the Fourier transforms of c and c^*, respectively (each contains the same phase information $\varphi(x_H, y_H)$), while $A\left(f_{x_H}, f_{yH}\right)$ is the low-frequency background illumination, and $\left(f_{x_H}, f_{yH}\right)$ is the frequency coordinate at the Fourier space. The spectrum of the three terms in the last equation can be distinctively seen in the Fourier space, this thanks to the introduction of the spatial carriers (the tilt given to the reference beam). Then it becomes a simple task to choose the A and C^* terms to be removed through a bandpass filter $\left(B\left(f_x x_H, f_y y_H\right)\right)$:

$$C\left(f_{x_H}, f_{yH}\right) = B\left(f_x x_H, f_y y_H\right) \otimes \mathscr{F}\{I(x_H, y_H)\}. \quad (6.28)$$

Then the remaining C term is frequency centered and its inverse Fourier transform is applied to yield

$$c(x_H, y_H) = \mathscr{F}^{-1}\left\{C\left(f_{x_H}, f_{yH}\right)\right\}. \quad (6.29)$$

Hence the optical phase can be obtained by

$$\varphi(x_H, y_H) = \arctan\frac{\mathrm{Im}\,c(x_H, y_H)}{\mathrm{Re}\,c(x_H, y_H)}, \quad (6.30)$$

where $\mathrm{Im}\,c(x_H, y_H)$ and $\mathrm{Re}\,c(x_H, y_H)$ are the imaginary and real parts of the Fourier inverted hologram of the C term.

Figure 6.9(a) shows the recorded hologram as captured by the sensor and figure 6.9(b) shows the result of the 2D FFT applied to the hologram: it contains the three terms given in equation (6.27), particularly its results in two angularly spaced sidelobes separated from the low-frequency background illumination at the center, representing a complex-conjugate pair of identical holograms.

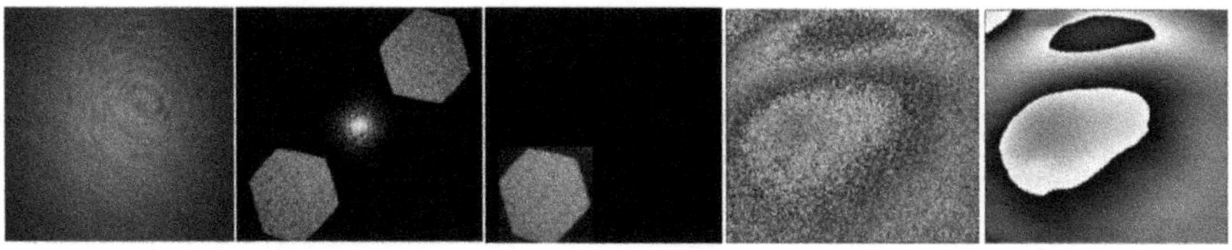

Figure 6.9. (a) Hologram, (b) Fourier, (c) bandpass filter, (d) optical phase, and (e) wrapped phase transform map.

To quantitatively evaluate the optical phase one of the lobes is chosen using a bandpass filter to retain it while eliminating the other, as shown in figure 6.9(c). After this bandpass is applied, the inverse FFT is applied to the filtered lobe giving the complex amplitude of the object beam $c(x_H, y_H)$ from which the optical phase $\Delta\varphi(x_H, y_H)$ is retrieved, as shown in figure 6.9(d)—notice the speckle noise contained in the image. In figure 6.9(e) the wrapped phase is depicted after a filtering process carried out on the image in figure 6.9(d).

Upon subtracting the two phases for the base (φ_{ud}) and deformed (φ_d) object states, a wrapped phase map is obtained showing the relative phase difference ($\Delta\varphi$) calculated with

$$\Delta\varphi(x_H, y_H) = \varphi_{ud}(x_H, y_H) - \varphi_d(x_H, y_H), \tag{6.31}$$

equation (6.31) can be expressed in terms of $c_{ud}(x_H, y_H)$ and $c_d(x_H, y_H)$ before and after the object's deformation as follows:

$$\Delta\varphi(x_H, y_H) = \frac{\mathrm{Re}\{c_{ud}\}\,\mathrm{Im}\{c_d\} - \mathrm{Re}\{c_d\}\mathrm{Im}\{c_{ud}\}}{\mathrm{Re}\{c_{ud}\}\,\mathrm{Re}\{c_d\} + \mathrm{Im}\{c_{ud}\}\mathrm{Im}\{c_d\}}, \tag{6.32}$$

$\Delta\varphi(x_H, y_H)$ is contained in a so-called wrapped map because its values range between $-\pi$ and π due to the discontinuous nature of the tangent function [17, 18].

DHI has been successfully applied in many applications in most areas of knowledge. These many applications meant that DHI could evolve to become a trusted tool to find and quantify variables such as surface deformation, surface shape, refractive index, and object thickness, and characterize the vibration behavior of objects and rotation information from the object, among many other applications.

References

[1] Trolinger J D and Mansoor M M 2022 History and metrology applications of a game-changing technology: digital holography *J. Opt. Soc. Am.* A **39** A29–43

[2] Gabor D 1948 A new microscopic principle *Nature* **161** 777–8

[3] Leith E N and Upatnieks J 1964 Wavefront reconstruction with diffused illumination and three-dimensional objects *J. Opt. Soc. Am.* **54** 1295–301

[4] Goodman J W and Lawrence R W 1967 Digital image formation from electronically detected holograms *Appl. Phys. Lett.* **11** 77–9

[5] Kronrod M A, Merzlyakov N S and Yaroslavskii L P 1972 Reconstruction of holograms with a computer *Sov. Phys.—Tech. Phys.* **17** 333–4
[6] Schnars U 1994 Direct phase determination in hologram interferometry with use of digitally recorded holograms *J. Opt. Soc. Am.* **A11** 2011–5
[7] Schnars U, Kreis T and Jüptner W 1996 Digital recording and numerical reconstruction of holograms: reduction of the spatial frequency spectrum *Opt. Eng.* **35** 977–82
[8] Goodman J W 1996 *Introduction to Fourier Optics* 2nd edn (New York: McGraw-Hill)
[9] Kreis T, Adams M and Jüptner W 1997 Methods of digital holography: a comparison *Proc. SPIE* **3098** 224–33
[10] Kreis T and Jüptner W 1997 The suppression of the dc-term in digital holography *Opt. Eng.* **36** 2357–60
[11] Kreis T, Jüptner W and Geldmacher J 1998 Principles of digital holographic interferometry *Proc. SPIE* **3478** 45–54
[12] Sharpe W N (ed) 2008 *Springer Handbook of Experimental Solid Mechanics* (New York: Springer)
[13] Hernández-Montes M D S, Mendoza-Santoyo F, Flores Moreno M, de la Torre-Ibarra M, Silva Acosta L and Palacios-Ortega N 2020 Macro to nano specimen measurements using photons and electrons with digital holographic interferometry: a review *J. Eur. Opt. Soc.: Rapid Publ.* **16** 1–31
[14] Takeda M, Ina H I and Kobayashi S 1982 Fourier-transform method of fringe-pattern analysis for computer-based topography and interferometry *JOSA.* **72** 156–60
[15] Mendoza Santoyo F, Kerr D and Tyrer J R 1989 Manipulation of the Fourier components of speckle fringe patterns as part of an interferometric analysis process *J. Mod. Opt.* **36** 195–204
[16] Ferraro P, Del Core C, Miccio L, Grilli S, De Nicola S, Finizio A and Coppola G 2008 Phase map retrieval in digital holography: avoiding the under-sampling effect by a lateral shear approach *Proc. SPIE* **6995** 699506
[17] Hernandez-Montes M D S, Mendoza Santoyo F, Muñoz S, Perez C, de la Torre M, Flores M and Alvarez L 2015 Surface strain-field determination of tympanic membrane using 3D-digital holographic interferometry *Opt. Lasers Eng.* **71** 42–50
[18] Hernandez-Montes M D S, Mendoza Santoyo F, Perez C, Muñoz Solís S and Esquivel J 2011 Digital holographic interferometry applied to the study of tympanic membrane displacements *Opt. Lasers Eng.* **49** 698–702

IOP Publishing

Full Field Optical Metrology and Applications

Fernando Mendoza-Santoyo, Manuel De la Torre-Ibarra, María del Socorro Hernández-Montes and Jorge Mauricio Flores Moreno

Chapter 7

Digital speckle pattern interferometry (DSPI)

In what follows, some fundamental optical interferometric methods based on the speckle phenomena are presented. All have important applications in a wide variety of fields such as design and testing, control, aerospace, and biomedicine, among many others.

7.1 Background

Speckle pattern interferometry (SPI) was probably the start of it all. It renders non-destructive measurements with a whole field of view of the object under study. Data are gathered with high sensitivity, namely around the order of the laser wavelength. Double exposure SPI is used for displacement measurements, where two object states are required, and the recording is made on photographic film. The speckled fringes appear after chemically processing the film and illuminating it with a laser. In the late 1960s photographic films were substituted by electronic cameras with frame acquisition storage. This technological change introduced the pioneering technique called electronic speckle pattern interferometry (ESPI), an optoelectronic system that eliminates the chemical photographic process and the time-consuming qualitative and quantitative measurements as the fringes are displayed in real time on a TV monitor. The ESPI system was first reported in the UK with a speed of 25 frames per second (the standard TV frame rate in that part of the world), although there were other researchers working on similar devices at the same time. ESPI also allows whole-field-of-view object measurements requiring two beams. The first is the reference beam, a smooth wavefront (typically diverging) superimposed on the object beam backscattered from the surface. The result is an interference speckled fringe pattern or specklegram stored in the electronic system frame grabber (today, many different storage devices can be used). The fringes are formed by electronic

doi:10.1088/978-0-7503-3027-5ch7

correlation and video processing of the speckle patterns by subtracting a surface-primary, object base state, speckle pattern on the frame grabber from the subsequently acquired frames, with the resulting fringe pattern shown in real time on a television monitor. The size of the speckle must be made large enough to fit within at least one pixel of the camera sensor to obtain high-contrast fringe patterns. The fringe patterns are produced when a phase change (change in optical path lengths) is present between the primary speckle pattern and the subsequent ones, i.e. the phase change is due to object displacement (or rigid body motion). ESPI can provide displacement measurements on the order of half the wavelength of light [1, 2].

7.2 DSPI technique

The fast and continuous hardware advances associated with ESPI made it possible to develop digital speckle pattern interferometry (DSPI). Examples of these advances include manufacturing and new technologies in optics and electronics (such as charge-coupled devices (CCDs), electron-multiplying CCDs (EMCCDs), complementary metal–oxide semiconductors (CMOSs), scientific CMOSs (sCMOS), and back-illuminated sCMOSs) in order to provide the appropriate speed, sensitivity, resolution, frame rate, and field of view for the object under study. More extensive and faster data storage computers and algorithms designed to acquire and manipulate data have also been developed [3, 4].

DSPI yields a fringe pattern viewed in real time on a computer screen. As mentioned, these fringes are the result of two object correlated speckle patterns captured at two different states: the reference (I_{ud}) can be considered when the object is at rest or immediately after it suffered a change and the modified state (I_d) when the object has undergone a displacement or deformation. These two object states are reflected in alterations of the speckle pattern along the x-, y-, and z-directions. In ESPI and DSPI, it is customary to correlate these two speckle patterns by subtraction (although addition or multiplication of these two patterns is also possible), resulting in a fringe pattern, see figure 7.1. The process for correlation by subtraction is described by

$$I_d - I_{ud} = 2u_o u_r[\cos\varphi_{o-r} - \cos(\varphi_{o-r} + \Delta\varphi)],$$

$$= 4u_o u_r \sin(\varphi_{o-r} + \Delta\varphi/2)(\sin\Delta\varphi/2), \tag{7.1}$$

where the terms I_{ud} and I_d are given by

$$I_{ud}(x, y) = |u_r(x, y)\mathrm{e}^{(\mathrm{i}\varphi_r)} + u_o(x, y)\mathrm{e}^{(\mathrm{i}\varphi_o)}|^2,$$

$$I_d(x, y) = |u_r(x, y)\mathrm{e}^{(\mathrm{i}\varphi_r)} + u_o(x, y)\mathrm{e}^{(\mathrm{i}\varphi_o + \Delta\varphi)}|^2 \tag{7.2}$$

with $\varphi_{o-r} = \varphi_o - \varphi_r$. $\Delta\varphi$ is the optical phase difference corresponding to the object displacement and is the variable we want to measure. As the object is displaced/deformed during the speckle pattern recording, the optical path length of the backscattered light from the object varies, resulting in $\Delta\varphi$, which is observed as

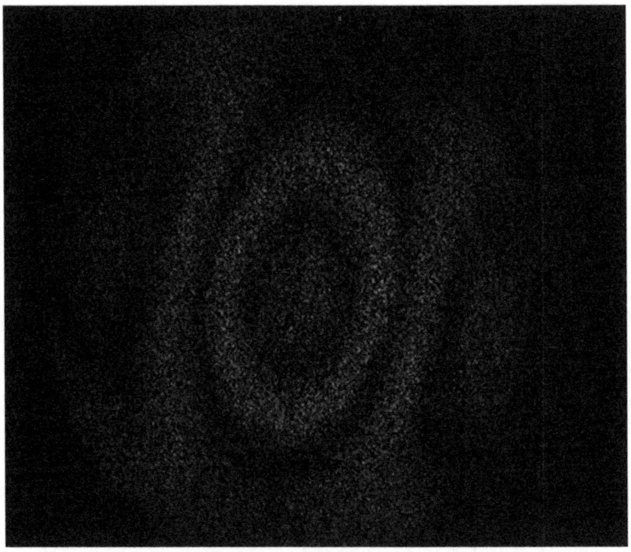

Figure 7.1. A typical ESPI/DSPI fringe pattern.

fringe patterns. When two speckle patterns are fully correlated, $I_d = I_{ud}$ or $\Delta\varphi = 2m\pi$, dark fringes are created, and when the two speckle patterns are uncorrelated, $I_d \neq I_{ud}$ or $\Delta\varphi = (2m + 1)\pi$ bright fringes are produced, with m an integer.

For completeness, the correlation by addition is given by

$$I_d + I_{ud} = 2(u_o^2 + u_r^2) + 2u_O u_R[\cos\varphi_{o-r} + \cos(\varphi_{o-r} + \Delta\varphi)]$$

$$= 2(u_o^2 + u_r^2) + 4u_O u_R \cos(\varphi_{o-r} + \Delta\varphi/2)\cos(\Delta\varphi/2). \tag{7.3}$$

The background random speckle noise added by $2(u_o^2 + u_r^2)$ dramatically decreases the fringe pattern contrast. In this case, when two speckle patterns are fully correlated or $\Delta\varphi = 2m\pi$, $\cos(\Delta\varphi/2) = 1$, bright fringes are created; and when $\Delta\varphi = (2m + 1)\pi$, $\cos(\Delta\varphi/2) = 0$, dark fringes are created, and the two speckle patterns are uncorrelated, with m an integer.

The fringe pattern carries information on the changes experiences by the test object. The number and direction of the fringes are determined by the object's magnitude and the direction of its displacement.

Finally, to quantify the displacement, it is necessary to assess the optical phase $\Delta\varphi$ between two speckle patterns captured at different states

$$\Delta\varphi = \varphi_d - \varphi_{ud}, \tag{7.4}$$

where φ_d and φ_{ud} are the phases of the modified and reference states. $\Delta\varphi$ is determined from the recorded fringe patterns where this variable is embedded [5, 6], and its quantification needs the analysis of some variables, a task dealt with in the next section.

7.3 The sensitivity vector, the displacement vector, and the optical phase difference

The sensitivity vector, the displacement vector, and the optical phase difference are variables that we must have for the quantitative analysis of interferometric techniques.

Let us start with the case of object displacements along its out-of-plane direction, i.e. displacements along the normal/perpendicular direction to its surface. The sensitivity vector (***S***) illustrated in figure 7.2 specifies the deformation direction in which the experimental set-up will be most sensitive. ***S*** bisects the unitary vectors $\boldsymbol{s}_i$ and $\boldsymbol{s}_o$ along with the illumination and observation directions.

The Cartesian components of $\boldsymbol{s}_i$ and $\boldsymbol{s}_o$ are and

$$\boldsymbol{s}_i = s_{ix}\hat{i} + s_{iy}\hat{j} + s_{iz}\hat{k} \text{ and } \boldsymbol{s}_o = s_{ox}\hat{i} + s_{oy}\hat{j} + s_{oz}\hat{k}. \quad (7.5)$$

Figure 7.2 shows the illumination point (I_p), e.g. the laser spot, O_p is the detector's observation position, and ***P*** is a point arbitrarily located on the object; the positions of these points are defined by $\mathbf{r}_1$, $\mathbf{r}_2$, and $\mathbf{r}_p$ in that order and are expressed in Cartesian coordinates as

$$\mathbf{r}_n = r_{nx}\hat{i} + r_{ny}\hat{j} + r_{nz}\hat{k} \quad n = 1,2, \ldots P, \quad (7.6)$$

where r_x, r_y, and r_z represent the coordinates of these vectors, and $\hat{\imath}$, $\hat{\jmath}$, and $\hat{k}$ are the unitary vectors in the direction of the x-, y-, and z-axes. Now, $\boldsymbol{s}_i$ and $\boldsymbol{s}_o$ can be rewritten in terms of the position vectors as

$$\boldsymbol{s}_i = k\frac{r_p - r_1}{\left|r_p - r_1\right|} \text{ and } \boldsymbol{s}_o = k\frac{r_2 - r_p}{\left|r_2 - r_p\right|}, \quad (7.7)$$

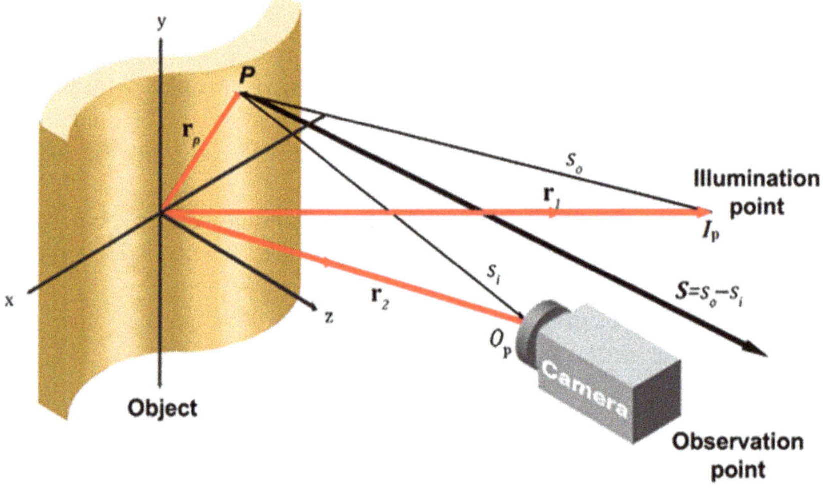

Figure 7.2. Schematic showing the geometry to determine the sensitivity vector ***S***. From the figure, $\mathbf{r}_1$, $\mathbf{r}_2$, and $\mathbf{r}_p$ are the position vectors for the illumination, observation, and a point ***P*** on the object surface, respectively. $\boldsymbol{S}_i$ and $\boldsymbol{S}_o$ are unitary vectors along with the illumination and the observation directions.

where k (the wavenumber, as defined in chapter 1) is the magnitude of $\boldsymbol{s}_i$ and $\boldsymbol{s}_o$, and is associated with the wavelength (M) of the illumination beam, $|\boldsymbol{s}_i| = |\boldsymbol{s}_o| = k = 2\pi/\lambda$. Then the sensitive vector $\boldsymbol{S}$ can be calculated as

$$\boldsymbol{S} = \boldsymbol{s}_o - \boldsymbol{s}_i. \tag{7.8}$$

The scalar product of the sensitivity vector $\boldsymbol{S}$ times the displacement vector $\boldsymbol{d}$ is related to the measurable optical phase $\Delta\varphi$ as

$$\Delta\varphi = \boldsymbol{S} \cdot \boldsymbol{d}. \tag{7.9}$$

Recall that the displacement $\boldsymbol{d} = d_x\hat{i} + d_y\hat{j} + d_z\hat{k}$, refers to the object change from the reference to the modified state (the change being static or dynamic). The components d_x and d_y are the in-plane displacements, while d_z represents the out-of-plane object displacement. These object displacement components are often referred to as $\boldsymbol{d} = [u, v, w]$, a notation that we will use later—this avoids using the x-, y-, and z-axes employed as coordinates in the observation plane, namely the camera sensor.

The optical phase difference can be rewritten relative to the optical path difference (OPD) as

$$\Delta\varphi = 2\pi/\lambda(\text{OPD}), \tag{7.10}$$

where $\text{OPD} = n(\boldsymbol{s}_o - \boldsymbol{s}_i) \cdot \boldsymbol{d}$, where we introduced n as the refractive index of the medium (for most practical purposes, experiments are conducted in solid objects surrounded by laboratory environmental conditions, with $n = 1$ corresponding to air).

In practice, optical layouts are built to be either sensitive for measurements on (a) object in-plane displacement components along with the x- and y-directions or (b) object out-of-plane displacements along the z-direction [7, 8]. It is worth pointing out here that many optical layouts can simultaneously measure displacements along the three coordinates x, y, and z. In addition, some of these layouts can also measure the object contour (shape), a feature that completes the quantitative object displacement on a point-by-point basis over the object contour data (this subject will be treated in chapter 10).

7.4 In-plane and out-of-plane sensitive set-ups

Typical DSPI optical set-ups are based on a two-beam interferometer that detects object displacements in different directions. The x- and y- displacement components (plane parallel to the object surface) are detected with an in-plane set-up. An out-of-plane set-up detects the z-displacement component (perpendicular to the object surface).

7.4.1 Optical set-up for in-plane displacements sensitivity

The optical set-up depicted in figure 7.3 has a sensitivity for in-plane displacement measurements along the x- and y-axes. As shown, for the system to be sensitive to in-plane displacement components, the optical axis, the imaginary axis that goes from the center of the camera sensor to the geometric center of the object surface

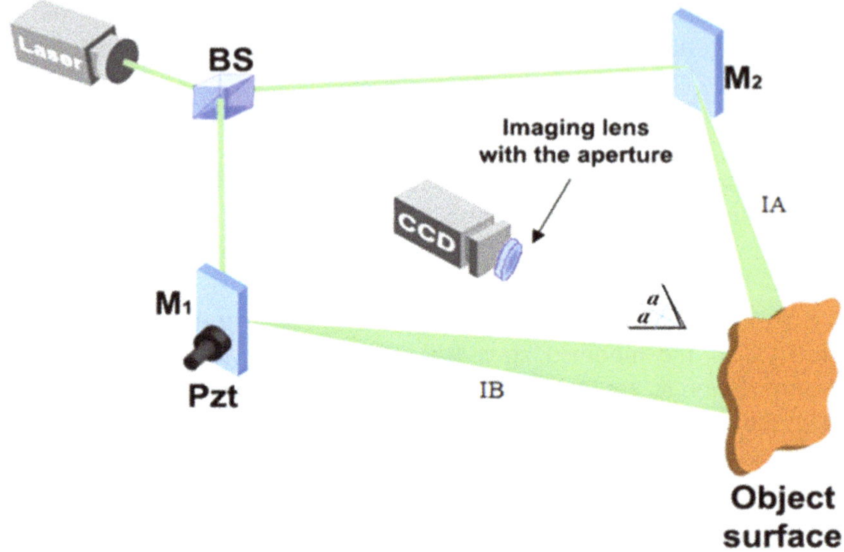

Figure 7.3. Set-up for in-plane displacement measurements.

(or indeed the center of the object area under inspection), must be along the normal to the object surface. This interferometer employs two illumination beams (IA and IB, shown as expanding beams, although not a necessary condition) that can be situated on the x–z or the y–z plane. An important feature is that each beam illuminates the object's surface at equal angles ($\pm a$) to the surface normal, i.e. a is half of the angle formed from IA to IB. The imaging optics includes an achromatic lens and an aperture. They are positioned in front of the camera, collecting most of the surface's backscattered light. A speckle pattern is produced as a result of the two beams interfering. The piezoelectric transducer (Pzt) fixed to the backside of the mirror (M_1) as it appears in figure 7.3, acts as an optical phase stepper and is used to retrieve the optical phase (via the phase-stepping method, see chapter 3).

We now move to figure 7.4, where object displacements at a point P are considered: OPD_A and OPD_B for the beams IA and IB give

$$OPD_A = d_z(1 + \cos a) + d_x \sin a.$$

$$OPD_B = d_z(1 + \cos a) - d_x \sin a. \tag{7.11}$$

Then the optical phase difference $\Delta\varphi$ results in

$$\Delta\varphi = 2\pi/\lambda(OPD_A - OPD_B) = 4\pi/\lambda d_x \text{sin}a, \tag{7.12}$$

$$\Delta\varphi = 4\pi/\lambda d_x \text{sin}a. \tag{7.13}$$

or

$$d_x = \frac{\lambda}{4\pi}\frac{\Delta\varphi}{\text{sin}a}. \tag{7.14}$$

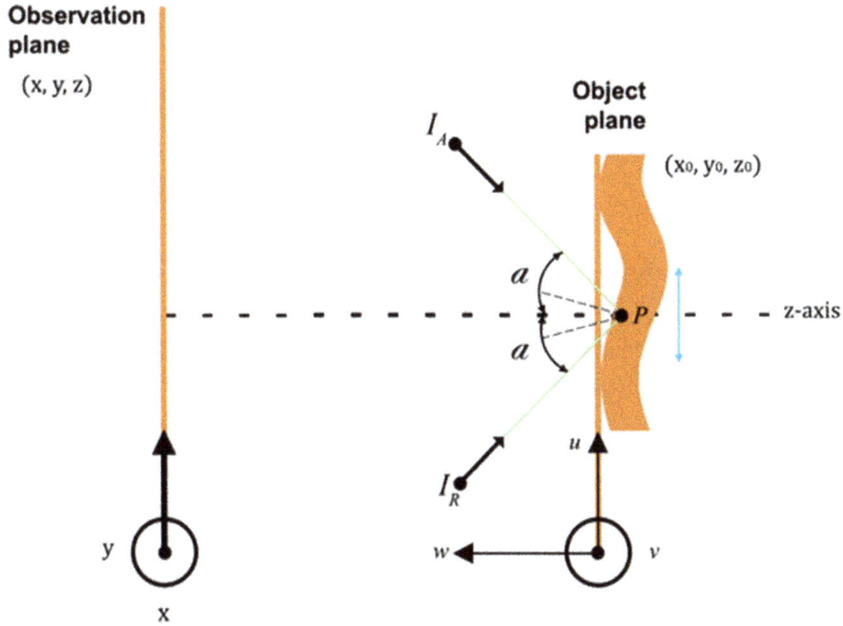

Figure 7.4. Geometry of in-plane sensitive set-up. The displacement components $\boldsymbol{d} = [u,\ v,\ w]$ indicate d_x, d_y, and d_z, respectively.

For bright fringes we need to use $\Delta\varphi = 2m\pi$ in equation (7.14), giving a displacement of

$$d_x(x,\ y) = \frac{m\lambda}{2\text{sin}a}. \tag{7.15}$$

For dark fringes, $\Delta\varphi = \pi(2m + 1)$, giving

$$d_x(x,\ y) = \frac{(2m + 1)\lambda}{4\text{sin}a}, \tag{7.16}$$

where $m = 0, 1, 2, \ldots$. If now the optical set-up is such that $\boldsymbol{S}$ is parallel to the y-axis, i.e. the set-up is sensitive to d_y, the phase difference results in

$$\Delta\varphi = 4\pi/\lambda d_y\text{sin}a, \tag{7.17}$$

$$d_y = \frac{\lambda}{4\pi}\frac{\Delta\varphi}{\text{sin}a}. \tag{7.18}$$

The fringes are formed by image correlation of a reference and modified object state (refers to equations (7.1) and (7.2)). The subtraction of these two images results in

$$\Delta I(x,\ y) = 2I_b(x,\ y)\gamma(x,\ y)\cdot\ \sin\left(\varphi_{o-r} + 2\pi/\lambda\cdot d_x(x,\ y)\sin a\right)\cdot\ \sin\left(2\pi/\lambda\cdot d_x(x,\ y)\sin a\right). \tag{7.19}$$

The first sinusoidal term corresponds to a high-frequency speckled background modulated by the second low-frequency sinusoidal term, which corresponds to the fringe pattern influenced by $d_x(x, y)$. The a-angle ranges from 0 to $\pi/2$, where higher sensitivity to in-plane displacements is obtained when the angle value is large [9, 10]. $I_b(x, y)$ is the background intensities and $\gamma(x, y)$ is the visibility of the fringes defined in chapter 4.

7.4.2 Optical set-up for out-of-plane displacements sensitivity

A smooth reference beam is combined with the speckled object beam at the camera sensor for the out-of-plane optical set-up. The out-of-plane interferometer is widely employed to measure the w-component of the displacement. A commonly employed out-of-plane DSPI system is illustrated in figure 7.5, where the laser beam is divided in two via a beam splitter (BS): the object beam (OB) and the reference beam (RB). The OB is reflected on the mirror (M_1) and is made divergent by a lens or a microscope objective (L_1) to fill the object's surface with light. The OB illuminates the object surface at an angle α_1 to the optical axis (as defined previously). The imaging arrangement includes a lens and an aperture, and it serves to collect the backscattered object light and convey a focused image on the camera sensor. The RB undisturbed, thus considered smooth, is directed to the camera's sensor using a diverging lens (L_2), a mirror (M_2), and a beam combiner (BC). In this way, both the RB and OB are overlapped on the camera sensor, interfering.

This optical set-up is shown as an example among several reported arrangements. For example, the set-up can be more compact and straightforward when optical fibers launch the laser beam in the object, and reference beams, say, coupling them independently via single-mode optical fibers. In any optical set-up, we must match

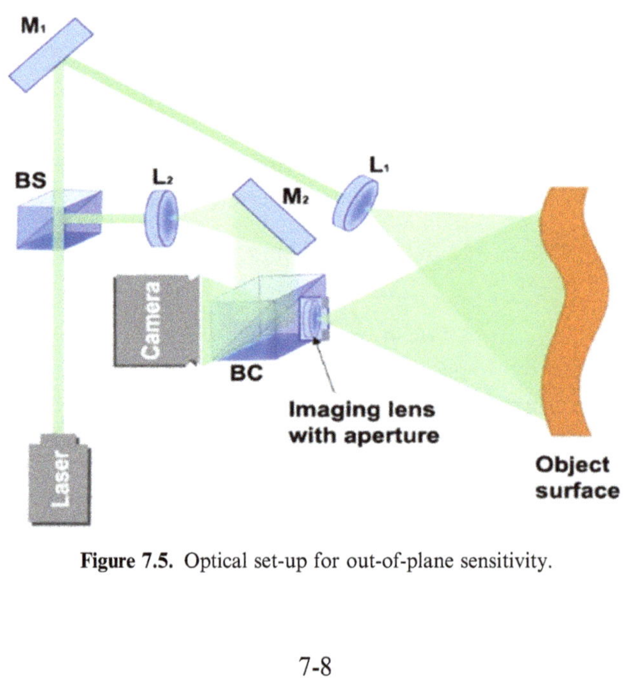

Figure 7.5. Optical set-up for out-of-plane sensitivity.

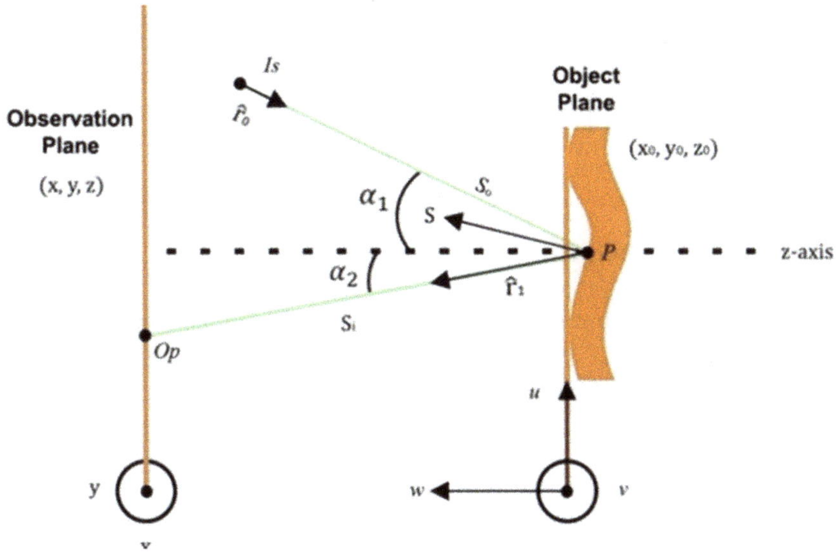

Figure 7.6. Geometry for an out-of-plane sensitive set-up. An object plane at (x_o, y_o, z_o) is illuminated by I_s with a direction represented by the $\boldsymbol{S}_o$ vector. The backscattered light reaches the observation plane at (x, y, z), with an arbitrary direction represented by the $\boldsymbol{S}_i$ vector. The w-displacement specifies d_z.

the optical path lengths of the two beams or have them with a mismatch that falls within the laser coherence length—the latter assures that interference will occur between the RB and OB.

See figure 7.6, where α_1 is the angle of object illumination and α_2 is the angle of observation, both with respect to the object's surface normal (in this case the z-axis). Let us also assume that (a) the object beam and the observation directions are very close to the z-direction, i.e. α_1 and α_2 are very small, and (b) we can thus neglect the contribution of the displacement components along x and y (this is experimentally achieved by subjecting the object to a *pure* out-of-plane displacement). Then the optical path difference is

$$\mathrm{OPD} = d_z(\cos\alpha_1 + \cos\alpha_2). \tag{7.20}$$

If we may consider an observation angle $\alpha_2 = 0$, then $\mathrm{OPD} = d_z(\cos\alpha_1 + 1)$. Thus the measurable phase difference $\Delta\varphi$ is

$$\Delta\varphi = \frac{d_z 2\pi(\cos\alpha_1 + 1)}{\lambda}, \tag{7.21}$$

or

$$d_z = \frac{\Delta\varphi\lambda}{2\pi(\cos\alpha_1 + 1)}. \tag{7.22}$$

As before, let us see two particular cases: when $\Delta\varphi = 2m\pi$ bright fringes appear or d_z fringes are formed in integer multiples of (λ).

Now assume that the object illumination beam is on the z-axis, then $\alpha_1 = 0$ and $\Delta\varphi$ becomes

$$\Delta\varphi = 4\pi/\lambda(d_z), \tag{7.23}$$

or

$$d_z = \lambda\Delta\varphi/4\pi. \tag{7.24}$$

If $\Delta\varphi = 2\pi$ equation (7.24) gives $d_z = \lambda/2$, which indicates the constitutive sensitivity of the fringe correlation.

When we use the subtraction by correlation process between the base and deformed object state, I_d and I_{ud}, (referring to equations (7.1) and (7.2)), the resulting intensity equation is given by

$$\Delta I(x, y) = 2I_b(x, y)\gamma(x, y) \cdot \sin(\varphi_{o-r} + 2\pi/\lambda(d_z)) \cdot \sin(2\pi/\lambda(d_z)). \tag{7.25}$$

The first term represents the background intensity, the $\sin(\varphi_{o-r} + 2\pi/\lambda(d_z))$ term corresponds to the high-frequency noise due to the randomness of the imaged speckles. The low-frequency interference term $\sin(2\pi/\lambda(d_z))$ corresponds to the fringe pattern directly related to the object deformation whose phase difference is $\Delta\varphi$ [11, 12]. $I_b(x, y)$ is the background intensities and $\gamma(x, y)$ is the fringe's visibility defined in chapter 4.

To retrieve the optical phase difference $\Delta\varphi$, we can use either the phase-stepping techniques in interferometry (PSI), or the Fourier transform method (FTM) discussed in chapter 6. Let us look more closely at PSI.

7.5 Phase-stepping interferometry

The PSI method retrieves the phase by shifting the frequency of a beam in the interferometer. It requires a phase change among the reference and object beams which is commonly achieved by stepping a mirror, a mirror driven through a piezoelectric transducer (Pzt), moving a grating, using an acousto-optic modulator, or employing a tunable laser in an interferometer.

A sequence of n-intensity patterns is recorded as needed for the selected algorithm to achieve phase retrieval in the phase-stepping procedure. The intensity equation has three unknowns, so at least three linearly independent equations are needed to solve $\Delta\varphi(x, y)$; the three images must be captured, called a three-step phase-stepping algorithm. The standard algorithms are 3, 4, 5, and 7 steps (1 step = 1 intensity image). The central concept is distributing the 2π modules into 3, 4, 5, or 7 parts to recover the optical phase. The optical phase sign of ambiguity is solved with this procedure.

The intensity of the nth interferograms for the reference state can be depicted as

$$I_{ud(n)}(x, y) = I_b(x, y) + I_m(x, y)\cos(\varphi_{o-r}(x, y) + \theta(t)_n). \tag{7.26}$$

For the modified state this is

$$I_{d(n)}(x, y, t) = I_b(x, y) + I_m(x, y)\cos(\varphi_{o-r}(x, y) + \Delta\varphi(x, y) + \theta(t)_n), \tag{7.27}$$

where $\theta(t)$ is the phase shift introduced, and n is the number of steps according to the algorithm used.

Using a piezoelectric transducer (Pzt) glued to a tilting mirror usually located at some point along the reference beam path generates the steps in the optical phase that are required for phase retrieval. The Pzt must be calibrated; one way is taking five images, that is, an image at 0, $\pi/2$, π, $3\pi/2$, and 2π, where the last image returns to the initial state, at 0, $(2\pi = 0)$; it may be concluded that the piezo went accurately through a 2π cycle, keeping the object in the static state. The Pzt receives a DC voltage that makes the mirror move a calculated and specific distance, producing a small variation of the optical phase and introducing an optical path length difference. If the light beam strikes at 45° and is reflected at 45° (concerning the mirror's normal), a Pzt–mirror displacement of $\lambda/8$ is enough to produce a phase shift of $\pi/2$ where the four steps give a total of 2π.

The phase difference is calculated between the two states of the object by $\Delta\varphi = \varphi_{ud} - \varphi_{d}$.

By using $n = 4$ in equations (7.26) and (7.27), with a phase shift introduced of $\pi/2$ between each intensity image, the usual four-step algorithm produces

$$\Delta\varphi = \tan^{-1}\left[\frac{(I_{ud1} - I_{ud3})\cdot(I_{d2} - I_{d4}) - (I_{ud2} - I_{ud4})\cdot(I_{d1} - I_{d3})}{(I_{ud1} - I_{ud3})\cdot(I_{d1} - I_{d3}) + (I_{ud2} - I_{ud4})\cdot(I_{d2} - I_{d4})}\right]. \tag{7.28}$$

The I_m is determined by

$$I_m = \sqrt{(I_{d2} - I_{d4} + I_{ud2} - I_{ud4})^2 + (I_{d1} - I_{d3} + I_{ud1} - I_{ud3})^2}. \tag{7.29}$$

The resulting $\Delta\varphi$ is an image known as a wrapped phase map that contains the information values codified between $-\pi$ and π or $|2\pi|$, from black to white fringes passing for gray levels. In order to obtain the measurand, the $|2\pi|$ phase discontinuities must be eliminated to obtain the unwrapped phase where great advances have been achieved in this issue. Chapter 3 discusses this topic.

PSI is being developed continuously due to new devices in cameras, computers, and steppers. In this way, fewer or more phase-stepped images are required than for the usual algorithms [13, 14].

7.6 Vibration modes obtained with DSPI

Vibrational analysis has become a valuable tool for measuring the mechanical properties and behavior of objects subjected to different kinds of loads. We can find these vibrational studies in a wide range of fields, from civil, aeronautical, automotive, and product design engineering, and testing of archaeological buildings, to the biomedical field. DSPI detects an object's state change while it is subjected to an external static or dynamic stimulus. In particular, the latter can be achieved by harmonic vibration (sinusoidal wave), non-sinusoidal vibration, sinusoidal sweep, transient vibration, environmental disturbances, etc. This type of stimulus is usually induced by an external source at the back of the object, close to or in contact, i.e. a loudspeaker, a shaker, or a Pzt device, all driven by an electronic function generator.

This type of excitation-stimulus is known as forced vibration, where the object is originally at rest, and the vibration starts when the external force is applied to the object. In this way, a wide range of frequencies is scanned to study the object's response.

The DSPI out-of-plane set-up is more commonly used for vibrational and modal analysis of objects since its sensitivity is ideal for finding, for instance, the object resonant frequencies. DSPI for vibration analysis may be used in various forms such as the time-average, the high-speed, and the time-resolved methods, where the first two use continuous-wave lasers (CW) and the latter strobed or pulsed lasers. Let us look at these methods more closely.

7.6.1 Time-average DSPI method

Consider figure 7.5 for the measurement of object harmonic vibrations. In order to acquire complete information on the vibration, it is necessary to obtain information on the object's position-dependent amplitude, frequency, and phase. The DSPI time-average method renders the magnitude of the full field of view object's surface vibration as a grayscale map of a fringe pattern. As the amplitude of vibration increases, the number of fringes increases, and the fringe visibility decreases, so large vibration amplitudes are often difficult to measure [15–18]. The resulting fringe patterns follow a Bessel function intensity distribution. The drawback is that the time-average method does not provide phase information of the vibrating object (because the argument of the Bessel function does not contain the phase term) according to equation (7.34).

Consider an object vibrating at a sinusoidal frequency with a period $2\pi/\omega_m$, which is less than the period of camera frame integration, inducing z deformations. The amplitude of the vibration provokes a displacement of the object surface at time t that can be represented by

$$d_z(x, y, t) = z_{m0}(x, y, t)\sin[\varnothing_m(x, y, t) - \omega_m t], \tag{7.30}$$

where z_{m0} represents the amplitude of vibration, x and y are the surface spatial coordinates, $\varnothing_m$ is the mechanical phase, and ω_m is the angular frequency.

The backscattered wave complex amplitude of the object recorded by the camera sensor at t is

$$U_o(x, y, t) = u_o(x, y)e^{i(\varphi_o(x, y)+\frac{4\pi}{\lambda}d_z(x, y, t))}, \tag{7.31}$$

where λ is the wavelength of the laser.

Now, if the object is illuminated for a specific time interval whose duration is less than or equal to the camera frame integration period Δt, the intensity $I(x, y)$ produced by the optical interference between a wave reflected from an object in movement and a set reference wave acquired by the digital sensor is

$$I(x, y) = \frac{1}{\Delta t}\int_t^{t+\Delta t} I_o(x, y) + I_r(x, y) + 2\sqrt{I_o I_r}\cos[\varphi_{o-r}(x, y) + \Delta\varphi(x, y, t)]dt, \tag{7.32}$$

with

$$\Delta\varphi(x, y, t) = \frac{4\pi}{\lambda} d_z(x, y, t).$$

The cosenoidal function includes the optical phase difference $\Delta\varphi(x, y, t)$, plus the random phase $\varphi_{o-r}(x, y)$. Note that the former relates directly to object displacements.

If $2\pi/\omega_m \ll \Delta t$, the integral can be assessed for a complete time of vibrating cycles and using the integral representation of the Bessel function $\mathcal{J}_0(z)$ defined as

$$\mathcal{J}_0(z) = \frac{1}{2\pi} \int_0^{2\pi} \mathrm{e}^{\mathrm{i}z\cos\Theta} \mathrm{d}\Theta.$$

Equation (7.32) can be rewritten as

$$I(x, y) = I_o(x, y) + I_r(x, y) + 2\sqrt{I_o I_r} \cos(\varphi_{o-r}(x, y)) \mathcal{J}_0\left(\frac{4\pi}{\lambda} z_{m0}\right). \tag{7.33}$$

Bessel fringes represent the time-average vibration amplitude. The bright fringes of the resulting pattern correspond to regions of maximum and minimum contrast governed by $\mathcal{J}_0\left(\frac{4\pi}{\lambda} z_{m0}\right)$.

For the case of sinusoidal vibrations, the intensity pattern $I(x, y)$ at any point is proportional to the square of the zero-order Bessel function, producing $\mathcal{J}_0$ fringes related to the displacement amplitude $d_z(x, y)$ by

$$I(x, y) \propto \mathcal{J}_0^2\left[\frac{4\pi}{\lambda} d_z(x, y)\right]. \tag{7.34}$$

The zero-order bright fringe has the highest intensity, with subsequent fringe intensities decreasing in brightness; this feature agrees with the object surface displacement and defines iso-displacement contours whose magnitudes are given by equation (7.34). The fringes follow a shape given by the Bessel function instead of the cosine function, making the use of the phase-shifting technique difficult (see figure 7.7). However, the phase-stepping method may be used to produce a pattern that has an intensity-modulated I_m by a periodic function whose argument is $\Delta\varphi(x, y, t)$. Thus, the resulting intensity when using phase stepping for the nth step can be written as

$$\begin{aligned} I_n(x, y, t) = I_o(x, y) + I_r(x, y) + 2\sqrt{I_o I_r} \cos[\varphi_{o-r}\varphi(x, y) + \Delta \\ \varphi(x, y, t) + \theta(t)_n] \cdot \mathcal{J}_0\left[\frac{4\pi}{\lambda} d_z(x, y)\right]. \end{aligned} \tag{7.35}$$

Using a four-step algorithm on equation (7.35), recording with a 90° phase shift between each intensity pattern, $I_1(x, y, t), \ldots, I_4(x, y, t)$, the modulation intensity I_m is calculated by (figure 7.7)

$$I_m = \sqrt{(I_1 - I_3)^2 + (I_2 - I_4)^2}. \tag{7.36}$$

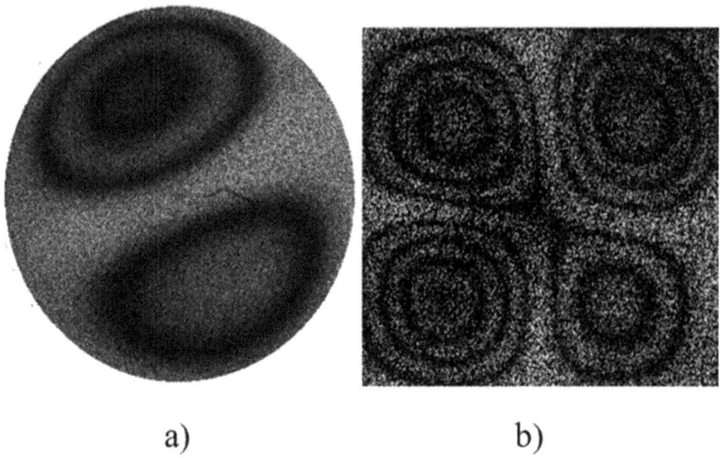

Figure 7.7. Time-averaged patterns of (a) a plastic membrane vibrating at 21 kHz and (b) a square aluminum test sample vibrating at 2.9 kHz.

Several methods have been introduced to evaluate Bessel fringes and obtain phase maps of the object's surface vibration. This measurement can be achieved by avoiding time averaging during the recording of fringe patterns, modulating the light source, shifting the bright irradiance to a region of large vibration amplitude using time speckle interferometry, and recording images at sufficiently short times using high-speed cameras, usually combined with temporal phase shifting. Another method suitable for measuring high-frequency vibrations is based on the use of stroboscopic illumination of the laser source that illuminates the object with short laser pulses synchronized with the oscillation of the object to 'freeze' the motion and convert the illumination beam into cosine fringes. Also, using computational procedures, for example, based on the use of the Hilbert transform, numerically adjusting the $\mathcal{J}_0$ function to the recorded light intensity or the consecutive images recorded in a region for an object vibrating with increasing amplitude.

7.6.2 Time-resolved method

The time-resolved method can be used to retrieve the optical phase map $\Delta\varphi(x, y, t)$. For this purpose, the object beam has to be strobe or pulsed, so the time-average Bessel fringes are changed to cosine fringes. The strobe light is obtained with a 'strobe switch' such as electro and optoacoustic modulators or mechanical shutters phase-locked to the stimulus. The pulsed illumination can be obtained with pulsed lasers, with pulse time widths shorter than the time of the vibrating object.

Let us work with pulsed illumination, figure 7.8, where the pulses are typical of a few tens of nanoseconds duration, meaning that they can *freeze* the object vibration motion. Then the two speckle patterns are gathered for two laser pulses at two different time instants, $I(t_1)$ and $I(t_1 + \tau)$, *freezing* two object displacements at $d_z(t_1)$ and $d_z(t_1 + \tau)$. These two laser pulses are synchronized at any two instants of the object's vibration cycle, and the image recording is synchronized with the vibration

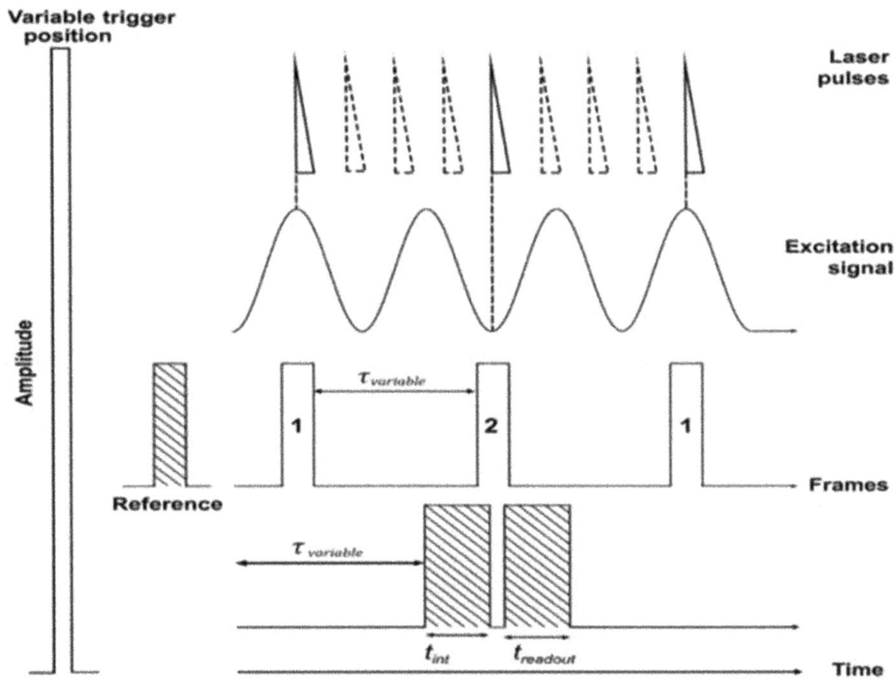

Figure 7.8. Pulsed illumination: the laser pulses sync with the excitation signal cycle phase and the recording frame rate.

stimuli. The image acquisition starts by giving an external trigger signal set by the stimuli period or implementing a time control to execute the measurements. The fringe patterns are obtained by subtracting these two speckled images. The first image may be with the object at rest (or at any other specific time in the cycle), and the second at any other time different from this first-base state. In practice, the laser pulse duration (P_t) must be shorter than the period of the vibration stimulus, (T_s), where 10% of the cycle has given a good result for vibration measurements,

$$P_t = 10\% \times T_s, \tag{7.37}$$

where $T_s = 1/f_s$. The pulse duration of the laser P_t can illuminate one T_s cycle n times of $\approx T_s/P_t$ of the vibration registered.

Let us now consider the case of stroboscopic illumination. Here, the object is illuminated by a train of pulses whose time duration depends on the hardware used (e.g. electro- or optoacoustic modulators). Again we can assume that the cycling movement of the object is frozen for each pulse, and is synchronized to specific locations of the vibration cycle, see figure 7.9. As mentioned before, stroboscopic illumination converts the time-averaged Bessel fringes into cosine fringes, making it possible to retrieve the optical phase. The camera records two speckled images acquired at different times to obtain the fringe patterns. The stroboscopic method is applied in principle for sinusoidal and non-sinusoidal vibrations with periodic motion.

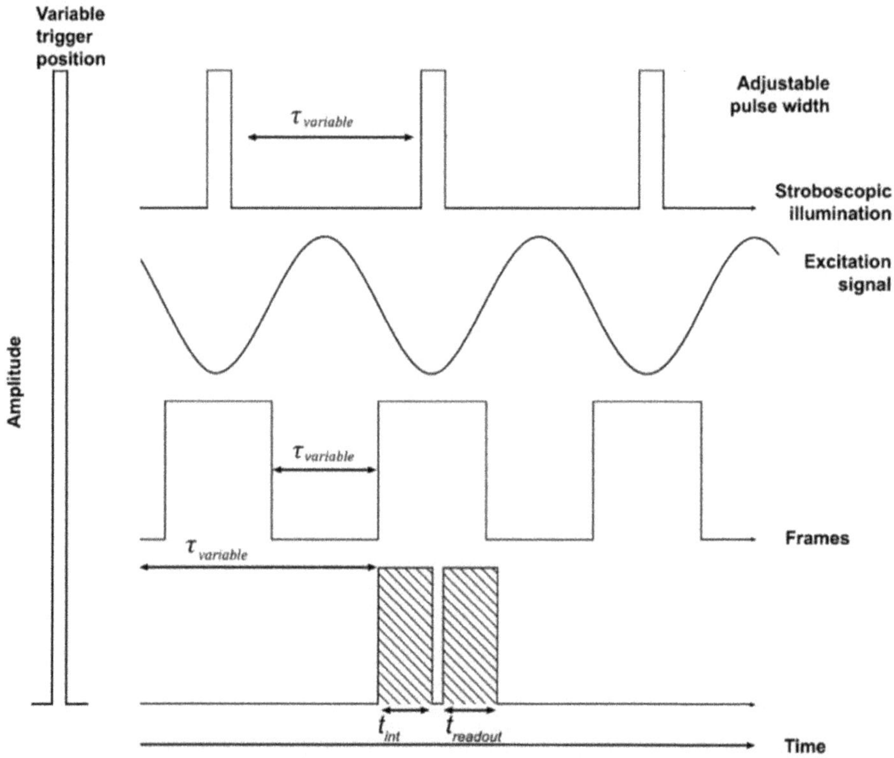

Figure 7.9. Stroboscopic illumination method. The illumination pulses are phase-locked with the excitation signal cycle and recording frame rate. The integration and readout times of the data are carried out by programming the camera, as illustrated in the lower part of the figure.

The vibrational analysis can be resolved if the object vibration event is synchronized as follows: the vibration stimulus on the object can be used as the input for triggering the electronics or for the user interface to perform the image acquisition at different parts of the vibration cycle. The period of the stimuli and the settings of the CCD camera, such as the image readout time, and the intrinsic delay plus the trigger delay, are fundamental to selecting the exposure time and calculating the period of the trigger. Managing the start of the external trigger or the time between pulses allows the study of the object movement evolution at different parts of the excitation signal. Another way to enhance the vibration analysis is to set some camera parameters (if possible), such as the exposure time chosen to be sufficiently short to freeze the object's motion [19–21].

The gathered images are processed using real-time subtraction. Subsequent frames can be subtracted in pairs. One form is by successive subtraction of frames, first – second, first – third, etc. Another form is continuously subtracting two consecutive frames, first – second, second – third, etc. The process renders fringe patterns depicting the object displacement. The modulation intensity I_m at each camera pixel results from processing the intensity images at different object states, namely I_{ud0} to I_{ud3} and I_{d0} to I_{d3}, and I_m can be calculated as

$$I_m = \sqrt{(I_{ud0} - I_{ud2} + I_{d0} - I_{d2})^2 + (I_{ud1} - I_{ud3} + I_{d1} - I_{d3})^2}. \tag{7.38}$$

7.6.3 High-speed time-average DSPI

With the development of digital and semiconductor technologies, it is possible to have CCD/EMCCD/CMOS/sCMOS sensors with high resolution, high dynamic range, and high acquisition speeds. The latter means that pulsed lasers or strobed object illumination are no longer must-have hardware to study vibrations. The sensor's resolution is based on the number of pixels and their size relative to the image size. The sensor format sets the high resolution; for EMCCD, it is considered from 0.25 to 1 Mpixel. For CMOS, it is about 3 Mpixels, and for sCMOS, it is about 4.19 or 5.5 Mpixels. These values depend on the pixel size and the detector size. The sCMOS detectors provide low noise at high frame rates and have high spatial resolution due to their small pixel size advantages over CCD sensors. EMCCDs are considered a solution for sCMOS camera systems as they have been developed to the degree that they can operate at 25 MHz and even maintain read noise equal to sub-electron noise. Commercially available fast cameras can be found at 5000 fps or higher at full resolution and with dynamic ranges well above 12 bits. Hence, such fast cameras mean that a continuous-wave laser can be used to study object vibrations. These cameras quantify the object deformation as vibration fringes with high spatial and temporal resolution: one can now record hundreds of object speckle patterns from a myriad of positions of the vibration cycle [22–24].

Here is the process for the quantitative analysis. As before, one image (I_{ud}) at time t_1 is captured of the object at rest (or at any other know position); then, the object is subjected to a stimulus, and a second image is recorded (I_d) at t_2. A fringe pattern is displayed by subtracting I_{ud} from I_d, showing the object response as vibration modes for the specific stimuli frequencies. Each fringe pattern characterizes the change undergone by the object relative to the reference state and any given state. The intensity pattern is described by

$$I_d - I_{ud} = 4u_o u_r \sin\varphi_{o-r} \sin\frac{\Delta\varphi}{2}, \tag{7.39}$$

where the random phase of the speckle is observed as a high-frequency pattern, while the low-frequency fringe pattern corresponds to the object surface displacement.

As was mentioned before, these types of fast cameras allow one to study object vibrations under certain conditions, namely when the object vibration frequency is lower than the camera frame rate ($f_{vib} < f_{cap}$), when the object vibration frequency is higher or equal to the camera frame rate ($f_{vib} \geqslant f_{cap}$), and when the camera frame rate is programmable, today a most desirable feature [25, 26].

Let us see the case when $f_{vib} < f_{cap}$ (see figure 7.10). A first object image is captured at a base state, then the camera is configured to acquire the object vibration at different parts of the cycle. Finally, the images are correlated by subtraction to yield the fringe patterns and then processed to quantify the resulting optical phase maps ($\Delta\varphi$). Note that it is unnecessary for the objects under study to be at their

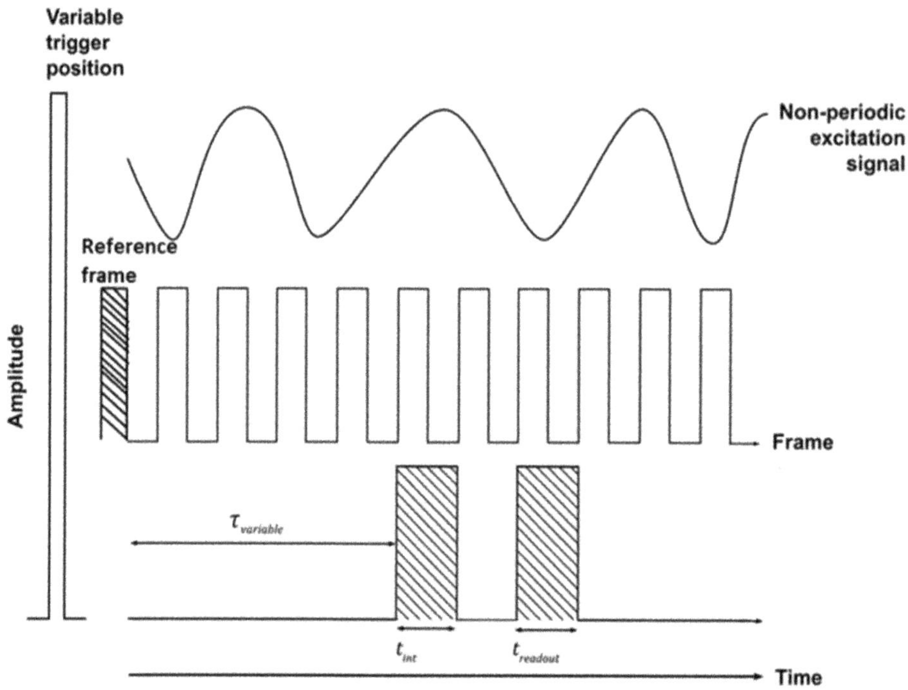

Figure 7.10. Sampling diagram of an object oscillating when a non-periodic excitation signal is applied to it, shown in the upper waveform. The sampling performed by the camera is seen in the middle part, and the integration time and reading of the data carried out by the camera are plotted in the lower graph.

resonant frequencies or be subjected to periodic vibrations for vibration analysis, i.e. all forms of induced object vibrations are allowed.

For the recording process, the camera can be triggered in an *external mode* where the trigger is induced by an external pulse (electrical signal) coming from an electronic timing control system, pulse generator, etc, or in an *internal mode* by a control software (programmed). The trigger's purpose is to control the start of the camera image acquisition. The frequency of the induced vibration allows the control of the acquisition frequency (acquisition frames) f_{rate}, and the integration time (exposure time) t_{int}, which is we assume is much shorter than the vibration period τ_{vib}, needed to acquire an image. The t_{int} is given as

$$t_{\text{int}} \leqslant 10\% \times \tau_{\text{vib}}. \tag{7.40}$$

This condition is linked to the recording rate, which is equivalent to freezing the object vibrating at the mean instant of t_{int} at which the recording process is made. During this time, the deformation is approximately constant, so the phase difference of the object is also constant. This condition allows the possibility of reconstructing the vibration from the acquired images free of any error. The period to acquire one image is calculated by

$$\tau = \frac{1}{f_{\text{rate}}}. \tag{7.41}$$

The quantity of captured images, $\#_{acq}$ for different positions of the vibration, the cycle may be determined by

$$\#_{acq} \cong \frac{\tau_{vib}}{t_{int}}. \tag{7.42}$$

These parameters also are affected by the laser power, the object size, working distance of the optical set-up, and the pixel surface.

Today's cameras avoid data averaging by handling the internal shutter function, whose opening time must be shorter than the exposure time.

Now, let us study the case when $f_{vib} \geqslant f_{cap}$, shown in figure 7.11. This condition is not suitable for recording many interferograms in one period of the vibration cycle, so it is necessary to find a resonance frequency to guarantee the study of the object vibration at different parts of its cycle. The high-speed camera and the excitation must be synchronized for regular cyclic vibrations by adjusting the camera's triggering at different parts of the vibration, which is also synchronized with the excitation signal. An image of the reference state is acquired (at rest or any chosen state of the vibration cycle, namely, if the excitation signal is well known, it is possible to capture the zero vibration amplitude of the object). Then, later is

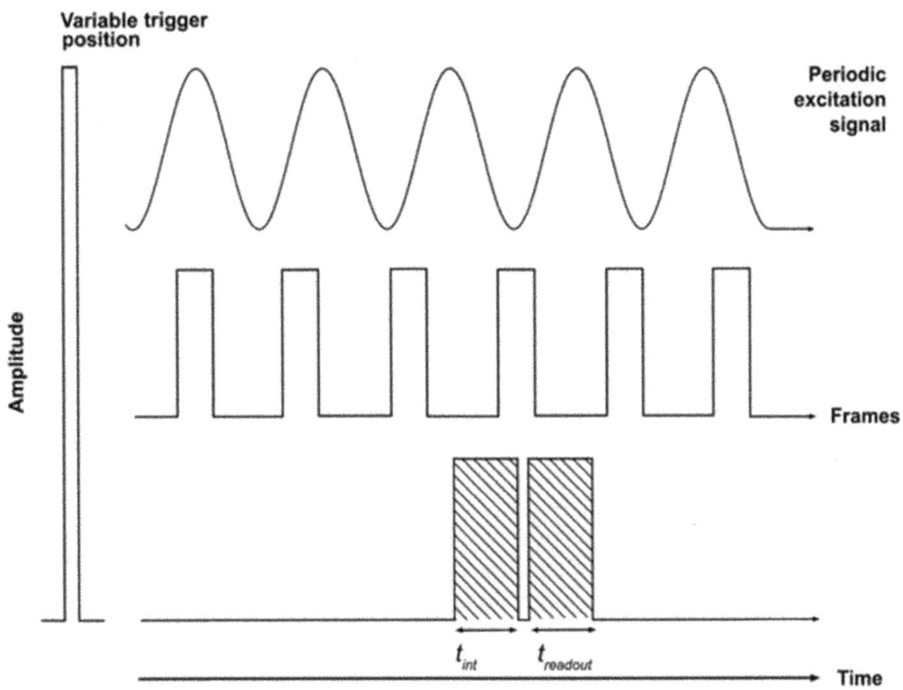

Figure 7.11. Sampling diagram for an object with vibration frequency $f_{vib} \geqslant f_{cap}$. The camera trigger is controlled to capture different parts of the object's vibration cycle. Periodic excitation signal (top image); sampling performed by the camera at different points in the cycle (middle image); and integration time and data readout from the camera (lower image).

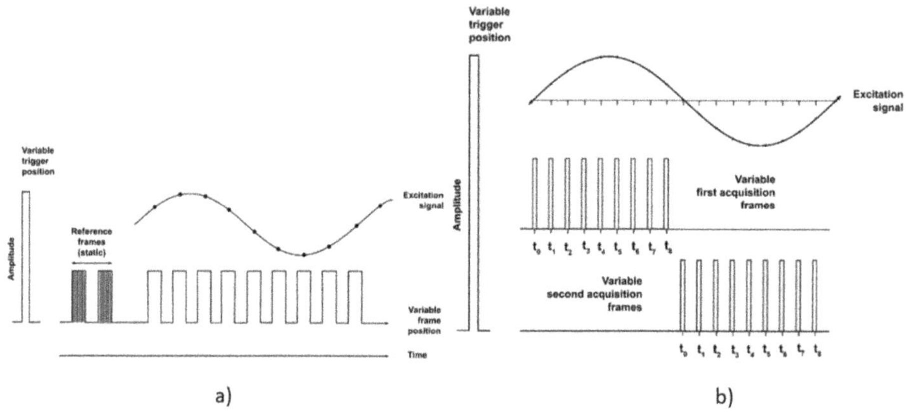

Figure 7.12. Recording of intensity images in two sample states either with (a) a static initial state or (b) at any given state of the external excitation. It is possible to record a sequence of images and record the object vibration evolution.

compared with a different object position: the vibrations can be determined from the method in figures 7.12(a) and (b).

The number of images that can be recorded is approximately

$$N_{\text{rec1}} \cong \frac{f_{\text{vib}}}{f_{\text{vib}} - f_{\text{rate}}}. \tag{7.43}$$

When the camera frame rate is programmable, i.e. the number of frames per second can be chosen, the optimal f_{rate} the setting for a vibration compromises the images recorded at different times of the vibration cycle with an effective integration t_{int} determined by equation (7.40). One way to freeze the vibration is to program the exposure time of each frame using the high-speed camera's internal electronic shutter and release the trigger to start the frame acquisition [27–29]. The programmable frame rate can be set by multiplying N times the vibration frequency, where N is an integer used to generate a multiple of the vibration frequency (see figure 7.13):

$$f_{\text{rate}} = Nf_{\text{vib}} \quad \text{or} \quad t_{\text{rate}} = t_{\text{vib}}/N. \tag{7.44}$$

For the case when f_{vib} exceeds the programmable frame rate of the camera, it can be determined by sub-multiples of the vibrating frequency. The number of recorded interferograms N_{rec} in a time interval τ_{int} during the object vibration is

$$N_{\text{rec}} \cong \frac{\tau_{\text{int}}}{\tau}, \tag{7.45}$$

where τ is the time interval to acquire an image.

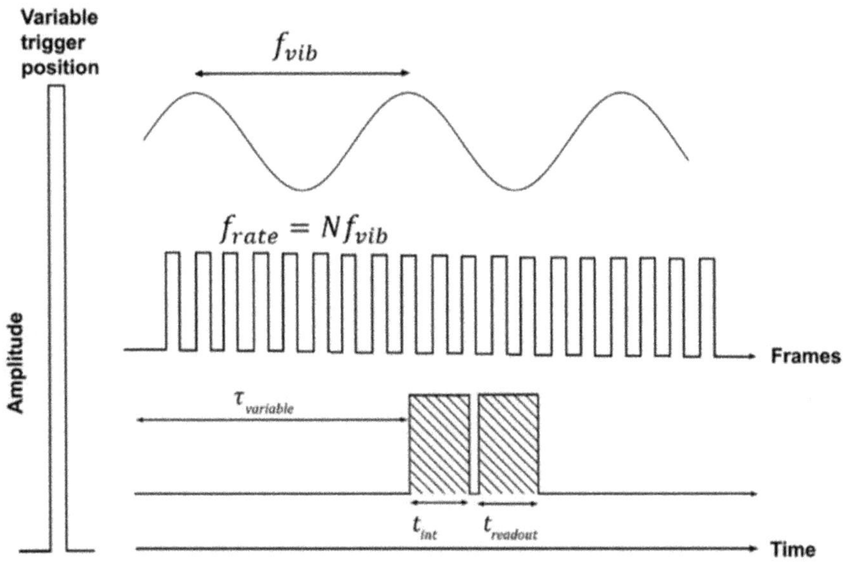

Figure 7.13. Sampling plot, when the camera frame rate is programmable, the vibration frequency determines the number of frames captured.

References

[1] Yang L, Xie X, Zhu L, Wu S and Wang Y 2014 Review of electronic speckle pattern interferometry (ESPI) for three dimensional displacement measurement *Chin. J. Mech. Eng.* **27** 1–13

[2] Patorski K 2001 Recent developments in electronic speckle pattern interferometry *Proc. SPIE* **4356** 124–32

[3] Tutt J H, Hall D J, Soman M R, Holland A D, Warren A J, Connolley T and Evagora A M 2014 Comparison of EM-CCD and scientific CMOS based camera systems for high resolution x-ray imaging and tomography applications *J. Instrum.* **9** P06017

[4] Types of camera sensor *Teledyne Photometrics* https://photometrics.com/learn/camera-basics/types-of-camera-sensor

[5] Dvoráková P, Bajgar V and Trnka J 2007 Dynamic electronic speckle pattern interferometry in application to measure out-of-plane displacement *Eng. Mech.* **14** 37–44

[6] Gong X L and Toyooka S 1999 Investigation on mechanism of plastic deformation by digital speckle pattern interferometry *Exp. Mech.* **39** 25–9

[7] Georges M P, Thizy C, Languy F, Zhao Y and Vandenrijt J F 2019 Digital holographic interferometry and speckle interferometry applied on objects with heterogeneous reflecting properties *Appl. Opt.* **58** G318–25

[8] Hansen R S 1997 Deformation measurements of specularly reflecting objects using holographic interferometry with diffuse illumination *Opt. Lasers Eng.* **28** 259–75

[9] Jallapuram R, Healy C, Mihaylova E and Toala V 2011 In-plane sensitive electronic speckle pattern interferometer using a diffractive holographic optical element *Am. J. Phys.* **79** 389–91

[10] Matham M V and Sujatha N 2004 Digital speckle pattern interferometry for deformation analysis of inner surfaces of cylindrical specimens *Appl. Opt.* **43** 2400–8

[11] Song C, Matham M V and Chan K H K 2015 Speckle referencing: digital speckle pattern interferometry (SR-DSPI) for imaging of non-diffusive surfaces *Proc. SPIE* **9524** 95242P
[12] Kemper B, Dirksen D, Kandulla J and von Bally G 2001 Quantitative determination of out-of-plane displacements by endoscopic electronic-speckle-pattern interferometry *Opt. Commun.* **194** 75–82
[13] Hernández-Montes M d S *et al* 2009 Optoelectronic holographic otoscope for measurement of nano-displacements in tympanic membranes *J. Biomed. Opt.* **14** 034023
[14] Li J, Zhong L, Liu S, Zhou Y, Xu J, Tian J and Lu X 2017 An advanced phase retrieval algorithm in *N*-step phase-shifting interferometry with unknown phase shifts *Sci. Rep.* **7** 44307
[15] Rosowski J J, Cheng J T, Ravicz M E, Hulli N, Hernandez-Montes M, Harrington E and Furlong C 2009 Computer-assisted time-averaged holograms of the motion of the surface of the mammalian tympanic membrane with sound stimuli of 0.4 to 25 kHz *Hear. Res.* **253** 83–96
[16] Moore T R and Skubal J J 2008 Time-averaged electronic speckle pattern interferometry in the presence of ambient motion. Part I. theory and experiments *Appl. Opt.* **47** 4640–8
[17] Creath K and Slettemoen G Å 1985 Vibration-observation techniques for digital speckle-pattern interferometry *J. Opt. Soc. Am.* A **2** 1629–36
[18] Acosta L S, Hernández M D S, Mendoza Santoyo F, De la Torre I M H, Flores Moreno J M, Frausto R G and Muñoz Solis S 2020 Study of skin rigidity variations due to UV radiation using digital holographic interferometry *Opt. Lasers Eng.* **126** 105909
[19] Tiziani H J and Pedrini G 1996 Digital double-pulse holographic interferometry for vibration analysis *Shock Vib.* 3 117–25
[20] Kumar U P, Somasundaram U, Mohan N K and Kothiya M P 2011 Comparative study of time average and stroboscopic illumination techniques for vibration fringe analysis *J. Opt.* **40** 184–92
[21] Trillo C, Doval A F, Hernández-Montes S, Deán-Ben X L, López-Vázquez J C and Fernández J L 2011 Pulsed TV holography measurement and digital reconstruction of compression acoustic wave fields: application to nondestructive testing of thick metallic samples *Meas. Sci. Technol.* **22** 025109
[22] What is a scientific CMOS camera? *Oxford Instruments* https://andor.oxinst.com/learning/view/article/scmos-technology-what-is-scmos
[23] https://andor.oxinst.com/products/scmos-for-physical-science-and-astronomy
[24] Trillo C, Doval Á F, Mendoza-Santoyo F, Pérez-López C, de la Torre-Ibarra M and Luis Deán J 2009 Multimode vibration analysis with high-speed TV holography and a spatiotemporal 3D Fourier transform method *Opt. Express* **17** 18014–25
[25] Meteyer E, Montresor S, Foucart F, Le Meur J, Heggarty K, Pezerat C and Picart P 2021 Lock-in vibration retrieval based on high-speed full-feld coherent imaging *Sci. Rep.* **11** 7026
[26] Pedrini G, Osten W and Gusev M E 2006 High-speed digital holographic interferometry for vibration measurement *Appl. Opt.* **45** 3456–62
[27] Razavi P, Tang H, Rosowski J J, Furlong C and Cheng J 2018 Combined high-speed holographic shape and full-field displacement measurements of tympanic membrane *J. Biomed. Opt.* **24** 031008
[28] Poittevin J, Picart P, Gautier F and Pezerat C 2013 Is high-speed digital holography the ultimate tool for vibration analysis? *MEDYNA 2013: 1st Euro-Mediterranean Conf. on Structural Dynamics and Vibroacoustics*
[29] Wu T, Jones J D C and Moore A J 2006 High-speed phase-stepped digital speckle pattern interferometry using a complementary metal-oxide semiconductor camera *Appl. Opt.* **45** 5845–55

IOP Publishing

Full Field Optical Metrology and Applications

Fernando Mendoza-Santoyo, Manuel De la Torre-Ibarra, María del Socorro Hernández-Montes and Jorge Mauricio Flores Moreno

Chapter 8

3D digital holographic interferometry

Digital holographic interferometry (DHI) is a highly versatile technique that can use different kinds of illumination sources, light guides, and configurations [1]. In this chapter, more elaborate set-ups will be presented where the use of more than one object illumination is described to retrieve more information about the object under study. Different illumination configurations are described, starting with the simplest one where the same laser is used to illuminate once at a time, i.e. simultaneously, up to a configuration where a different laser is used for each illumination position. Between these two configurations, a third one uses only one laser, but due to its coherence length it works as if several independent lasers were employed.

8.1 Principle

DHI uses light as the inspection tool that can be launched using optical fibers with diameters no bigger than a few tenths of a millimeter (typically mono-mode optical fibers are used, this helps to avoid speckled beams coming out of the fiber). This feature allows several possible illumination positions (IP) to cover the object [2]. As an example, in figure 8.1, to study object vibrations, four illumination positions (IP1–IP4) or n positions (IPn) may be used to illuminate it.

The number of these positions is related to the interferometer's space (x, y, z) sensitivity. The simplest case involves only one object illumination beam as the classical out-of-plane sensitive configuration, i.e. this configuration will only have the sensitivity along the z-axis [3]. However, there are applications where measuring the object displacement in two or three axes is required [4]. There are mainly two forms to do this: illuminating the object once at a time or simultaneously.

doi:10.1088/978-0-7503-3027-5ch8

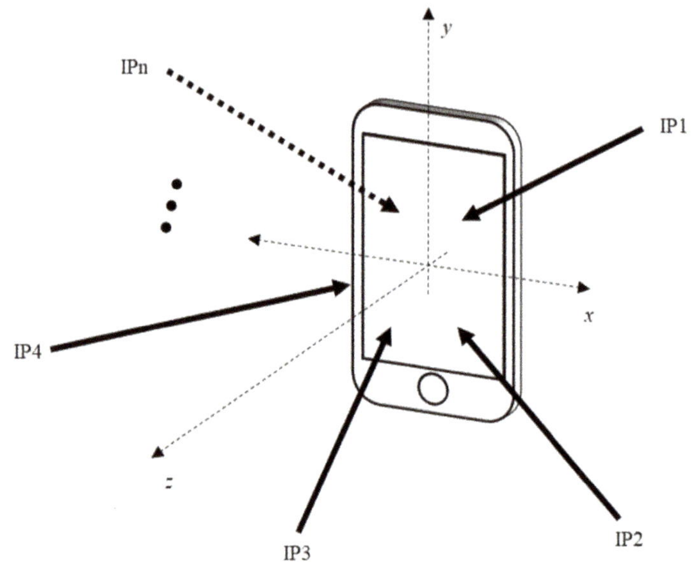

Figure 8.1. Object with *n* illumination positions.

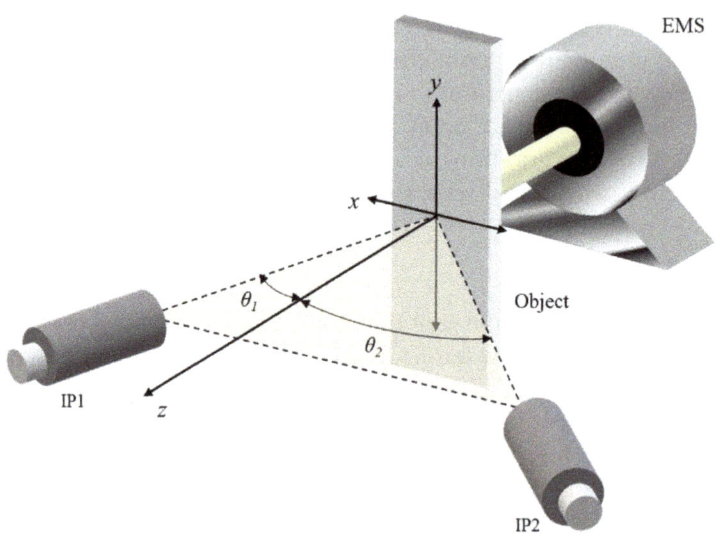

Figure 8.2. DHI system with a once at a time two-beam configuration with θ_1 and θ_2 as the illumination angles in the x–z plane. EMS = electromechanical shaker.

The first case is possible when well-controlled deformations are provoked on the object, and it is possible to apply them in continuous steps. This basic configuration could be extended to more illumination positions, obtaining an independent image hologram for each illumination. Let us consider the case for two illumination positions in a vibration test, as shown in figure 8.2.

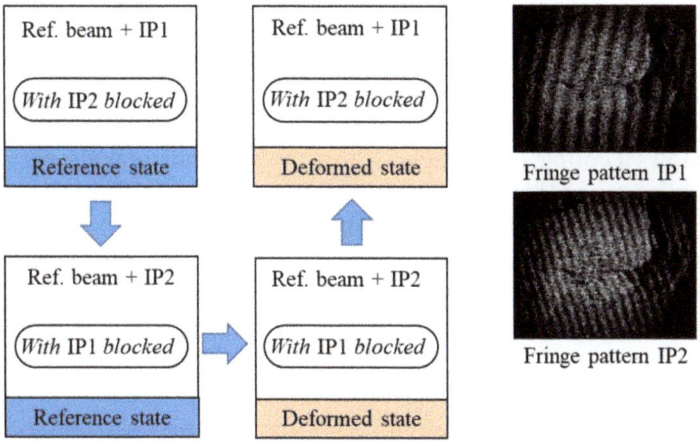

Figure 8.3. Steps for once at a time illumination. Their resulting fringe patterns are shown with different fringe inclinations.

The object under study is a cantilever deformed with a fixed resonant vibration mode. In this case, the interferometer records four image holograms: 1. IP1 in the reference state; 2. IP2 in the reference state; 3. IP1 with the deformed object; and 4. IP2 with the deformed object. The illumination not being used is blocked in every step, i.e. IP2 is blocked when IP1 is used [5]. Figure 8.3 shows a block diagram for this method.

At this point, both phase maps (exemplified in figure 8.3 with fringe patterns) are independent of each other (there is not a direct correlation), so it is not possible to relate them. However, it is possible to correlate these two-phase maps through the sensitivity vector [6]. To achieve this, let us first rewrite this equation in the general form:

$$\Delta\varphi = k\boldsymbol{S}.\ d, \tag{8.1}$$

where k is the wavenumber $2\pi/\lambda$ with λ as the wavelength of the illumination beam of the interferometer. The term d refers to the object displacement and $\boldsymbol{S}$ is the sensitivity vector that makes it possible to match the radians of the phase map with the distance units of the displacement map (normally several micrometers, but could be much less than this order of magnitude). For the case where one illumination is used, equation (8.1) is reduced to equation (8.2). The latter applies when an out-of-plane configuration is used and the sensitivity is aligned with the z-axis. This axis in the displacement field is expressed as w, thus

$$\Delta\varphi = 2\pi/\lambda\boldsymbol{S}.\ w. \tag{8.2}$$

However, the interferometer renders two displacement maps for the case with two object illumination directions. One is aligned between the plane created by the illumination directions, and the other is perpendicular to the object's surface (see figure 8.2). For this reason, the previous expression can be rewritten in a general form as

$$\begin{pmatrix} \Delta\varphi_1 \\ \Delta\varphi_2 \end{pmatrix} = 2\pi/\lambda \begin{pmatrix} S_{1\eta} & S_{1\xi} \\ S_{2\eta} & S_{2\xi} \end{pmatrix} \cdot \begin{pmatrix} d_1 \\ d_2 \end{pmatrix}. \tag{8.3}$$

where ξ and η are the resulting sensitivity axes due to the illumination position. Figure 8.4 shows a couple of examples where the perpendicular (in-plane) sensitivity axis η is modified according to the illuminating positions. In these cases, the object's axes (x and y) are also aligned with the observation plane (for instance, a recording medium such as the sensor in a digital camera parallel to the x–y plane of the object).

A common practice to avoid the complexity of having this new η-axis with information of the x- and y-axes combined, is to have the illumination directions aligned with the planes x–z or y–z. This condition simplifies the expression to uniquely have the contribution of the x- or the y-axis, as figures 8.5(a) and (b) show, respectively.

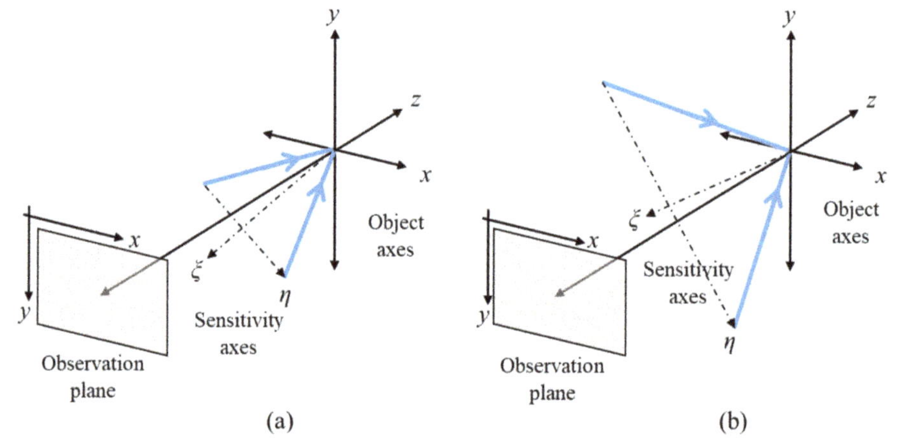

Figure 8.4. Examples of how the sensitivity η-axis changes according to the illumination direction.

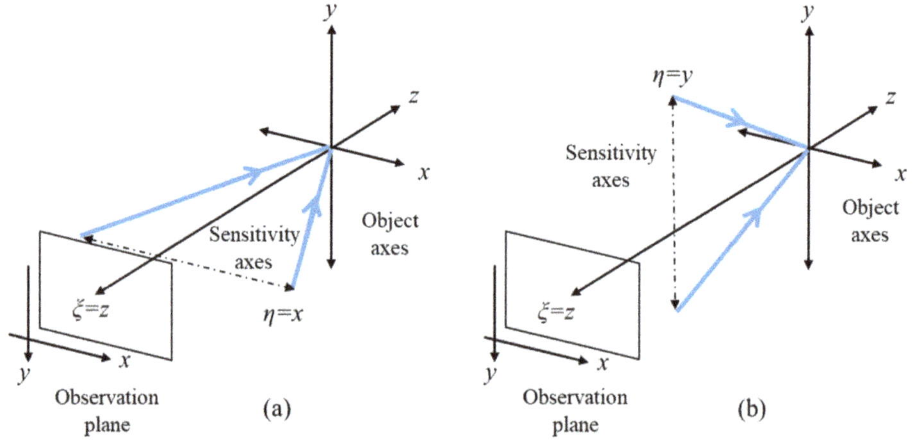

Figure 8.5. Examples of the sensitivity axes aligned with the object's and camera sensor's axes.

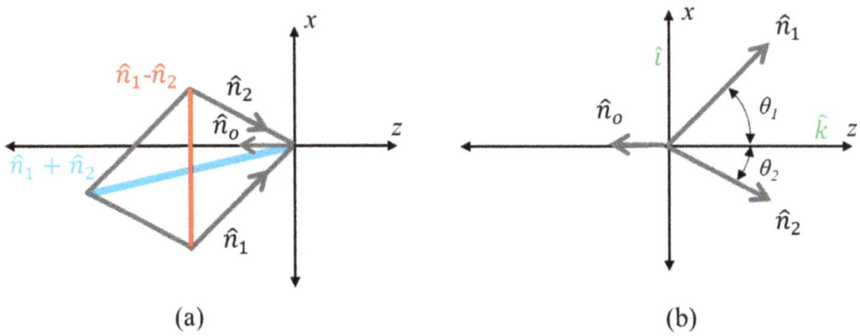

Figure 8.6. Schematic of (a) the resulting sensitivities and (b) their corresponding coordinate system.

By taking the illumination alignment shown in figure 8.5(a), the illumination ($\hat{n}_1$, $\hat{n}_2$) and observation ($\hat{n}_0$) directions are as shown in the schematic view of figure 8.6(a). Thus, translating into a coordinate system, as in figure 8.6(b), it is possible to determine the sensitivity vectors for each illumination direction.

Then equation (8.3) may be expressed as

$$\begin{pmatrix} \Delta\varphi_1 \\ \Delta\varphi_2 \end{pmatrix} = 2\pi/\lambda \begin{pmatrix} S_{1x} & S_{1z} \\ S_{2x} & S_{2z} \end{pmatrix} . \begin{pmatrix} u \\ w \end{pmatrix}. \tag{8.4}$$

With the terms S_1 and S_2 as the sensitivity vectors for the illumination position 1 and 2, respectively. The variables u and w represent the object displacements along the x- and z-axes. From figure 8.6(b) we have

$$\hat{n}_o = -\hat{k},$$
$$\hat{n}_1 = \cos\theta_1\hat{k} + \text{sen}\theta_1\hat{i},$$
$$\hat{n}_2 = \cos\theta_2\hat{k} - \text{sen}\theta_2\hat{i}.$$

Here, i and k are the unitary vectors along the x- and z-axes, respectively, then

$$S_1 = \hat{n}_1 - \hat{n}_o = (1 + \cos\theta_1)\hat{k} + \text{sen}\theta_1\hat{\imath}, \tag{8.5}$$

$$S_2 = \hat{n}_2 - \hat{n}_o = (1 + \cos\theta_2)\hat{k} - \text{sen}\theta_2\hat{\imath}. \tag{8.6}$$

Substituting equations (8.5) and (8.6) in (8.4),

$$\begin{pmatrix} \Delta\varphi_1 \\ \Delta\varphi_2 \end{pmatrix} = 2\pi/\lambda \begin{pmatrix} \text{sen}\theta_1 & (1 + \cos\theta_1) \\ -\,\text{sen}\theta_2 & (1 + \cos\theta_2) \end{pmatrix} . \begin{pmatrix} u \\ w \end{pmatrix}. \tag{8.7}$$

Now, solving the equation to retrieve the $u(\Delta\varphi_1 - \Delta\varphi_2)$ and $w(\Delta\varphi_1 + \Delta\varphi_2)$ displacements, it is found that

$$\Delta\varphi_1 - \Delta\varphi_2 = k[(\text{sen}\theta_1 + \text{sen}\theta_2)\hat{\imath} + (\cos\theta_1 - \cos\theta_2)\hat{k}].\ u, \tag{8.8}$$

$$\Delta\varphi_1 + \Delta\varphi_2 = k[(\text{sen}\theta_1 - \text{sen}\theta_2)\hat{\imath} + (2 + \cos\theta_1 + \cos\theta_2)\hat{k}].\ w. \tag{8.9}$$

These expressions are a general solution when two object illumination directions are used. However, as shown in figure 8.6(a), the out-of-plane vector (w) is not entirely aligned with the z-axis. The reason is that the angles of both illumination directions are different ($\theta_1 \neq \theta_2$), so it is not possible to eliminate the term $\hat{\imath}$ from equation (8.9) and the term $\hat{k}$ in equation (8.8). However, if both angles are the same ($\theta = \theta_1 = \theta_2$), these expressions are simplified as in common experimental practice:

$$\Delta\varphi_1 - \Delta\varphi_2 = 2k(\text{sen}\theta)\hat{\imath}.\ u, \tag{8.10}$$

$$\Delta\varphi_1 + \Delta\varphi_2 = 2k(1 + \cos\theta)\hat{k}.\ w. \tag{8.11}$$

8.2 Illuminating once at a time

To simplify the experimental conditions, when a once at a time illumination is used, the illumination angles are set equal and aligned to the same plane as figure 8.7(a) shows; an equivalent configuration using optical fibers is presented in figure 8.7(b). Here, each image hologram is recorded once at a time using the light blockers B1 and B2 (according to the illumination applied). If the deformation is controlled, stepped, or occurs slowly, B1 and B2 could be achieved by blocking the beams with a non-transparent material. If the deformation is a fast event such as a vibration, both B1 and B2 could be electronically controlled, one at a time.

In this form, equations (8.10) and (8.11) are valid, and the displacements can be calculated through

$$u = \frac{\lambda(\Delta\varphi_1 - \Delta\varphi_2)}{4\pi\text{sen}\theta}, \tag{8.10}$$

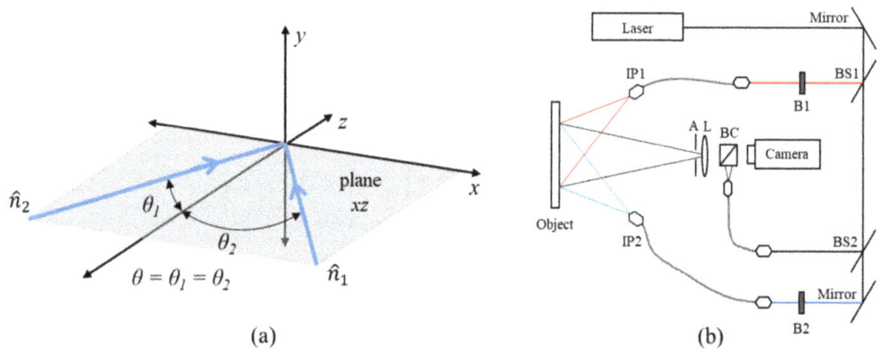

Figure 8.7. (a) Configuration with two object illumination used once at a time by blocking the beams with either B1 or B2; and (b) an example set-up where BS1 and BS2 are beam splitters, A and L are the aperture and lens, respectively, and B1 and B2 are beam blockers.

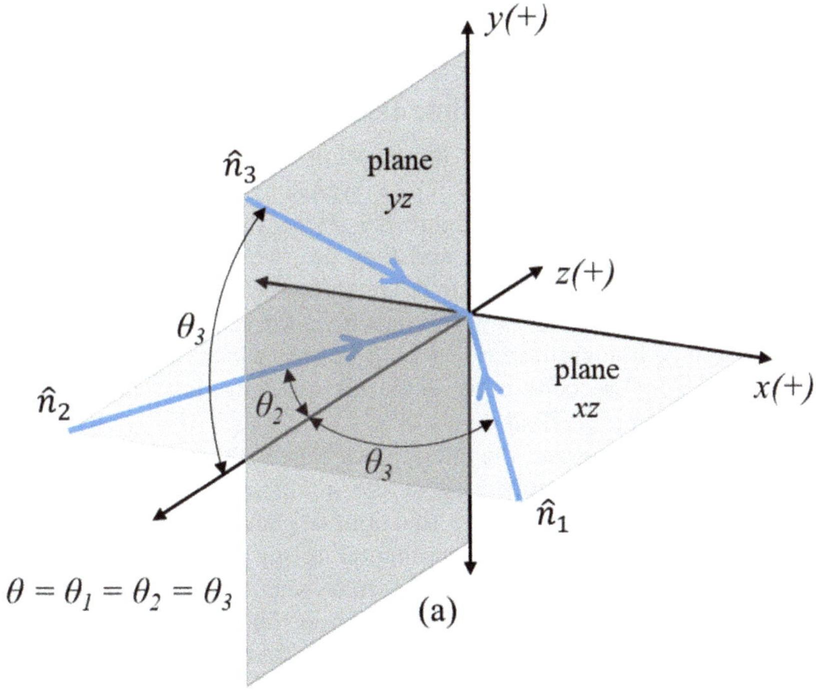

Figure 8.8. Configuration with three object illuminations used once at a time.

$$w = \frac{\lambda(\Delta\varphi_1 + \Delta\varphi_2)}{4\pi(1 + \cos\theta)}. \tag{8.11}$$

Therefore, this model could be extended to use three illumination positions to retrieve the full displacement components u, v, and w along the x-, y-, and z-axes. In this case, the third illumination should give the v component without introducing terms in the rest of the equations. For this reason, the configuration to be used is the one presented in figure 8.8. Here, the same condition of the illumination angle is used, having: $\theta = \theta_1 + \theta_2 + \theta_3$.

The displacement terms u and w are obtained in the usual manner using equations (8.10) and (8.11). But now, including that of the y-axis,

$$v = \left(\frac{\lambda}{2\pi}\right)\left(\frac{\Delta\varphi_1 + \Delta\varphi_2 - 2\Delta\varphi_3}{2\text{sen}\theta}\right). \tag{8.12}$$

As this is a once at a time configuration, there will be six image holograms that, once correlated, will give the three-phase maps $\Delta\varphi_1$, $\Delta\varphi_2$, and $\Delta\varphi_3$. The advantage is that the same illumination source could be used in any position. However, this is most suitable for slow and controlled deformations. High-speed deformations will require faster-blocking devices and high output power illumination sources to deal with the shorter exposure times needed at high-speed camera acquisition. As an alternative to this, there is a configuration where independent illumination sources may be used to avoid these latter issues.

8.3 Illuminating at the same time (simultaneously)

Each illumination position uses an illumination source with a different output wavelength (e.g. a solid-state laser) [7]. A typical configuration for two object illumination positions is shown in figure 8.9. Here, it is possible to see that the illumination directions are different, and they use two lasers with different wavelengths. However, the difference rests in the reference beams, coming from each laser simultaneously illuminating the camera's sensor via the beam combiner (BC). In this form a single image hologram retrieves both beams, which are spatially separated from each other. Before continuing with this configuration, let us explain the hologram formation and its data.

In a typical image hologram, the interference between the object's backscattering and the reference beam is embedded in the hologram. As DHI uses a spatial frequency carrier method to retrieve the optical phase [8], the geometry of the aperture (or pupil) used in the experimental set-up is imaged in the Fourier spectrum. The size and purpose of this aperture were explained in chapter 6, but let us focus on its location. For example purposes, figure 8.10 shows a gray level image hologram and its corresponding Fourier spectrum. Note the scale differences between the axes in figures 8.10(a) and (b): the first is a spatial *xy* image, while the second is expressed in frequency units for both axes.

Now, when more than one object illumination is used simultaneously, the Fourier spectrum is as the one shown in figure 8.11, where three object illumination directions are used.

Each reference beam requires a different spatial carrier to avoid overlapping among them. In this example, the effect of using different laser wavelengths is evident due to the size differences among the pupils. The pupil or aperture of the

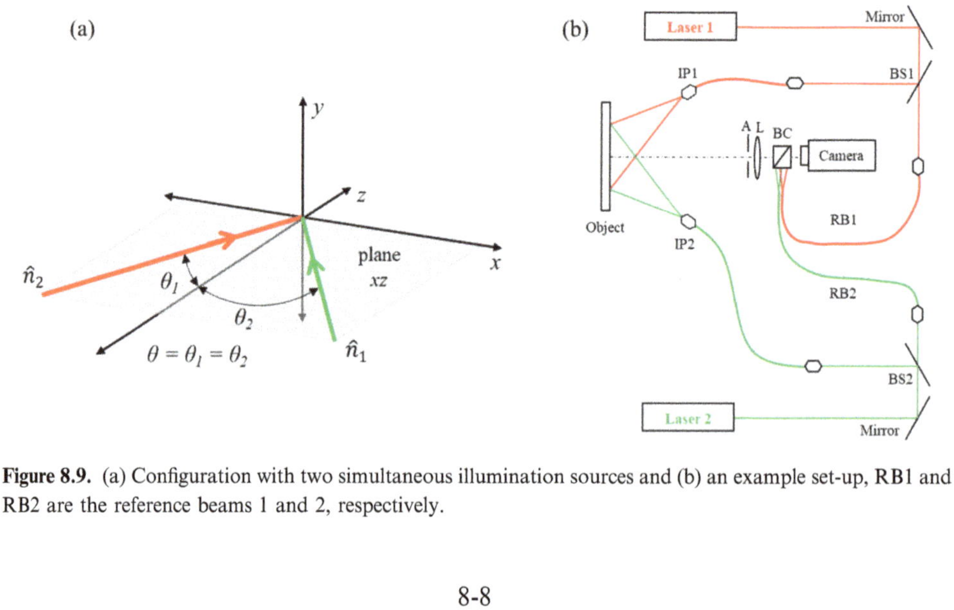

Figure 8.9. (a) Configuration with two simultaneous illumination sources and (b) an example set-up, RB1 and RB2 are the reference beams 1 and 2, respectively.

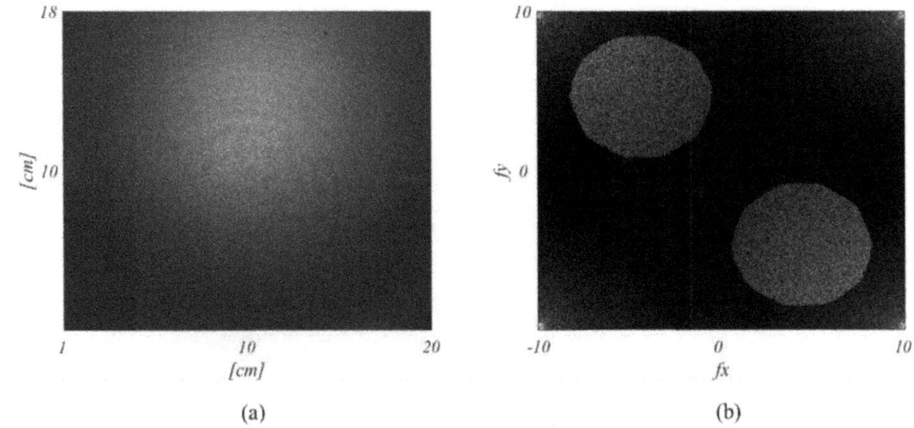

Figure 8.10. (a) Image hologram with one illumination and (b) its Fourier spectrum showing the aperture image and its corresponding complex conjugate term.

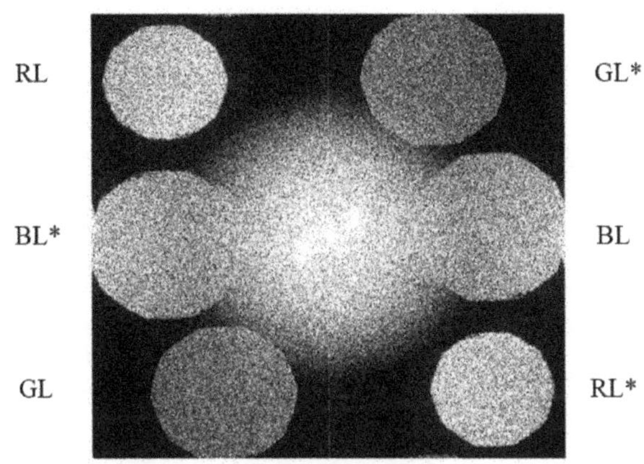

Figure 8.11. Fourier spectrum resulting from three simultaneous object illumination directions, * indicates the complex conjugate terms. RL, BL, and GL are the apertures from a red, blue, and green laser, respectively.

optical system is the same, but the speckle size changes according to the pupil's size and the wavelength. For this reason, the red laser's aperture (RL) looks smaller than the aperture for the blue laser (BL). The wavelength effect is also observed in the optical phase when the same deformation introduces fewer fringes for the larger wavelength (red laser) with respect to the shorter wavelength (i.e. the green laser). Figure 8.12 compares wrapped phase maps for the same deformation on a plate.

As can be seen, the wrapped phase map from the green (532 nm) laser has more fringes than that of the red (632.8 nm) laser. The latter implies that the unwrapped phase maps from both lasers have different values (radians). However, the expressions used so far to determine the displacements are modified to scale each phase map with their corresponding wavelength. After this adjustment, it is possible

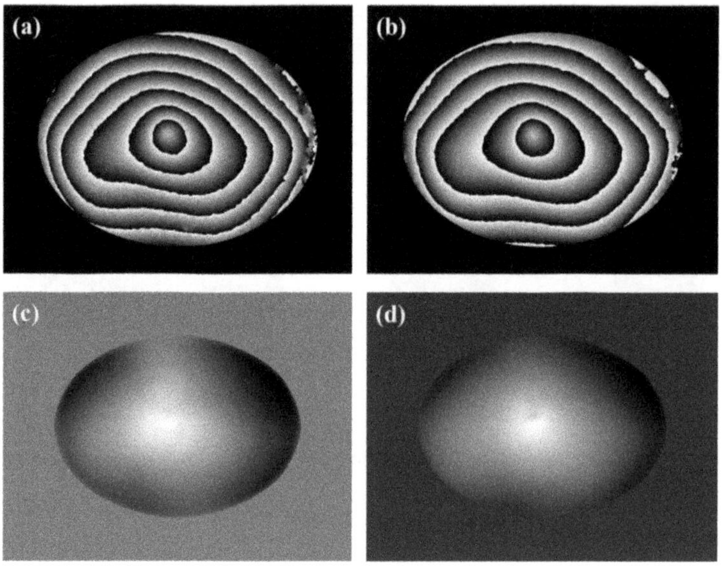

Figure 8.12. Wrapped phase maps for the same deformation with (a) green and (b) red lasers and their corresponding, (c) and (d), unwrapped phase maps.

to match the displacement information. The wavelength scaled equations are as follows:

$$u = \frac{(\lambda_1 \Delta\varphi_1 - \lambda_2 \Delta\varphi_2)}{4\pi \mathrm{sen}\theta}, \tag{8.13}$$

$$v = \left(\frac{1}{2\pi}\right)\left(\frac{\lambda_1 \Delta\varphi_1 + \lambda_2 \Delta\varphi_2 - \lambda_3 2\Delta\varphi_3}{2\mathrm{sen}\theta}\right), \tag{8.14}$$

$$w = \frac{(\lambda_1 \Delta\varphi_1 + \lambda_2 \Delta\varphi_2)}{4\pi(1 + \cos\theta)}. \tag{8.15}$$

Figure 8.13 shows an example of three lasers using three different wavelengths registered by the same monochromatic sensor and their corresponding fringe patterns and phase maps during a thermal deformation in a polymer.

If it is impossible to have several lasers illuminating simultaneously to gather all the holograms, a third configuration for multiple object illumination positions is feasible. This configuration acquires the desired information with only one laser: taking advantage of the laser coherence length and the optical paths, it is possible to illuminate all three directions with no interference among the individual illumination directions. Let us consider the example in figure 8.14; IP1 and RB1 form the first illumination position pair. Then, an optical fiber (OF) is used to increase the optical path length of the laser beyond its coherence length for the second illumination pair IP2 and RB2, such that this will not interfere with the first illumination position pair.

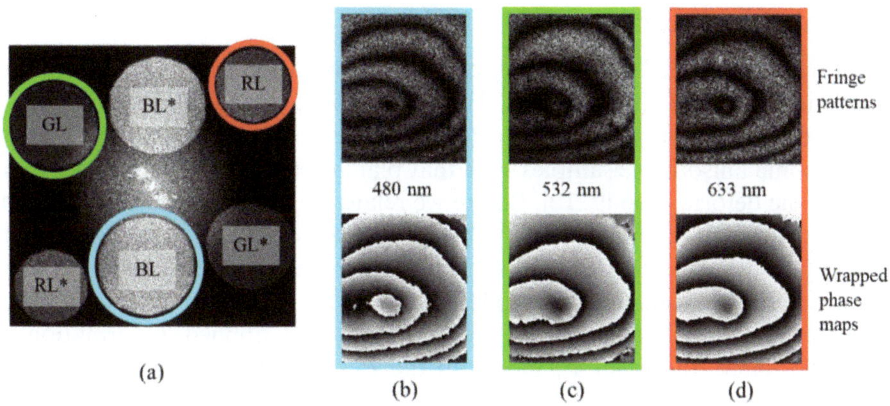

Figure 8.13. Simultaneous measurement example with (a) three lasers in (b) blue, (c) green, and (d) red.

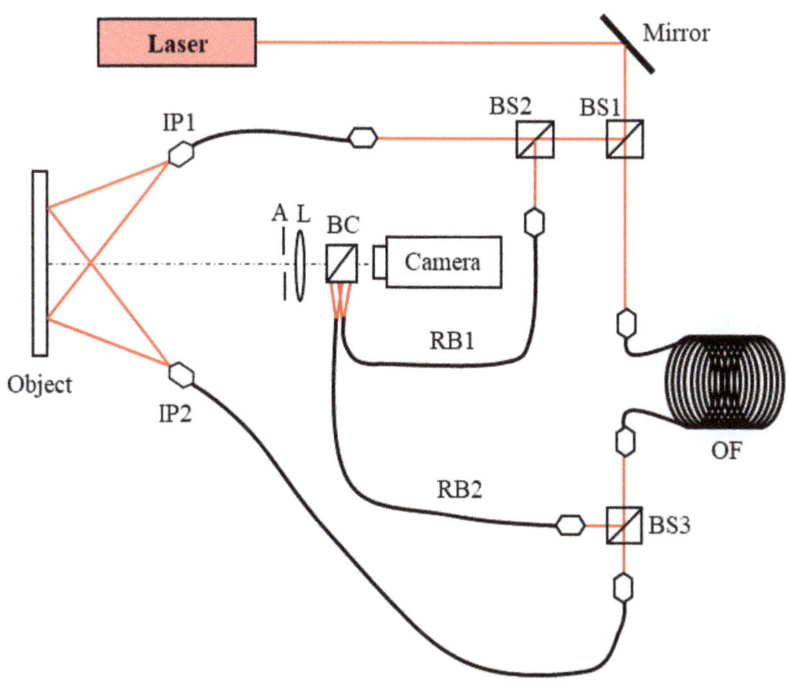

Figure 8.14. Optical set-up for two directions of object illumination using the same laser (ideally with a few meters' coherence length).

Also, the laser's coherence length should be considered. The pair of object and reference beams for the *n*th illumination position matches their optical paths within this coherence length. The rest of the interferometer is similar with beam splitters (BS1, BS2, and BS3), a beam combiner (BC), and a lens aperture (L and A) array in front of the camera.

8.4 Strain measurement using DHI

In several mechanical studies, the object strain information gives more detail about its behavior than the displacement data. This conversion is usually applied for isotropic media. However, the applied deformation is restricted to low rates and values in some anisotropic samples such that planes remain planes and lines remain lines after the deformation [9, 10]. Before we relate the displacement to strain, let us consider that the object under study has two allowed movements, one as a whole and a second relative to a point's position. It is this second movement that one uses to determine the displacement–strain relation. In order to exemplify this, a volume inside the sample is analyzed before and after the applied deformation (see figure 8.15). Here, the point P_1 moves to the P_1* position with a change in u, v, and w for the x-, y-, and z-axes, respectively. The remaining corners of this cube are also deformed to different positions that may be the same at the P_1* point or different (i.e. u*, v*, and w* associated with the P_3* point).

These general displacement terms could be expressed in terms of a Taylor expansion series as follows:

$$u^* = u + \frac{\partial u}{\partial x}\Delta x + \frac{\partial u}{\partial y}\Delta y + \frac{\partial u}{\partial z}\Delta z + \dots, \tag{8.16}$$

$$v^* = v + \frac{\partial v}{\partial x}\Delta x + \frac{\partial v}{\partial y}\Delta y + \frac{\partial v}{\partial z}\Delta z + \dots, \tag{8.17}$$

$$w^* = w + \frac{\partial w}{\partial x}\Delta x + \frac{\partial w}{\partial y}\Delta y + \frac{\partial w}{\partial z}\Delta z + \dots. \tag{8.18}$$

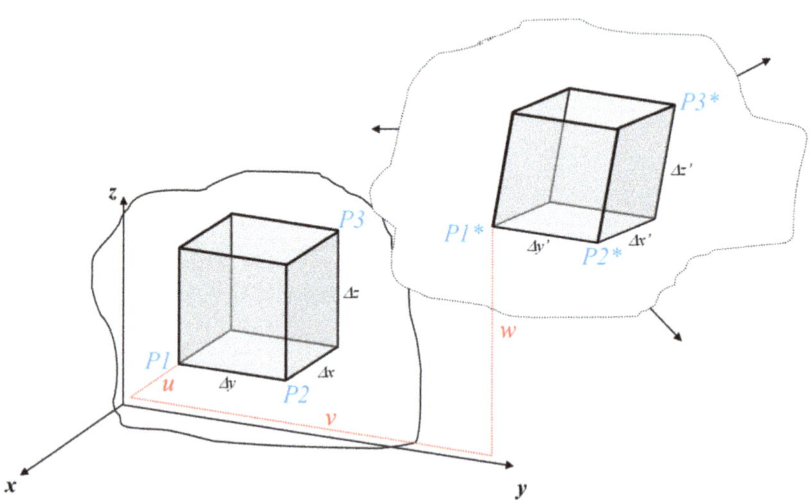

Figure 8.15. Schematic example employed to relate displacement with strain.

The previous expressions are valid if the cube size is small enough to neglect the higher-order terms of the series. Let us consider the line segment P_1–P_2 (Δy), which is aligned to the y-axis. After the deformation, its displacement terms are

$$u^* = u + \frac{\partial u}{\partial y}\Delta y, \tag{8.19a}$$

$$v^* = v + \frac{\partial v}{\partial y}\Delta y, \tag{8. 19b}$$

$$w^* = w + \frac{\partial w}{\partial y}\Delta y. \tag{8. 19c}$$

Defining the normal strain for the deformed line segment $\Delta y'$ as

$$\varepsilon_{yy} = \frac{\Delta y' - \Delta y}{\Delta y}, \tag{8.20}$$

and rewriting

$$\Delta y' = (1 + \varepsilon_{yy})\Delta y, \tag{8.21}$$

from figure 8.15, the line segment $\Delta y'$ could be expressed as

$$(\Delta y')^2 = \left(\frac{\partial u}{\partial y}\Delta y\right)^2 + \left[\left(1 + \frac{\partial v}{\partial y}\right)\Delta y\right]^2 + \left(\frac{\partial w}{\partial y}\Delta y\right)^2. \tag{8.22}$$

By substituting equation (8.22) in (8.21),

$$(1 + \varepsilon_{yy})\Delta y = \sqrt{\left(\frac{\partial u}{\partial y}\Delta y\right)^2 + \left[\left(1 + \frac{\partial v}{\partial y}\right)\Delta y\right]^2 + \left(\frac{\partial w}{\partial y}\Delta y\right)^2}$$

$$(1 + \varepsilon_{yy}) = \sqrt{\left(\frac{\partial u}{\partial y}\right)^2 + \left[\left(1 + \frac{\partial v}{\partial y}\right)\right]^2 + \left(\frac{\partial w}{\partial y}\right)^2},$$

$$\varepsilon_{yy} = \sqrt{\left(\frac{\partial u}{\partial y}\right)^2 + 1 + 2\frac{\partial v}{\partial y} + \left(\frac{\partial v}{\partial y}\right)^2 + \left(\frac{\partial w}{\partial y}\right)^2} - 1. \tag{8.23}$$

In a similar form, the strain components ε_{xx} and ε_{zz} are expressed as follows:

$$\varepsilon_{xx} = \sqrt{1 + 2\frac{\partial u}{\partial x} + \left(\frac{\partial u}{\partial x}\right)^2 + \left(\frac{\partial v}{\partial x}\right)^2 + \left(\frac{\partial w}{\partial x}\right)^2} - 1, \tag{8.24}$$

$$\varepsilon_{zz} = \sqrt{\left(\frac{\partial u}{\partial z}\right)^2 + \left(\frac{\partial v}{\partial z}\right)^2 + 1 + 2\frac{\partial w}{\partial z} + \left(\frac{\partial w}{\partial z}\right)^2} - 1. \tag{8.25}$$

The quadratic terms could be neglected if tiny deformations are applied as they are too small. Hence, these expressions are reduced to their well-known engineering strain–displacement form [11].

$$\varepsilon_{xx} = \frac{\partial u}{\partial x}, \tag{8.26}$$

$$\varepsilon_{yy} = \frac{\partial v}{\partial y}, \tag{8.27}$$

$$\varepsilon_{zz} = \frac{\partial w}{\partial z}. \tag{8.28}$$

The normal strain components define the length change of a line segment divided by the original line length. The reduction is applied to the shear strain terms. A shear strain indicates an angular change between two line segments of the illustrated cube in figure 3.18.

$$\gamma_{xy} = \frac{\partial v}{\partial x} + \frac{\partial u}{\partial y}, \tag{8.29}$$

$$\gamma_{yz} = \frac{\partial w}{\partial y} + \frac{\partial v}{\partial z}, \tag{8.30}$$

$$\gamma_{zx} = \frac{\partial u}{\partial z} + \frac{\partial w}{\partial x}. \tag{8.31}$$

When a material is tested, the stress–strain curve is an essential tool to evaluate its mechanical behavior; this curve is also related to the optical signal coming from an optical method such as DHI. In addition, it is possible to associate these two values by employing a constant known as Young's modulus (E) or elasticity modulus,

$$\sigma = E.\, \varepsilon. \tag{8.32}$$

If the material is isotropic and known, it is possible to retrieve the stress information (σ) through equation (8.32), considering the strain definitions of the previous section.

8.5 Deformation in 3D measurements

Let us write some relationships with data taken from the experimental set-up to find the object deformation. Consider figure 8.16, where S_i and S_o are unity vectors along with the object illumination and observation (e.g. CCD sensor) directions, respectively, $P(x, y, z)$ is a point on the object's surface, $\mathrm{IP}_k(x, y, z)$ is the object source illumination position, and $C(x, y, z)$ is the CCD sensor position. Now recall equation (8.1),

$$\Delta\varphi = 2\pi/\lambda \boldsymbol{S}.\, d, \tag{8.33}$$

where the sensitivity vector $\boldsymbol{S} = S_i - S_o$, and it is associated with the space (x, y, z) sensitivity of the interferometer. Now consider the object being illuminated from three illumination directions (see figure 8.8), equation (8.33) will take the form

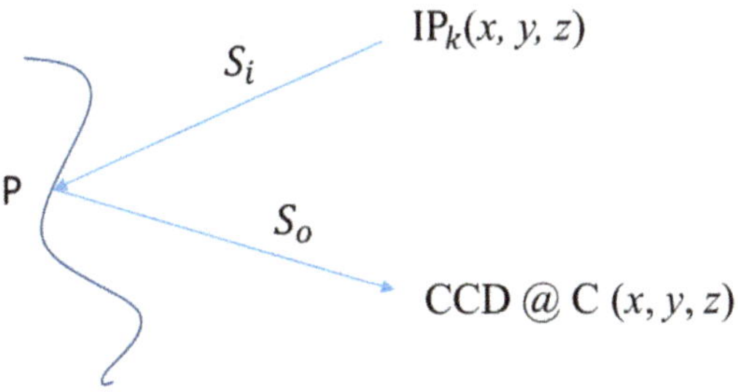

Figure 8.16. Illumination and observation direction coordinates. The object source of illumination position changes to three different points.

$$\Delta\varphi_k = 2\pi/\lambda(S_{ik} - S_o).\ d, \tag{8.34}$$

with $k(x, y, z) = 1, 2, 3$, corresponding to the direction of object illumination 1, 2, or 3 for the corresponding position of the light source. It follows that,

$$S_{ik} = P - \mathrm{IP}_k$$
$$S_o = C - P$$

and the sensitivity vector for each object illumination direction takes the form

$$\boldsymbol{S}_k = (S_{ik} - S_o) = P - \mathrm{IP}_k - C + P = 2P - \mathrm{IP}_k - C$$

or for instance

$$\boldsymbol{S}_{1x} = 2P - \mathrm{IP}_{1x} - C.$$

From equation (8.34),

$$\Delta\varphi_1 = 2\pi/\lambda\left(S_{1x}d_x + S_{1y}d_y + S_{1z}d_z\right) \tag{8.35}$$

$$\Delta\varphi_2 = 2\pi/\lambda\left(S_{2x}d_x + S_{2y}d_y + S_{2z}d_z\right) \tag{8.36}$$

$$\Delta\varphi_3 = 2\pi/\lambda\left(S_{3x}d_x + S_{3y}d_y + S_{3z}d_z\right) \tag{8.37}$$

These relations represent a system of equations where the three unknowns d_x, d_y, d_z (or as using the notation in the last section for the displacement field u, v, w along the axes x, y, z) can be found. To achieve this and for simplicity, let us define

$\frac{\lambda}{2\pi}\Delta\varphi_n = \varphi_n$.

After some algebraic manipulation, the solution is

$$d_z = \frac{(S_{2x}\varphi_3 - S_{3x}\varphi_2)(S_{1x}S_{2y} - S_{1y}S_{2x}) - (S_{1x}\varphi_2 - S_{2x}\varphi_1)(S_{3y}S_{2x} - S_{2y}S_{3x})}{(S_{2x}S_{3z} - S_{2z}S_{3x})(S_{1x}S_{2y} - S_{1y}S_{2x}) - (S_{1x}S_{2z} - S_{1z}S_{2x})(S_{3y}S_{2x} - S_{2y}S_{3x})} \tag{8.38}$$

$$d_y = \frac{(S_{1x}\varphi_2 - S_{2x}\varphi_1) - (S_{1x}S_{2z} - S_{1z}S_{2x})d_z}{(S_{1x}S_{2y} - S_{1y}S_{2x})} \tag{8.39}$$

$$d_x = \frac{\varphi_1}{S_{1x}} - \frac{S_{1y}}{S_{1x}}d_y - \frac{S_{1z}}{S_{1x}}d_z. \tag{8.40}$$

Experimentally the variables φ_n are obtained as the phase maps from each individual object illumination direction, as happens with all the sensitivity vectors $\boldsymbol{S}_k$. Hence, equations (8.38), (8.39), and (8.40) can be written in an algorithm whose output will calculate the object displacement along the x-, y-, and z-axes (deformation along u, v, w).

8.6 Amplitude and phase calculation for harmonic excitation with pulsed object illumination: an example

Let us now consider an object subjected to vibration, particularly when it is externally driven, so its surface undergoes a harmonic excitation. The surface displacement between two object states at times t and $t + \Delta t$, with Δt being the light source pulse separation, i.e. if the light source is a laser with a continuous wave (CW), it can be made as if it was pulsed utilizing simply a light rotating chopper, or a more sophisticated optomechanical device; if the laser is pulsed Δt is the pulse separation. The absolute deformation may be written as

$$\vec{d}(x, y, z, t) = \vec{D}(x, y, z)\cos(\omega t + \varphi_0), \tag{8.41}$$

where $\vec{D}$ is a vector whose amplitude is the object displacement due to the harmonic vibration, ω the angular frequency, and the starting phase is defined as

$\varphi_0 = \varphi_0(x, y, z, t_0)$.

The measured deformation between these two object positions is

$\Delta\vec{d}(x, y, z, t) = \vec{D}(x, y, z)[\cos(\omega(t + \Delta t) + \varphi_0) - \cos(\omega t + \varphi_0)]$ or

$$\begin{aligned}\Delta\vec{d}(x, y, z, t) = \vec{D}(x, y, z)[&\cos(\omega t + \varphi_0)\\ &(\cos\omega\Delta t - 1) - \sin(\omega t + \varphi_0)\sin\omega\Delta t].\end{aligned} \tag{8.42}$$

Using equation (8.41) the velocity of the vibrating surface is

$$\vec{v}(x, y, z, t) = \frac{\mathrm{d}\vec{d}}{\mathrm{d}t} = -\omega\vec{D}(x, y, z)\sin(\omega t + \varphi_0). \tag{8.43}$$

Then substituting in equation (8.42)

$$\Delta\vec{d}(x, y, z, t) = \vec{d}(x, y, z, t)(\cos\omega\Delta t - 1) + \frac{1}{\omega}\vec{v}(x, y, z, t)\sin\omega\Delta t \tag{8.44a}$$

or

$$\frac{1}{\omega}\vec{v}(x, y, z, t) = \frac{\Delta\vec{d}(x, y, z, t) - \vec{d}(x, y, z, t)(\cos\omega\Delta t - 1)}{\sin\omega\Delta t}. \tag{8.44b}$$

The measured deformation at time $t + \tau$ is (using equation (8.42))

$$\Delta\vec{d}(x, y, z, t + \tau) = \vec{D}(x, y, z)[\cos(\omega t + \varphi_0 + \omega\tau)(\cos\omega\Delta t - 1) - \sin(\omega t + \varphi_0 + \omega\tau)\sin\omega\Delta t] \tag{8.45}$$

that after some algebraic manipulation yields

$$\Delta\vec{d}(x, y, z, t + \tau) = \left[\vec{d}(x, y, z, t)(\cos(\omega(\tau + \Delta t)) - \cos\omega\tau) + \frac{1}{\omega}\vec{v}(x, y, z, t)(\sin(\omega(\tau + \Delta t)) - \sin\omega\tau)\right]. \tag{8.46}$$

Substituting equation (8.44b) into equation (8.46) gives

$$\Delta\vec{d}(t + \tau) = \left[\begin{array}{l}\vec{d}(t)(\cos(\omega(\tau + \Delta t)) - \cos\omega\tau) + \dfrac{\Delta\vec{d}(t) - \vec{d}(t)(\cos\omega\Delta t - 1)}{\sin\omega\Delta t}\\ (\sin(\omega(\tau + \Delta t)) - \sin\omega\tau)\end{array}\right], \tag{8.47}$$

where for simplicity, the dependency on the coordinates x, y, and z was dropped and is implied henceforth.

This equation can be simplified to

$$\Delta\vec{d}(t + \tau) = \alpha\Delta\vec{d}(t) - \beta\vec{d}(t), \tag{8.48}$$

where

$$\alpha = \frac{\sin\omega\tau}{\sin\omega\Delta t}(\cos\omega\Delta t - 1) + \cos\omega\tau \tag{8.49}$$

$$\beta = \frac{\sin\omega\tau}{\sin\omega\Delta t}[(\cos\omega\Delta t - 1)^2 + \sin^2\omega\Delta t]. \tag{8.50}$$

From equation (8.48)

$$\vec{d}(t) = \frac{\alpha}{\beta}\Delta\vec{d}(t) - \frac{1}{\beta}\Delta\vec{d}(t + \tau) \tag{8.51a}$$

or using the symbol i to refer to a particular coordinate x, y, or z

$$d_i(t) = \frac{\alpha}{\beta}\Delta d_i(t) - \frac{1}{\beta}\Delta d_i(t + \tau). \tag{8.51b}$$

From equation (8.41)

$$d_i(x, y, z, t) = D_i(x, y, z)\cos(\omega t + \varphi_0), \tag{8.51c}$$

so for instance

$$d_x(x, y, z, t) = D_x(x, y, z)\cos(\omega t + \varphi_0), \text{ etc}$$

and from equation (8.43)

$$v_i(x, y, z, t) = -\omega D_i(x, y, z)\sin(\omega t + \varphi_0), \tag{8.51d}$$

namely

$$v_x(x, y, z, t) = -\omega D_x(x, y, z)\sin(\omega t + \varphi_0), \text{ etc}$$

hence

$$\tan(\omega t + \varphi_0) = -\frac{1}{\omega}\frac{v_i(t)}{d_i(t)}. \tag{8.52}$$

Using equations (8.44b) and (8.51b) to substitute into equation (8.52) gives

$$\tan(\omega t + \varphi_0) = \frac{(\cos\omega\Delta t - 1)}{\sin\omega\Delta t} + \frac{\beta}{\sin\omega\Delta t}\left(\frac{1}{\alpha - \frac{\Delta d_i(t+\tau)}{\Delta d_i(t)}}\right). \tag{8.53}$$

Let us now substitute (8.51b) into (8.44b)

$$\begin{aligned} v_i(x, y, z, t) = \frac{\omega}{\sin\omega\Delta t}\Bigg(&\Delta d_i(t)\left(1 - \frac{\alpha}{\beta}(\cos\omega\Delta t - 1)\right) \\ &+ \Delta d_i(t+\tau)\frac{(\cos\omega\Delta t - 1)}{\beta}\Bigg). \end{aligned} \tag{8.54}$$

Finally, from equations (8.41) and (8.43) it follows that the amplitude of the object displacement is given by

$$D_i(x, y, z) = \sqrt{d_i^2(x, y, z, t) + \frac{1}{\omega^2}v_i^2(x, y, z, t)}, \tag{8.55}$$

where $d_i(x, y, z, t)$ is the absolute deformation given by equation (8.51c) and $v_i(x, y, z, t)$ by equation (8.54)—these are variables found from the experimental data embedded and from the fringe and phase maps. In particular, the experimental absolute deformation variable $d_i(x, y, z, t)$ is given by equations (8.38), (8.39), and (8.40).

From all the above, let us recall some critical variables and parameters:

$\Delta d_i(x, y, z, t)$ and $\Delta d_i(x, y, z, t + \tau)$

ω = angular frequency = $2\pi f$

τ = time at second measurement; Δt = (double) pulse separation.-

And, for convenience, define the following variable, $\widetilde{v_i}$:

$\widetilde{v}_i = \frac{v_i}{\omega}$; and $\cos(\omega t + \varphi_0) = \cos\phi(x, y, z) = \frac{d_i(x, y, z, t)}{D_i(x, y, z)}$.

As a final example, let us think that at a time, say $t = 0$, the object is at the position one and has an absolute deformation of

$d_1(x, y, z, t) = D(x, y, z)\cos(\phi)$and at time $t + \tau = t + T/4$, the absolute object deformation at position two is

$d_2(x, y, z, t + \tau) = D(x, y, z)\cos(\phi + \pi/2) = -D(x, y, z)\sin\phi$,which implies that

$$\sqrt{d_1^2 + d_2^2} = D(x, y, z) \tag{8.56}$$

and

$$\tan\phi = -\frac{d_2}{d_1} \tag{8.57}$$

hence

$$\phi = \arctan\left(-\frac{d_2}{d_1}\right). \tag{8.58}$$

The phase given by equation (8.58) is directly related to the measured phase from the experiment.

References

[1] De la Torre-Ibarra M, Mendoza-Santoyo F, Flores Moreno J M and del Socorro Hernandez-Montes M 2022 Gabor's holography: seven decades influencing optics *Appl. Opt.* **61** B225–36

[2] De la Torre-Ibarra M, del Socorro Hernández-Montes M, Flores Moreno J M and Mendoza-Santoyo F 2016 Laser speckle based digital optical methods in structural mechanics: a review *Opt. Laser Eng.* **87** 32–58

[3] Pedrini G, Osten W and Gusev M E 2006 High-speed digital holographic interferometry for vibration measurement *Appl. Opt.* **45** 3456–62

[4] Schedin S, Pedrini G, Tiziani H J and Mendoza Santoyo F 1999 Simultaneous three-dimensional dynamic deformation measurements with pulsed digital holography *Appl. Opt.* **38** 7056–62

[5] De la Torre-Ibarra M D L, Mendoza-Santoyo F, Pérez-López C and Saucedo-A T 2005 Detection of surface strain by three-dimensional digital holography *Appl. Opt.* **44** 27–31

[6] Javidi B *et al* 2021 Roadmap on digital holography *Opt. Express* **29** 35078–118

[7] Saucedo-A T, De la Torre-Ibarra M H, Mendoza Santoyo F and Moreno I 2010 Digital holographic interferometer using simultaneously three lasers and a single monochrome sensor for 3D displacement measurements *Opt. Express* **18** 19867–75

[8] Takeda M, Ina H and Kobayashi S 1982 Fourier-transform method of fringe-pattern analysis for computer-based topography and interferometry *J. Opt. Soc. Am.* **72** 156–60

[9] Dally J W and Riley W F 1991 *Experimental Stress Analysis* 3rd edn (New York: McGraw-Hill)

[10] Hertzberg R W 1996 *Deformation and Fracture Mechanics of Engineering Materials* (New York: Wiley)

[11] Sciammarella C A and Gilbert J A 1973 Strain analysis of a disk subjected to diametral compression by means of holographic interferometry *Appl. Opt.* **12** 1951–6

Fernando Mendoza-Santoyo, Manuel De la Torre-Ibarra, María del Socorro Hernández-Montes and Jorge Mauricio Flores Moreno

Chapter 9

Digital holographic microscopy for surface profilometry

Contributing author: Myung K Kim

A review of digital holographic microscopy as a tool for surface profilometry is presented in this chapter. Quantitative phase microscopy with digital holography provides nanometric surface information and optical thickness profiles in a direct and straightforward manner. With the complete amplitude and phase information of the optical field, the numerical representation of the optical field can be manipulated and processed in ways not possible in real space optics. We will briefly describe several techniques with some examples given. These techniques include optical phase unwrapping, low coherence interferography, digital interference holography, and spatial phase scanning digital holography.

9.1 Introduction

Surface profilometry is an essential tool in many fields to measure the surface roughness, step size, curvature, and flatness of microstructures. It can also play important roles in biomedical applications to reveal cellular and tissue structures, and some information on the physiology of the object under study. Traditional profilometers are based on the stylus or other probes raster-scanned over a surface. They provide nanometric and sub-nanometric resolution, but they can be prone to surface damage and distorted measurements. Non-contact optical profilometers have many advantages, such as being non-destructive, fast, and highly precise. Much of the current technology of optical profilometry is based on white-light interferometry, structured light illumination microscopy, and confocal microscopy. Confocal and white-light systems usually require point detection and raster scanning, while structured light is a full field and fast approach. However, it has limitations in lateral and axial resolutions. Digital holographic microscopy is a new

doi:10.1088/978-0-7503-3027-5ch9

alternative approach that provides fast, high-resolution, and versatile capabilities in optical profilometry [1].

From previous chapters we know that digital holography records both the amplitude and phase of the optical field. The phase field directly reflects the surface height profile of the reflective surface or the optical thickness of the transparent dielectric structure. The full field acquisition of digital holography is fast compared to raster scanning systems, and the quantitative phase microscopy by digital holography gives high axial resolution on the nanometric or sub-nanometric scale. Many new applications of digital holographic microscopy are advancing in biomedical microscopy, materials characterization, and microstructure fabrication and offer a host of versatile capabilities in surface metrology [2].

In what follows we aim to provide a brief overview of the development of digital holographic microscopy and its potential applications in surface metrology. In section 9.2 a short history and basic principles of digital holography are given, together with some representative examples of quantitative phase microscopy by digital holography. In section 9.3 the process of optical phase unwrapping that allows height measurements beyond the illumination light wavelength is presented. Multi-wavelength holography removes the limitation of optical phase measurements, and examples will be given of surface structure measurements of almost macroscopic dimensions up to millimeters high. Section 9.4 describes the use of low coherence illumination light for digital holography that removes the persistent problem of coherent noise and significantly simplifies the optomechanical system. Concluding remarks are given at the end.

9.2 Digital holographic microscopy

Digital holography acquires the interference between the object and optical reference fields, which is then processed computationally to reconstruct both the amplitude and phase profiles of the original object's optical field. Replacing the photochemical processes of analog holography with numerical processing greatly simplifies the hologram acquisition and reconstruction and the optomechanical system requirements. Furthermore, digital holography leads to many imaging and processing capabilities that are difficult or impossible in analog holography [2].

Quantitative phase microscopy by digital holography is possible because the phase information is directly available as soon as the complex optical field is calculated by digital holography, as was demonstrated by Depeursinge's group in 1999 [3]. One of the major application areas of digital holographic quantitative phase microscopy has been cellular microscopy [4–7]. The mostly transparent cellular structure is challenging to observe with standard bright field microscopy. Conventional phase-contrast microscopy transforms the phase profile to intensity profile in a highly nonlinear manner and cannot be used for quantitative phase imaging. The various types of fluorescence microscopy represent a highly developed and powerful technology but require elaborate preparation and are subject to significant limitations such as toxicity of the staining process. Digital holography is ideally suited to generating quantitative phase contrast without any sample preparation that can hinder natural cellular physiology [4]. Digital holographic microscopy has been demonstrated to reveal microbiological processes in ways that have been inaccessible with conventional imaging methods, such as the growth of live neurons [8], dry cell mass measurement in yeast proliferation [6], and cellular response to drugs [7], as well as surface topography of microstructures [9]. Several companies now commercialize digital holographic microscopes for biological and materials science applications, such as LynceeTec, Ovizio, PhiOptics, and TomoCube.

In traditional analog holography, it is possible to generate multiple holograms acquired with different parameters, such as fluid flow or deformation, and combine them for holographic interferometry by carrying out additional interference experiments. However, in digital holography, combining holograms simply entails numerical addition, or other manipulations, of arrays of complex numbers that represent the holograms [10]. These processes led to many powerful techniques for materials characterization and morphological reconstruction. By using two or more wavelengths, the phase profile of an object can be obtained with a synthetic wavelength that can be much larger than the original wavelengths [11]. Furthermore, by combining holograms with the wavelengths scanned over a range, one can obtain an effective coherence length much shorter than the laser source, which can then be used for generating tomographic or topographic profiles of the object [12]. The amplitude and phase profiles of digital holography represent the complete knowledge of the object's optical field. This fact allows one to emulate many different methods of classical microscopy, such as darkfield, various types of phase contrast, and differential interference contrast [13].

Holography is a coherent imaging method, and various types of lasers are typically used as the illumination source. In many applications, the coherent noise of laser can be problematic, and the requirement of optomechanical stability of conventional interferometers can add to the complexity of the system. Digital holography using incoherent sources has been a significant direction of development. An important approach is to generate the reference by spatial filtering of the object beam, such as in optical diffraction tomography [14], diffraction phase microscopy [15], and spatial light interference microscopy [16]. Of course, coherence is a relative concept—a low coherence source has a finite coherence length, but the field is coherent within the coherence length. The latter means a low coherence source such as LED can carry out a holographic experiment as long as the path difference between the reference and the object does not exceed the coherence length. This significantly benefits from suppressing stray coherent interference between unintended optical surfaces, thus yielding final images with high fidelity [17, 18]. A new holographic interferometer approach is now available, called Fresnel incoherent holography [19] or self-interference digital holography [20], which can acquire holographic images under the illumination of completely incoherent light such as a natural sky light or ambient indoor light. Other significant trends in recent years have been the incorporation of modern signal processing concepts such as compressive sensing and deep learning to solve some of the problems that have been enigmatic in this field, such as phase unwrapping, phase retrieval without reference or insufficient information, and coherent noise removal [14, 21, 22].

There are two basic approaches to acquiring holograms: off-axis holography and phase-shifting holography [2]. In both, the goal is to acquire the phase profile of the object field, which is not normally available in non-holographic imaging. There are other methods to retrieve the phase information non-holographically, namely various phase retrieval methods [23]. These methods include transport of intensity equation [24], ptychography [25], and coherent diffraction tomography [26]. The heavy computational load of these methods is becoming a less critical issue with ever-increasing computational power, and some of these methods may very well become more commonplace.

We now describe the basic holographic processes. In off-axis holography, figure 9.1, the object phase information is encoded as modulation of a high-frequency interference fringe between the object and reference, tilted at a certain angle—hence the name off-axis. The Fourier transform of the fringe pattern, filtered with a mask, then eliminates the dc and twin-image terms. The remaining term is then multiplied with a propagation factor followed by an inverse Fourier transform, which yields a term proportional to the object's complex—amplitude and phase–optical field. Succinctly,

$$E(x, y; z) \propto \mathscr{F}^{-1}\left\{\mathscr{F}\{I(x, y; 0)\}\left[k_x, k_y\right] \cdot M\left(k_x, k_y\right) \cdot \exp\left(\mathrm{i}z\sqrt{k^2 - k_x^2 - k_y^2}\right)\right\}[x, y]. \quad (9.1)$$

Here $\mathscr{F}\{I(x, y; 0)\}\left[k_x, k_y\right]$ is the Fourier transform of the input field with respect to the angular spatial frequencies $\left(k_x, k_y\right)$, $M\left(k_x, k_y\right)$ is the mask applied to the spectrum to select one of the twin-image terms, $\exp\left(\mathrm{i}z\sqrt{k^2 - k_x^2 - k_y^2}\right)$ is the propagation factor, and $\mathscr{F}^{-1}\{\}[x, y]$ is the inverse Fourier transform. This particular method of calculating the diffraction is known as the angular spectrum method, which has constant lateral (x, y) resolution [27].

With phase-shifting digital holography [28], the object and reference fields are parallel or on-axis, and three or more interference patterns $I_n(x, y)$ are recorded

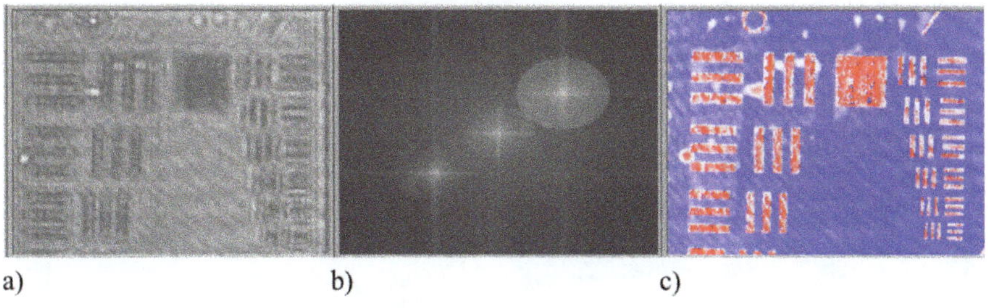

Figure 9.1. Digital holographic microscopy in the off-axis configuration: (a) camera acquisition of holographic interference; (b) Fourier transform showing the dc and twin-image terms, as well as the mask to select one of the twin terms; and (c) the phase image obtained using the angular spectrum method. The image area is $369 \times 302\ \mu\mathrm{m}^2$ with 1024×768 pixels.

while the phase of the reference is shifted in steps of $2\pi/N$, where N is the number of steps. The complex optical field is obtained by

$$E(x, y) = \frac{1}{N}\sum_{n=0}^{N-1} I_n(x, y)\exp\left(2\pi i\frac{n}{N}\right). \tag{9.2}$$

The field can be propagated as with the angular spectrum method. The resulting object field retains the full pixel resolution of the camera, while in the off-axis method, the resolution is reduced by the masking process. On the other hand, the off-axis configuration requires a single exposure, while the phase-shifting method requires multiple exposures. The amount of required information is fundamentally equivalent between the two methods.

The holographic configurations can be further classified in terms of the conjugate plane of the camera sensor. In the image hologram, the object plane is focused onto the camera plane so that the reconstructed holographic image is in the focus of the object plane. This procedure has the convenience of direct monitoring and reconstruction of the object image without the need for numerical propagation. If the camera plane is conjugate to the Fourier plane, i.e. the focal plane of the objective lens, the Fourier hologram needs to be converted to an object image by a Fourier transform. This can be useful for filtering and processing the reconstructed image. In general, if the camera is focused to an arbitrary distance, the configuration is referred to as a Fresnel hologram, and an appropriate numerical propagation is necessary to form a focused image of the object.

Quantitative phase microscopy by digital holography was first demonstrated by Depeursinge's group [3] and since then many researchers have developed techniques and applications in diverse areas. High-resolution, high fidelity images of various cell types have been obtained, such as cancer cells [4], live neurons [8], red blood cell membrane fluctuation [5], cellular response to the anticancer drug [29], and laser microsurgery [30], to name a few examples.

Figure 9.2 is an example of cellular microscopy by digital holography, showing the quantitative phase profiles of epithelial cells from a mouth swab. Panels (a) and (b) are while the sample is still wet; (c) and (d) are after the sample has dried in exposure to air. In the former, the relative index of refraction of the cell is small compared to the background water, and the optical phase variation is correspondingly tiny. After drying, the cellular refraction index is much larger than the

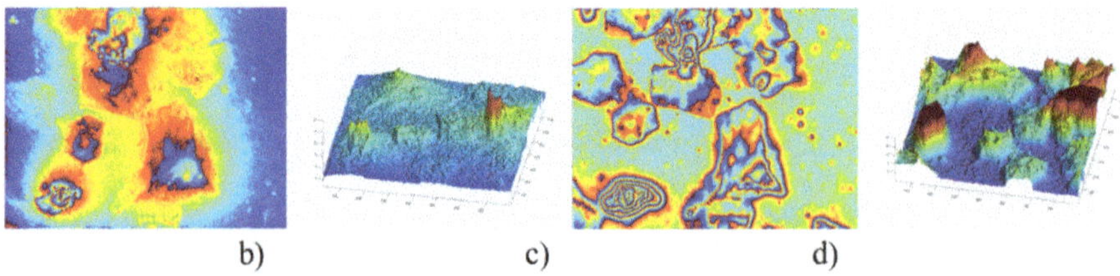

a) b) c) d)

Figure 9.2. Quantitative phase microscopy by digital holography of mouth epithelial cells: (a) and (b) are with the wet sample; (c) and (d) are after the sample is dried; (b) and (d) are after numerical phase unwrapping.

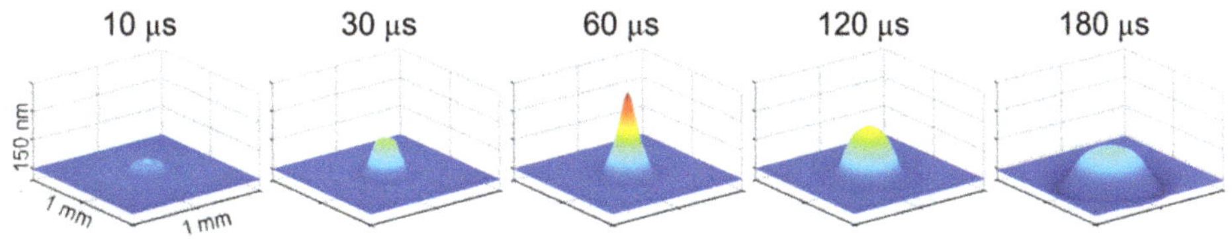

Figure 9.3. Time-dependent surface response to the optical impulse for deionized water at various time steps [31].

background air. The 3D plots in (b) and (d) are after numerically unwrapping the corresponding phase profiles.

In figure 9.3 a pulsed laser impinging on the surface of pure deionized water causes an impulsive response on the surface, not unlike the ripples formed after a pebble drops in the water. The surface profile is measured with digital holographic quantitative phase microscopy at various delays from the laser pulse [31]. The laser pulse is 5 ns long and contains much less than 1 mJ when it hits the water. Note the large difference in the lateral and axial scales and that the maximum height of the ripple is approximately 100 nm, at about 60 μs after excitation. From the measurement of the speed of ripple propagation, the surface tension of water can be determined in a non-contact non-invasive manner.

9.3 Optical phase unwrapping by multi-wavelength digital holographic microscopy

As seen above, nanometric height or optical thickness profiles can be generated using digital holography. On the other hand, the measurable height without ambiguity is limited by the wavelength used in generating the hologram. This so-called wrapped phase problem is familiar to most phase measuring methods and has been a subject of much research. Most numerical methods are computationally intense and subject to erroneous identification of phase discontinuity. A common approach is to apply any number of available numerical algorithms, mostly based on identifying the location of phase discontinuity and determining the strategy to stitch the phase discontinuity. New advances in deep learning-based phase unwrapping methods may bring new exciting tools to this problem [22].

What we call optical phase unwrapping has been introduced by applying multi-wavelength interferometry to digital holography [11, 32]. Suppose two holograms are acquired of an object surface using two wavelengths λ_1 and λ_2. If the height profile of the object is $z(x, y)$, then the hologram's phase profiles are $\varphi_1(x, y) = \text{mod}\left(\frac{2\pi z}{\lambda_1}, 2\pi\right)$ and $\varphi_2(x, y) = \text{mod}\left(\frac{2\pi z}{\lambda_2}, 2\pi\right)$. The modulo of the difference $\Phi_{12} = \text{mod}\,(\varphi_1 - \varphi_2, 2\pi)$ produces another phase profile corresponding to an effective or synthetic wavelength $\Lambda_{12} = \frac{\lambda_1\lambda_2}{\lambda_1 - \lambda_2}$: $Z_{12} = \Lambda_{12}\frac{\Phi_{12}}{2\pi}$. Using two wavelengths with a small difference can make the synthetic wavelength sufficiently large for the necessary height range. The optical phase unwrapping method has

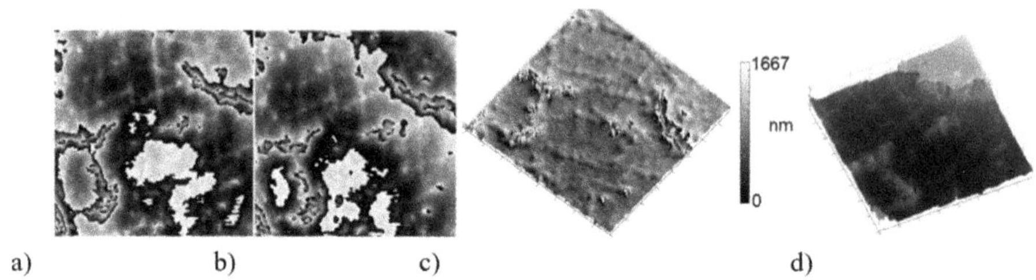

Figure 9.4. Surface profile of polished porous coal particle. Panels (a) and (b) are phase images at $\lambda_1 = 633$ nm and $\lambda_2 = 532$ nm, respectively. Panel (c) is the result of a two-wavelength optical phase unwrapping with synthetic wavelength $\Lambda_{12} = 3.334\ \mu\text{m}$. Panel (d) is the result of numerical phase unwrapping. The image area is $98 \times 98\ \mu\text{m}^2$ [33].

several advantages over numerical methods. The computational load is very light, consisting of simple subtraction and modulo operations. It is deterministic and not subject to algorithmic strategy as with numerical unwrapping.

An example of optical phase unwrapping is shown in figure 9.4 for quantitative phase microscopy of the polished surface of a porous coal particle [33]. Two holographic phase images are acquired using the wavelengths 633 nm of a HeNe laser and 532 nm of a diode-pumped solid-state laser. The modulo difference phase profile, adequately scaled, is shown in (c), displaying the nanometric scratch patterns of the polishing process and the presence of several pores. On the other hand, the numerically unwrapped profile in (d) has gross errors in the upper right corner because of the algorithmic misinterpretation of the phase discontinuity near that area.

In order to cover a larger height range, a smaller wavelength difference is required. For example, a combination of $\lambda_1 = 532$ nm and $\lambda_2 = 633$ nm gives $\Lambda_{12} = 3.334\ \mu\text{m}$, while $\lambda_1 = 632$ nm and $\lambda_2 = 633$ nm gives $\Lambda_{12} = 400\ \mu\text{m}$. On the other hand, it can be shown that the large synthetic wavelength is accompanied by proportionately amplified phase noise. If the hologram phase profile contains phase noise $\varepsilon \cdot 2\pi$ radians, this corresponds to height profile noise $\varepsilon \cdot \lambda$. The noise in the synthetic wavelength phase profile Φ_{12} also has the same level of phase noise $\varepsilon \cdot 2\pi$ radians, but now the noise in the height profile is $\varepsilon \cdot \Lambda_{12}$. There can be situations where this amplified noise is so large as to make the synthetic phase profile unusable. Here the noise is from any fluctuation or variation of phase that does not correlate with the wavelength change, such as detector noise, in particular from dark areas or sharp corners of an object, and speckle noise from diffuse areas or stray reflections.

As long as the synthetic phase noise $\varepsilon \cdot \Lambda_{12}$ is still small compared to λ_1, it is possible to use the Φ_{12} phase profile as a guide for unambiguously counting integer multiples of the original wavelength λ_1 and stitch the φ_1 phase profiles according to $\overline{\Phi}_{12} = \text{int}\left[\dfrac{Z_{12}}{\lambda_1}\right] \cdot 2\pi + \varphi_1$ and $\overline{Z}_{12} = \lambda_1 \dfrac{\overline{\Phi}_{12}}{2\pi}$. On the other hand, if the condition $\varepsilon \cdot \Lambda_{12} \ll \lambda_1$ cannot be met, a third wavelength λ_3 can be used such that, $\varepsilon \cdot \Lambda_{12} \ll \Lambda_{13}$ and $\varepsilon \cdot \Lambda_{13} \ll \lambda_1$, with $\Lambda_{13} = \dfrac{\lambda_1 \lambda_3}{\lambda_1 - \lambda_3}$. The process can go on hierarchically to

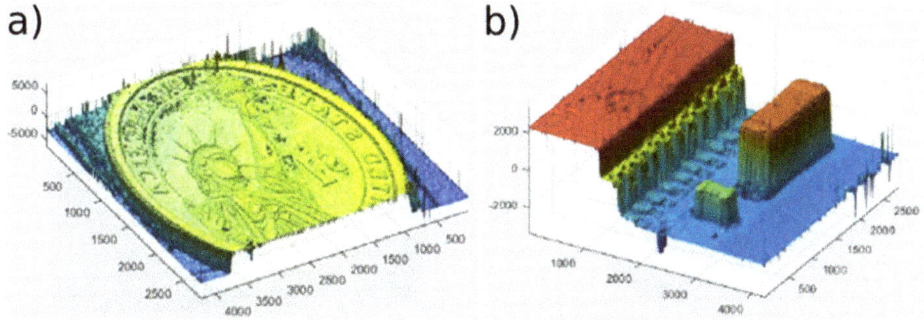

Figure 9.5. Surface profiles from MWOPU using four wavelengths: $\lambda_1 = 0.639615\ \mu$m, $\lambda_2 = 0.639585\ \mu$m, $\lambda_3 = 0.639426\ \mu$m, and $\lambda_4 = 0.637935\ \mu$m . The image volume is 30.5 mm × 20.4 mm × 6.8 mm.

generate synthetic wavelengths that are progressively smaller, starting from Λ_{12} and reduce to the noise in a multi-step process [34, 35].

Figure 9.5 displays the surface profiles of a US dollar coin and components on a printed circuit obtained with optical phase unwrapping using four wavelengths at $\lambda_1 = 0.639615\ \mu$m, $\lambda_2 = 0.639585\ \mu$m, $\lambda_3 = 0.639426\ \mu$m, and $\lambda_4 = 0.637935\ \mu$m. These values correspond to the synthetic wavelengths $\Lambda_{12} = 13\,640\ \mu$m, $\Lambda_{13} = 2164\ \mu$m, and $\Lambda_{14} = 242.9\ \mu$m. Because of the reflection geometry, the measured height is half of the synthetic wavelength, and up to 6.8 mm of height profile is generated with a typical noise level of less than 1 μm. In these examples, the object surface has many sharp edges and dark crevices, making it necessary to apply median filtering and other filtering techniques to extract the high fidelity images.

9.4 Digital holography using low coherence sources

Holography, in general, is based on the coherent interference of the object and reference fields, and therefore the coherence of the light source is an important parameter. Typically, various types of lasers are required for holography, particularly for macroscopic imaging objects. On the other hand, the concept of coherence is a relative one, and all light sources have finite coherence lengths greater than zero and less than infinity. One only needs to ensure that the coherence length is large enough for the maximum optical path difference between reference and object. For microscopic imaging, the object dimensions are typically micrometers or millimeters at most, meaning that light sources of fairly short coherence length can be used for holographic imaging. Such light sources may include LEDs, filtered incandescent lamps, or arc lamps [17]. The short coherence length also gives the added benefit of reduced stray interference between unintended reflections from optical surfaces, which is a significant cause of coherent noise in digital holography using lasers. In order to reduce such coherent noise, the long coherence length of lasers can be reduced by a rotating diffuser, whose time average of fluctuating spatial coherence gives the effect of short coherence length and reduced coherence noise [36]. Examples of quantitative phase microscopy by digital holography using LEDs as the light source are shown in figure 9.6. In particular, the height variation of chrome

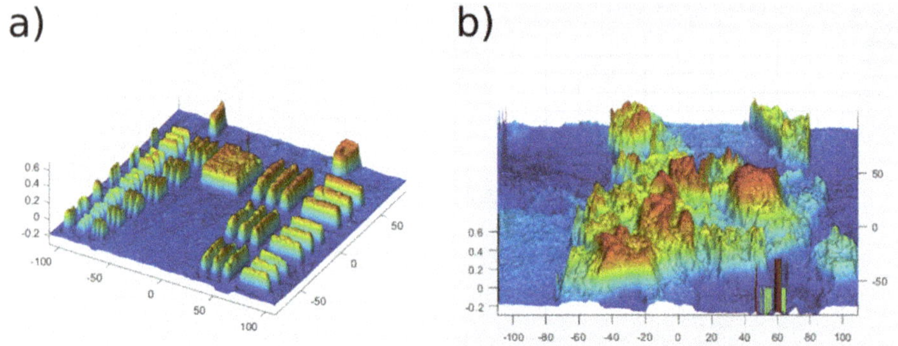

Figure 9.6. Quantitative phase microscopy by LED-based DHM. (a) A resolution target, showing the area of groups 6 and 7 and the patterned chrome film height ranging from 50 to 100 nm. (b) QPM image of mouth epithelial cells. Cell optical thickness ranges up to 400 nm relative to the water background.

film pattern on the resolution target with smallest patterns, group 7, is accurately and clearly displayed, thanks to the low coherent noise of LED.

Using a low coherence source entails a much more stringent requirement of optical alignment precision to maintain overlap of reference and object fields within the spatial and temporal coherence range. A practical approach is to spatially filter a part of the object field to generate a reference field. Instead of providing a separate reference field, this results in a typical path configuration between the reference and object and, therefore, very robust mechanical stability and operational simplicity. Various techniques have been developed with such names as diffraction phase microscopy [15] and spatial light interference microscopy [16]. By phase shifting, the reference, holographic phase microscopy is achieved with high fidelity phase images.

The preceding paragraphs describe the use of low coherence sources in effect as a coherent source with short coherence length. Quantitative phase profiles are obtained as long as the object height range is within the coherence length. On the other hand, the short coherence length can also generate contours of an object's height profile corresponding to a height relative to the reference mirror distance. The latter is the basis of white-light interferometry.

The typical configuration of white-light interferometry as a surface profiler is to detect the interference maximum of a point on an object's surface by axial scanning a Mirau-type microscope objective to construct a two-dimensional profile of surface height variations [37]. With phase-sensitive detection of the interference, the height resolution can be a few nanometers or better, and the obtained profile image can be of very low noise and high fidelity without distortion. The raster scanning can limit the speed and throughput of the system.

The principle can be extended to two dimensions by acquiring interference images on a two-dimensional sensor, e.g. a CCD or CMOS camera. The method may then be referred to as interferography or wide-field optical coherence tomography [38]. Only axial scanning is required, and therefore the process can be much faster in principle but is often limited by the rate of the camera image transfer and scanning of a mechanically bulkier object or camera. For example, figure 9.7 illustrates the formation of a 3D surface profile from the white-light interferometer. The camera view in (a) contains interference at contour points of the object. A phase-shifting method is used to extract the contour points of the object, as shown in (b). Axial scanning of the reference mirror builds up the 3D profile of the fingerprint over an area of 15×12 mm^2 and a maximum height of 0.70 mm, shown in (c) and (d) [39].

More examples of white-light interferography or wide-field optical coherence tomography (WFOCT) are shown in figure 9.8. Here a more straightforward method is used to extract the contour points, namely, taking the difference between frames as the z-distance of the reference mirror is slowly scanned and taking the locus of maximum interference along the z-axis as the z-location of the surface. A relatively noise-free surface map over large areas is obtained with an axial resolution of 10 or better in each of these examples. Compared with the laser-based multi-wavelength holographic method of figure 9.5, one may note the significant lack of the spiky coherent noise, although the axial resolution is somewhat larger, 10–20 μm, than the laser-based images, a few microns.

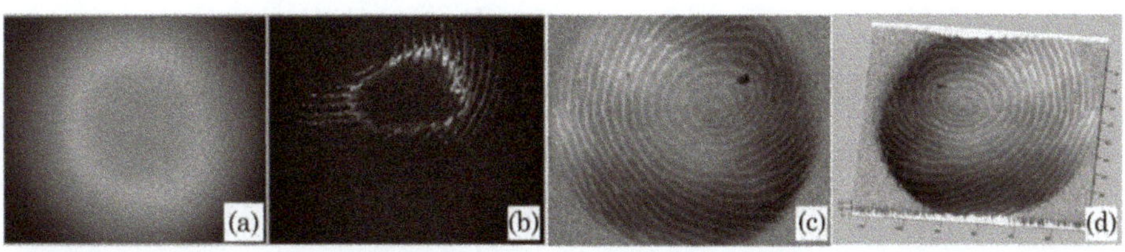

Figure 9.7. Topographic imaging of a plastic fingerprint by low coherence interferography: (a) an interferogram as captured by the camera, (b) a coherent interference envelope extracted by the phase-shifting method, (c) a two-dimensional map of the fingerprint surface topography with the grayscale representing the height of the surface at each pixel, and (d) a three-dimensional perspective rendering of (c). The image volume is 15 mm $\times$ 12 mm $\times$ 0.70 mm [39].

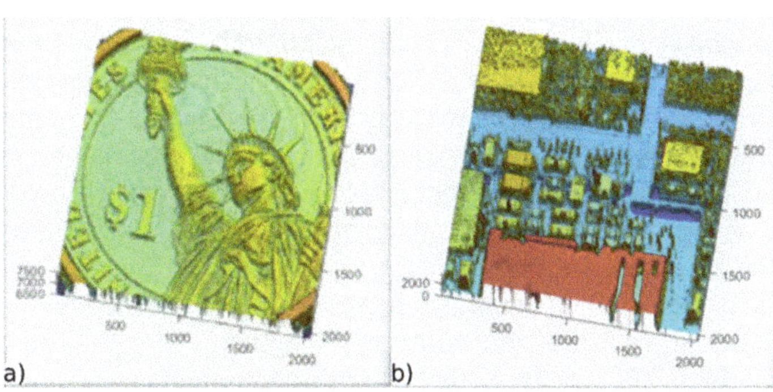

Figure 9.8. Examples of WFOCT. (a) Surface of a US dollar coin. (b) Surface of a printed circuit with electronic components. The image area is 20×20 mm^{2}, and the scan height is 1000 μm for (a) and 2000 μm for (b).

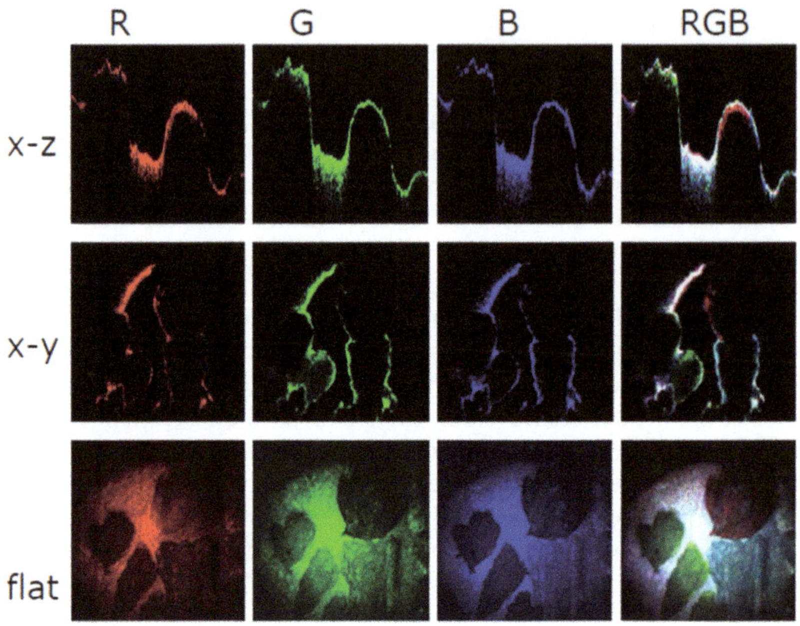

Figure 9.9. Color WFOCT images of a painted coin surface. The top, middle, and bottom rows represent the *x–z* cross-section, *x–y* cross-section, and flat view, respectively. The first three columns are the red, green, and blue channels combined to generate the full-color images on the right column. The image volume is 7.2 mm × 7.2 mm × 335 μm [38].

In figure 9.9 full-color three-dimensional tomography of a painted coin surface is obtained by repeating the interferography experiment three times using red, green, and blue LEDs and combining the resulting three sets of image data to compose an RGB type full-color image [38]. The experiment can be done with a single interferometry scan if an RGB color camera is used and the three color channels are separately processed. Also, note the tomographic nature of the 3D image, indicating reflections from layers below the surface due to light penetration into the paint material.

The low coherence of a light source is characterized by its broad-spectrum, i.e. the coherence length and the spectral width are in a Fourier inverse relationship. This relation implies that the low coherence can be synthesized by combining many holograms using a wide range of wavelengths. This process is referred to as wavelength scanning digital interference holography and can be used to generate tomographic or topographic images similar to white-light interferography [39, 40]. In each of the examples shown in figure 9.10, 30 holograms are acquired using wavelengths starting from 575.0 nm and incrementing by 1.06 nm, corresponding to a spectral width of 19.2 nm and coherence length of 17 μm. The result is the topography of the coin's surface or the fire ant's compound eye. The method is also applied to tomographic imaging of fingerprints.

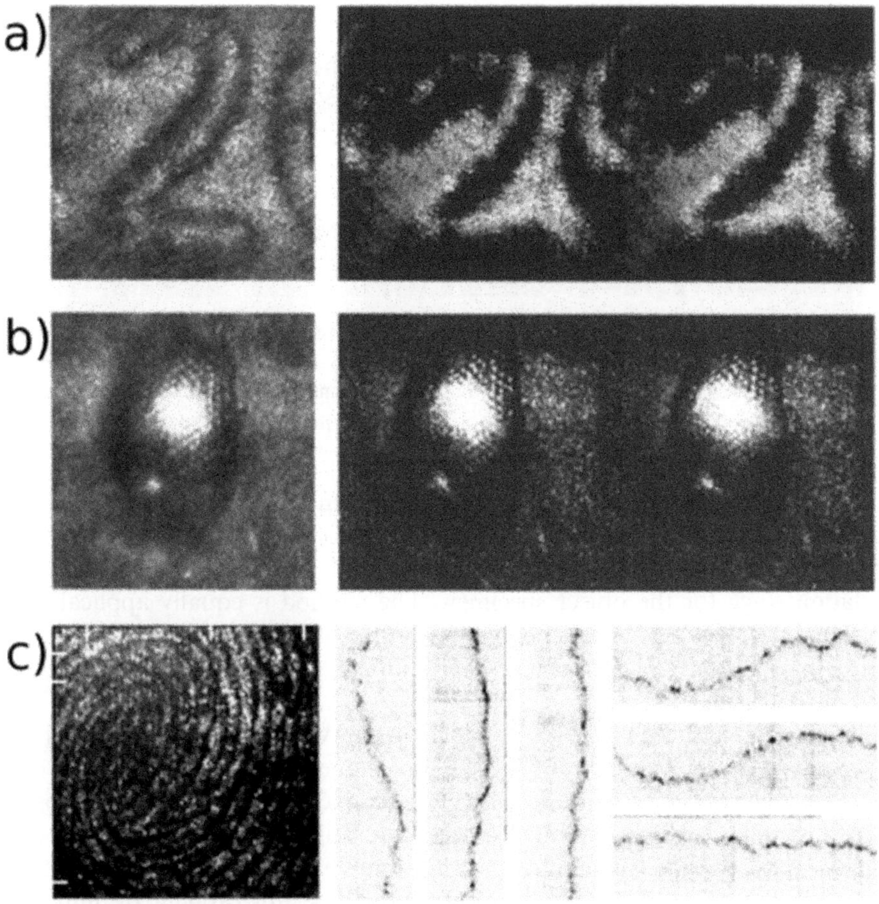

Figure 9.10. (a) Flat and stereoscopic images of a numeral 2 in the 2000 mintage mark of a penny. The area is 1×1 mm^2. (b) Flat and stereoscopic images of a fire ant's compound eye. The area is 1×1 mm^2. (c) Tomographic image of a plastic fingerprint, with the image volume of 10.4 mm × 10.4 mm × 0.213 mm [39, 40].

For many surface profilometry applications, acquiring high-resolution images over a large area requires special consideration to optimize acquisition and processing speed, size of the data set, and system complexity. A straightforward and common approach is to translate the object one field of view (FOV) at a time and stitch together the acquired image frames. When applied to phase-shifting digital holography, multiple exposures of each FOV require a stop-and-go motion that can cause problems in the system's speed, stability, and complexity. In a new technique named spatial phase scanning digital holographic microscopy (SDHM), the object is scanned across 2π phase variation of the tilted reference field instead of shifting the reference phase itself, as shown in figure 9.11(a). This obviates phase shifting by a piezo-modulator and significantly reduces optomechanical complexity, and allows continuous and efficient scanning of a large area [41]. Figure 9.11(b)

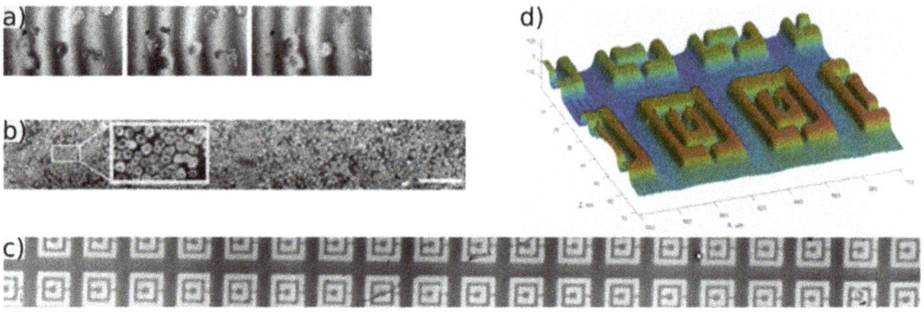

Figure 9.11. (a) Spatial phase shifting by translating the object across the tilted reference field. (b) Example of an SDHM phase image of blood smear across an area of 1090 μm × 160 μm. The scale bar is 100 μm, and the inset shows a 40 μm × 70 μm area detail. (c) SDHM of a THz antenna metastructure across an area of 960 μm × 71 μm. (d) A detailed 3D rendering of the height profile of an 127 μm × 71 μm area of the THz antenna [41].

shows a quantitative phase microscopy scan of a human blood smear covering an area of 1090 μm × 160 μm, with 7310 × 1080 pixels. The optical system, in this case, is based on a Mach–Zehnder configuration with the addition of a motorized translation stage for the object specimen. The method is equally applicable for a reflective object using Michelson configuration, as shown in figures 9.11(c) and (d). An area of 960 μm × 71 μm of a terahertz antenna metastructure is imaged, with 14561 × 1080 pixels.

Summing up, in this chapter we provided an overview of the capabilities of digital holography for applications in surface profilometry. These are based on the important characteristics of quantitative phase microscopy by digital holography that is direct and of high nanometric precision. Highly detailed surface structures can be measured and manipulated using simple optomechanical systems and straightforward numerical processing. Optical phase unwrapping by multi-wavelength holography can provide versatile and efficient tools to profile an area with significant variations in height. A more efficient method to process and minimize phase decorrelation noise is still under investigation. New methods from deep learning may have something significant to offer in this respect. Low coherence holography is another important development direction because of the significant reduction in coherent noise problems and system complexity. The new method of spatial phase scanning digital holographic microscopy has the potential to improve the efficiency and throughput of digital holography-based profilometry significantly. Surface profilometry is an essential tool in many scientific and industrial applications. The contribution from optical profilometry in general and digital holographic profilometry, in particular, will undoubtedly continue to grow.

References

[1] Geng J 2011 Structured-light 3D surface imaging: a tutorial *Adv. Opt. Photonics* **3** 128–60

[2] Kim M K 2011 *Digital Holographic Microscopy: Principles, Techniques, and Applications* (Berlin: Springer)

[3] Cuche E, Marquet P and Depeursinge C 1999 Digital holography for quantitative phase contrast imaging *Opt. Lett.* **24** 291–3
[4] Mann C J, Yu L F, Lo C M and Kim M K 2005 High-resolution quantitative phase-contrast microscopy by digital holography *Opt. Express* **13** 8693–8
[5] Rappaz B, Barbul A, Charriere F, Kuhn J, Marquet P, Korenstein R, Depeursinge C and Magistretti P J 2007 Erythrocytes analysis with a digital holographic microscope *Proc. SPIE* **6631** 66310H
[6] Rappaz B, Cano E, Colomb T, Kuhn J, Depeursinge C, Simanis V, Magistretti P J and Marquet P 2009 Non-invasive characterization of the fission yeast cell cycle by monitoring dry mass with digital holographic microscopy *J. Biomed. Opt.* **14** 034049
[7] Kemper B and von Bally G 2008 Digital holographic microscopy for live cell applications and technical inspection *Appl. Opt.* **47** A52–61
[8] Marquet P, Rappaz B, Magistretti P J, Cuche E, Emery Y, Colomb T and Depeursinge C 2005 Digital holographic microscopy: a non-invasive contrast imaging technique allowing quantitative visualization of living cells with subwavelength axial accuracy *Opt. Lett.* **30** 468–70
[9] De Nicola S *et al* 2004 Surface topography of microstructures in lithium niobate by digital holographic microscopy *Meas. Sci. Tech.* **15** 961–8
[10] Kreis T 2004 *Handbook of Holographic Interferometry* (New York: Wiley)
[11] Gass J, Dakoff A and Kim M K 2003 Phase imaging without 2π ambiguity by multi-wavelength digital holography *Opt. Lett.* **28** 1141–3
[12] Kim M K 1999 Wavelength scanning digital interference holography for optical section imaging *Opt. Lett.* **24** 1693–5
[13] Liu C, Bae Y S, Yang W Z and Kim D Y 2008 All-in-one multifunctional optical microscope with a single holographic measurement *Opt. Eng.* **47** 087001
[14] Choi G, Ryu D H, Jo Y J, Kim Y S, Park W, Min H S and Park K Y 2019 Cycle-consistent deep learning approach to coherent noise reduction in optical diffraction tomography *Opt. Express* **27** 4927–43
[15] Bhaduri B, Edwards C, Pham H, Zhou R, Nguyen T H, Goddard L L and Popescu G 2014 Diffraction phase microscopy: principles and applications in materials and life sciences *Adv. Opt. Photonics* **6** 57–119
[16] Kim T, Zhou R, Mir M, Babacan S D, Carney P S, Goddard L L and Popescu G 2014 White-light diffraction tomography of unlabelled live cells *Nat. Photonics* **8** 256–63
[17] Dubois F and Yourassowsky C 2012 Full off-axis red–green–blue digital holographic microscope with LED illumination *Opt. Lett.* **37** 2190–2
[18] Isikman S O, Bishara W, Mavadadi S, Yu F W, Feng S, Lau R and Ozcan A 2011 Lens-free optical tomographic microscope with a large imaging volume on a chip *Proc. Natl Acad. Sci.* **108** 7296–301
[19] Rosen J and Brooker G 2007 Digital spatially incoherent Fresnel holography *Opt. Lett.* **32** 912–4
[20] Kim M K 2012 Adaptive optics by incoherent digital holography *Opt. Lett.* **37** 2694–6
[21] Zhang H, Liu S, Cao L and Brady D J 2020 Noise suppression for ballistic-photons based on compressive in-line holographic imaging through and inhomogeneous medium *Opt. Express* **28** 10337–49
[22] Wang K, Li Y, Kemao Q, Di J and Zhao J 2019 One-step robust deep learning phase unwrapping *Opt. Express* **27** 15100–15

[23] Fienup J R 2013 Phase retrieval algorithms: a personal tour *Appl. Opt.* **52** 45–56
[24] Gureyev T E, Roberts A and Nugent K A 1995 Partially coherent fields, the transport-of-intensity equation, and phase uniqueness *J. Opt. Soc. Am.* A **12** 1942–6
[25] Marrison J, Raety L, Marriott P and O'Toole P 2013 Ptychography—a label free, high-contrast imaging technique for live cells using quantitative phase information *Sci. Rep.* **3** 2369–75
[26] Putkunz C T, Pfeifer M A, Peele A G, Williams G J, Quiney H M, Abbey B, Nugent K A and McNulty I 2010 Fresnel coherent diffraction tomography *Opt. Express* **18** 11746–53
[27] Yu L F and Kim M K 2005 Wavelength-scanning digital interference holography for tomographic three-dimensional imaging by use of the angular spectrum method *Opt. Lett.* **30** 2092–4
[28] Yamaguchi I and Zhang T 1997 Phase-shifting digital holography *Opt. Lett.* **22** 1268–70
[29] Kemper B, Carl D, Schnekenburger J, Bredebusch I, Schafer M, Domschke W and von Bally G 2006 Investigation of living pancreas tumor cells by digital holographic microscopy *J. Biomed. Opt.* **11** 034005
[30] Yu L F, Mohanty S, Zhang J, Genc S, Kim M K, Berns M W and Chen Z P 2009 Digital holographic microscopy for quantitative cell dynamic evaluation during laser microsurgery *Opt. Express* **17** 12031–8
[31] Clark D C and Kim M K 2012 Noncontact single-pulse optical method to measure interfacial properties in intact systems *Opt. Lett.* **37** 5145–7
[32] Khmaladze A, Kim M and Lo C M 2008 Phase imaging of cells by simultaneous dual wavelength reflection digital holography *Opt. Express* **16** 10900–11
[33] Khmaladze A, Restrepo-Martinez A, Kim M, Castaneda R and Blandon A 2008 Simultaneous dual-wavelength reflection digital holography applied to the study of the porous coal samples *Appl. Opt.* **47** 3203–10
[34] Warnasooriya N and Kim M K 2007 LED-based multi-wavelength phase imaging interference microscopy *Opt. Express* **15** 9239–47
[35] Warnasooriya N and Kim M 2009 Quantitative phase imaging using three-wavelength optical phase unwrapping *J. Mod. Opt.* **56** 85–92
[36] Dubois F, Requena M N, Minetti C, Monnom O and Istasse E 2004 Partial spatial coherence effects in digital holographic microscopy with a laser source *Appl. Opt.* **43** 1131–9
[37] de Groot P 2015 Principles of interference microscopy for the measurement of surface topography *Adv. Opt. Photonics* **7** 1–65
[38] Yu L and Kim M K 2004 Full-color three-dimensional microscopy by wide-field optical coherence tomography *Opt. Express* **12** 6632–41
[39] Potcoava M C and Kim M K 2009 Fingerprint biometry applications of digital holography and low-coherence interferography *Appl. Opt.* **48** H9–15
[40] Dakoff A, Gass J and Kim M K 2003 Microscopic three-dimensional imaging by digital interference holography *J. Electronic Img.* **12** 643–7
[41] Kim M K 2020 Wide area quantitative phase microscopy by spatial phase scanning digital holography *Opt. Lett.* **45** 784–6

IOP Publishing

Full Field Optical Metrology and Applications

Fernando Mendoza-Santoyo, Manuel De la Torre-Ibarra, María del Socorro Hernández-Montes and Jorge Mauricio Flores Moreno

Chapter 10

Contouring

It is common practice to use optical non-invasive methods, such as DSPI or DHI, to measure displacements from flat objects or objects whose shape varies smoothly (for instance, a tiny radius of curvature, i.e. near a flat surface). Then the x, y, and z displacement components can be located pointwise over the flat or smooth surface and in this way one may *safely* claim that the object surface displacement was correctly measured and the data placed accurately on its surface. However, this is often not the case, and there is a wide variety of object shapes whose surface is irregular and/or presents geometric discontinuities, an issue that complicates the precise location of the x, y, and z displacement components. Thus finding the shape information of a non-flat object is an essential requirement to accurately solve and quantify object displacements/deformations that are the start data that enable us to find other physical and mechanical (and more) variables leading to measuring, for example, stress, strain, temperature, elasticity coefficients, etc.

A starting point to quantify the object shape rests in the knowledge of the sensitivity vector, previously described as one that specifies the deformation direction. The sensitivity vector depends on the wavelength, refractive index, and, more importantly, the illumination and observation directions referenced to object points. During the object shape assessment, the measured optical phase is a function of the varying sensitivity vector directly related to the object's geometric shape.

There are some full-field of view methods for shape (or, as it is often called, contouring) measurements. In what follows, we will discuss two: the two points of object illumination method based on double-exposure DSPI, and the two-wavelength or two-frequency contouring method where a sensitivity vector change is introduced between the recording of the fringe patterns.

doi:10.1088/978-0-7503-3027-5ch10

10.1 Two points of object illumination

An out-of-plane sensitive DSPI is shown in figure 10.1 and we will use it to measure the object's contour, where it will be possible to place the displacement data. The two points of the object illumination method are based on the double-exposure DSPI, i.e. it is an interferometric optical method. Experimentally, the optical arrangement has a physical reference plane tangential to the object's surface, say at its back or middle portion. Two object interferograms are captured, one for each object illumination position. The first serves as a reference and the second is recorded with a small tilt $\Delta\alpha$ of the object beam (so we have two object illumination positions). The intensity of these interferograms is given by equations (4.15) and (4.16) in chapter 4. This method renders a phase map between both interferograms related to constant elevation surface contours. The phase map is used to determine the contour of a 3D object that has an inclination produced by the angle of the object's beam relative to the optical axis. This tilt must be eliminated by subtracting the phase difference from the reference plane.

Also, in figure 10.1 we can see one of the illumination beams at an angle (α) with respect to the optical axis (from the geometrical center of the camera sensor to the object surface origin of coordinates, point P in the figure) and ($\Delta\alpha$) is the tilt angle or angular displacement of the two object illumination positions from one point I to another I'. $\vec{u}_{IP}$ is a unit vector from I to P, and $\vec{u}_{I'P}$ is the unit vector from I' to P. $\vec{r}$ is the vector from the point I to I' that represents a constant displacement on the x–z plane caused by the tilt ($\Delta\alpha$) of the object's illumination beam.

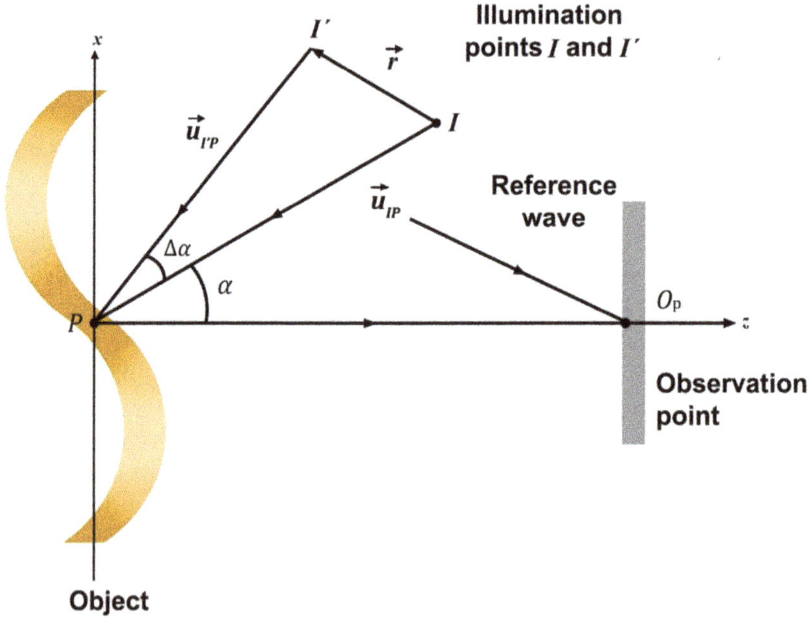

Figure 10.1. Geometry for shape measurement using two points of object illumination.

The unit vectors $\vec{u}_{IP}$ and $\vec{u}_{I'P}$ can be defined as

$$\begin{aligned}\vec{u}_{IP} &= (\sin\alpha, \cos\alpha)\\ \vec{u}_{I'P} &= (\sin(\alpha + \Delta\alpha), \ \cos(\alpha + \Delta\alpha)).\end{aligned} \tag{10.1}$$

Upon subtraction

$$\vec{u}_{IP} - \vec{u}_{I'P} = (-\Delta\alpha\cos\alpha, \ \Delta\alpha\sin\alpha). \tag{10.2}$$

Since the tilt $(\Delta\alpha)$ is very small, the vector $\vec{r} = (x_{I'} - x_I, z_{I'} - z_I)$ represents pointwise differences that are also very small so that we can define them as differentials

$$\vec{r} = (\Delta x, \Delta z), \tag{10.3}$$

where (Δz) is the change in the height of the object surface, a variable that can be found from the optical phase change $\Delta\varphi$, obtained from the phases φ_1 and φ_2 recorded before and after the small tilt $(\Delta\alpha)$ of the object illumination beam. This optical phase difference $\Delta\varphi$ at an object point $P(x, z)$ is

$$\Delta\varphi = \varphi_1 - \varphi_2 = \frac{2\pi}{\lambda}(\overrightarrow{u_{IP}} - \overrightarrow{u_{IP'}}) \cdot \vec{r}. \tag{10.4}$$

Substituting equations (10.2) and (10.3) in equation (10.4) results in

$$\Delta\varphi = \frac{2\pi}{\lambda}\Delta\alpha[-\Delta x\cos(\alpha) + \Delta z\sin(\alpha)]. \tag{10.5}$$

The phase difference in the last equation is said to be wrapped, so it needs to be unwrapped.

From equation (10.5) the equation for the object contour Δz is

$$\Delta z = \frac{\Delta\varphi\lambda}{2\pi\Delta\alpha\sin(\alpha)} + \frac{\Delta x\cos(\alpha)}{\sin(\alpha)}. \tag{10.6}$$

The first term in this equation corresponds to the object optical phase with a tilt or slope generated by the first term of the right-hand side in equation (10.5). The second term describes the linear part of the equation that depends only on x, and it is unrelated to the object contour [1–3]. Hence, the contribution along x can be discarded, and the height variations Δz in the z-direction can be calculated by

$$\Delta z = \frac{\Delta\varphi\lambda}{2\pi\Delta\alpha\sin(\alpha)}. \tag{10.7}$$

If $\Delta\varphi = 2\pi$, the maximum height sensitivity of the system $\Delta z_{\max}$ is attained, which gives the minimum measurement details that the system can resolve, i.e. the sensitivity is inversely proportional to the object beam tilt angle, $\Delta\alpha$,

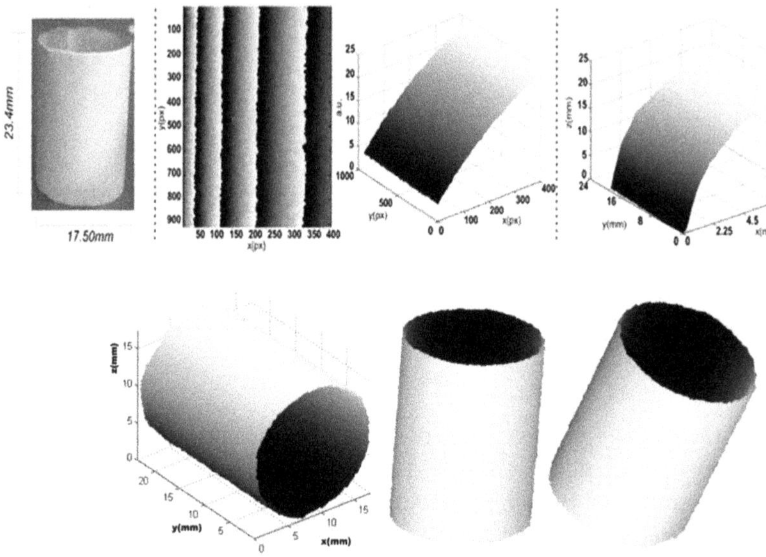

Figure 10.2. Surface contour reconstruction from the shape measurement of a cylinder. The original cylinder is at the top left-hand side, its wrapped phase and unwrapped phase maps are at the top center, the image at the top right-hand corner depicts the actual surface contour, and at the bottom we show three different views of the 3D reconstructed cylinder.

$$\Delta z_{\max} = \frac{\lambda}{\Delta\alpha \sin(\alpha)}. \tag{10.8}$$

Figure 10.2 shows random views of the surface contours of a cylinder reconstructed by the two points of the illumination contouring method.

10.2 Two-wavelength contouring

Changes in the light source wavelength modify the sensitivity vector. A two-wavelength contouring technique allows extending the shape measuring range since a longer equivalent synthetic wavelength is created. The object surface height variations are found as a function of the beating frequency between the two wavelengths. The shorter the synthetic wavelength, the higher the measurement resolution of the fringe patterns, and a higher sensitivity renders more refined features on the object surface. These two wavelengths may be obtained by using a tunable laser, or by driving the current or temperature of a light source made for that purpose, or by using two lasers with different wavelengths.

The process is as follows. Two consecutive object interferograms are recorded. The first is recorded with λ_1 giving a wave number $k = 2\pi/\lambda_1$. The second one is captured after the wavelength is changed to a new value λ_2 with a wave number $k = 2\pi/\lambda_2$. As mentioned, this can be achieved through two lasers with different wavelengths or with sequential illumination by successively changing the wavelength of a tunable laser [4–7].

The optical phases φ_1 and φ_2 obtained with λ_1 and λ_2 are subtracted from one another to obtain a fringe pattern corresponding to the contour of the object

$$\Delta\varphi(x, y) = \varphi_{\lambda 1} - \varphi_{\lambda 2} = 2\pi\frac{\text{OPL}_{\lambda_1}}{\lambda_1} - 2\pi\frac{\text{OPL}_{\lambda_2}}{\lambda_2} = 2\pi\left(\frac{\lambda_2\text{OPL}_{\lambda_1} - \lambda_1\text{OPL}_{\lambda_2}}{\lambda_1\lambda_2}\right), \quad (10.9)$$

where OPL stands for optical path length, and

$$\text{OPL}_{\lambda 1} = \Delta z(\cos\alpha_{\lambda 1} + \cos\alpha_2) + \Delta x\sin\alpha_{\lambda 1} \quad (10.10\text{a})$$

$$\text{OPL}_{\lambda 2} = \Delta z(\cos\alpha_{\lambda 2} + \cos\alpha_2) + \Delta x\sin\alpha_{\lambda 2}. \quad (10.10\text{b})$$

$\text{OPL}_{\lambda 1}$ and $\text{OPL}_{\lambda 2}$ are invariant optical path lengths, i.e. they are constant since the object remains at rest during the image recording and the illumination of the two wavelengths occurs perpendicular to the object surface. Δz is the object height variation in the z-direction, α_2 is the observation angle, and Δx indicates the transverse displacement, which can be neglected since it is not related to the object contour, thus

$$\Delta\varphi(x, y) = 2\pi\left(\frac{\lambda_2\text{OPL}_{\lambda_1} - \lambda_1\text{OPL}_{\lambda_2}}{\lambda_1\lambda_2}\right) = 2\pi\left(\frac{\lambda_2 - \lambda_1}{\lambda_1\lambda_2}\right)\text{OPL} = 2\pi\frac{\text{OPL}}{\Lambda}. \quad (10.11)$$

Λ is an equivalent synthetic wavelength defined by

$$\Lambda = \frac{\lambda_1\lambda_2}{\lambda_2 - \lambda_1}. \quad (10.12)$$

If we assume that $\lambda_2 > \lambda_1$, then $\lambda_2 = \lambda_1 + \Delta\lambda$, where $\Delta\lambda$ is the spectral gap between the two wavelengths. $\Delta\varphi(x, y)$ renders contours of constant elevation that correspond to the object shape along the z-axis, and it is directly related to the surface contour Δz as

$$\Delta z = \frac{\Delta\varphi(x, y)\Lambda}{2\pi(\cos\alpha + 1)}. \quad (10.13)$$

The sensitivity of the system when $\Delta\varphi(x, y) = 2\pi$ is

$$\Delta z = \frac{\Lambda}{(\cos\alpha + 1)}. \quad (10.14)$$

For example, figure 10.3 depicts random views of the surface shape of a skin tissue reconstructed by the two-wavelength contouring method.

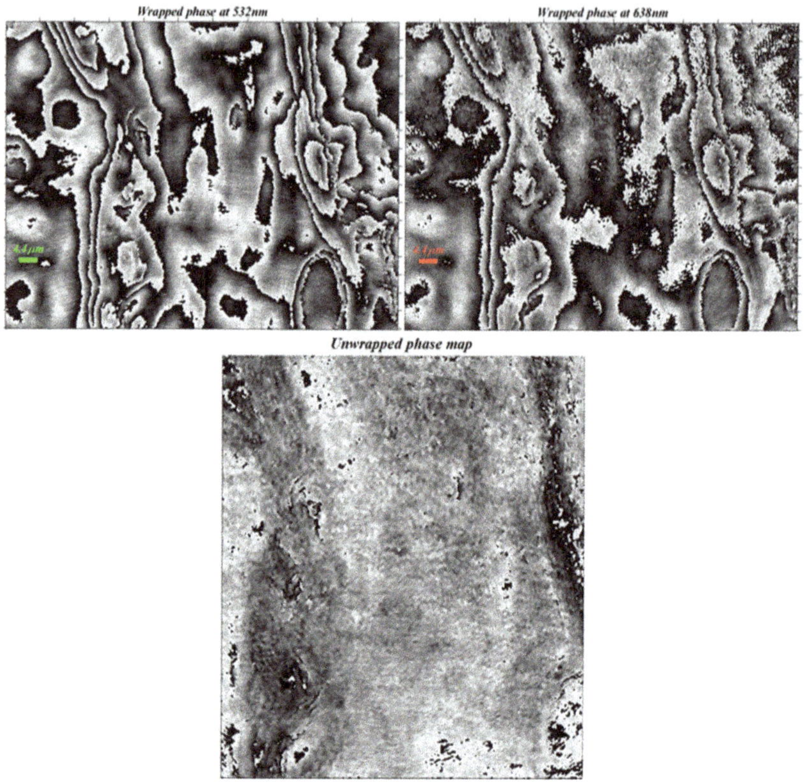

Figure 10.3. Wrapped phase maps showing the contour of a skin tissue sample obtained using the two-wavelength contouring method; φ_1 and φ_2 were found for $\lambda_1 = 532$ nm and $\lambda_2 = 638$ nm, respectively (top two images); the unwrapped phase map for a synthetic wavelength Λ of 3.20 μm is shown at the bottom.

10.3 Contouring from experimental data

To have the displacement components once at a time or simultaneously allows retrieving alternative measurements from these values. To precisely describe the object deformation, the object contour is needed so that the pointwise object deformation can be mapped on the object contour on a one-to-one point basis. The following describes the equations that lead to the expression that can be used to compute the object contour from experimental data obtained through the phase map and the coordinates of the experimental optical layout.

10.3.1 Contouring with one illumination

Consider figure 10.4.

$P(x, y, z)$ is an arbitrary point on the object's surface, $\hat{n}_i$ and $\hat{n}_o$ are unitary vectors along with the object illumination and observation directions, respectively, and l_s and l_c are the distances from the illuminating source to point P and from the coordinate origin to the CCD sensor, respectively.

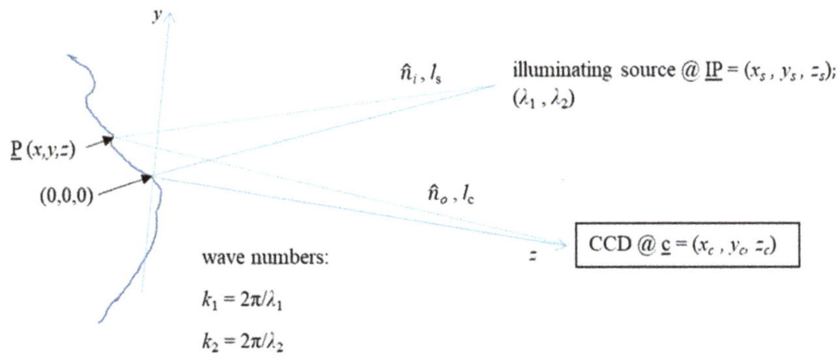

Figure 10.4. Illumination and observation direction coordinates; an arbitrarily shaped surface with two reference points; note that the light source represents two wavelengths, λ_1 and λ_2.

The wavefront from the illumination source reaching point P takes the form

$$U_p = u_o e^{ik_1 l_s}, \text{ for } \lambda_1$$
$$U_p = u_o e^{ik_2 l_s}, \text{ for } \lambda_2$$

and the wavefront at the CCD center is

$$U_c = u_o e^{ik_1(l_s + l_c)}, \text{ for } \lambda_1$$
$$U_c = u_o e^{ik_2(l_s + l_c)}, \text{ for } \lambda_2,$$

where the optical phase may be defined as $\varphi_1 = k_1(l_s + l_c)$ and $\varphi_2 = k_2(l_s + l_c)$, so that the phase difference $\varphi_1 - \varphi_2$ is

$$\Delta\varphi = (k_2 - k_1)(l_s + l_c). \tag{10.15}$$

This phase difference can be rewritten as

$$\frac{\Delta\varphi}{2\pi} = \frac{(\lambda_1 - \lambda_2)}{(\lambda_1\lambda_2)}(l_s + l_c), \tag{10.16}$$

where the so-called synthetic wavelength can be defined as

$$\Lambda = \frac{\lambda_1\lambda_2}{\lambda_2 - \lambda_1}, \tag{10.17}$$

so that

$$(l_s + l_c) = l = \frac{\Delta\varphi}{2\pi}\Lambda. \tag{10.18}$$

Now, let us consider the schematics in figure 10.5.

From this figure

$$l_s = \sqrt{(x - x_s)^2 + (y - y_s)^2 + (z - z_s)^2}, \tag{10.19}$$

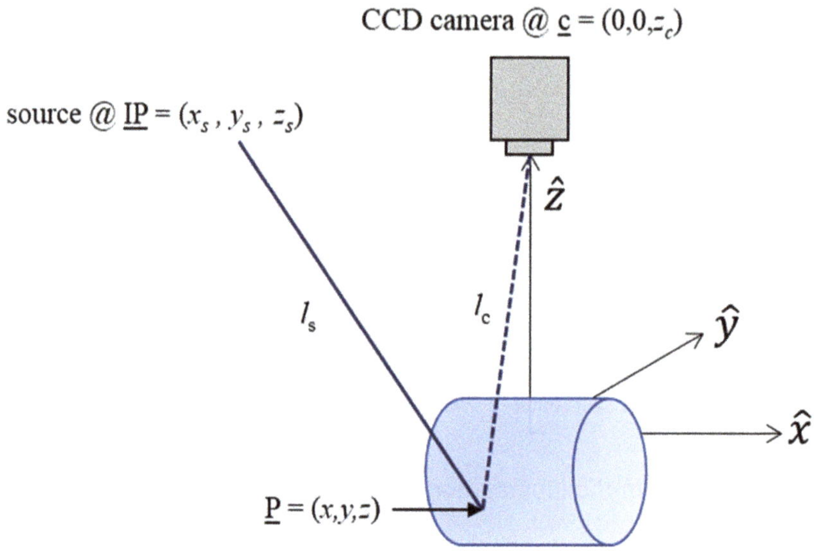

Figure 10.5. For simplicity, a regular shaped object is chosen with the observation position set at the coordinates shown.

and

$$l_c = \sqrt{(x)^2 + (y)^2 + (z - z_c)^2}. \tag{10.20}$$

In order to compute z (the object height) using all the known quantities, namely, all the coordinates (within the associated uncertainties while measuring these coordinates) and the phase $\Delta\varphi$ and the synthetic wavelength Λ; we are required to go through the following process.

Recall that $(l_s + l_c) = l$, upon squaring it, $l^2 - 2l_s l_c - l_s^2 - l_c^2 = 0$, and squaring this leads to

$$l^4 + (l_c^2 - l_s^2)^2 - 2l^2 (l_c^2 + l_s^2) = 0. \tag{10.21}$$

Using equations (10.19) and (10.20) to find the second and third terms of equation (10.21), and after some straightforward algebraic manipulation, it is found that

$$z = -\frac{B}{2} \pm \frac{1}{2}\sqrt{B^2 - D}, \tag{10.22}$$

where B is defined as

$$B = \frac{l^2 (z_s + z_c) + \left[\left(x^2 + y^2 + z_c^2\right) - ((x - x_s)^2 + (y - y_s)^2 + z_s^2)\right](z_s - z_c)}{(z_s - z_c)^2 - l^2}.$$

Moreover, D is defined as

$$D = \frac{l^4 - 2l^2\beta + \alpha^2}{(z_s - z_c)^2 - l^2}.$$

with

$$\alpha = (x^2 + y^2 + z_c^2) - ((x - x_s)^2 + (y - y_s)^2 + z_s^2),$$
$$\beta = (x^2 + y^2 + z_c^2) + ((x - x_s)^2 + (y - y_s)^2 + z_s^2).$$

So equation (10.22) gives the object contour through equation (10.18) and the use of all coordinates in the optical set-up.

10.3.2 Contouring with two illuminations

This section gives the equation that can be used to compute the object contour from experimental data obtained through the phase map and the coordinates of the optical layout. Consider figure 10.6.

In this figure $\hat{n}_{1,2}$ and $\hat{n}_o$ are unitary vectors along with the two object illumination and observation directions, respectively, and $l_{1,2}$ and l_c are the distances from the two illuminating sources to point P and from the coordinate origin to the CCD sensor, respectively.

The wavefront originating from the source at point P is:

- $U_p = u_o e^{ikl_1}$, for light coming from source 1;
- $U_p = u_o e^{ikl_2}$, for light coming from source 2.

The wavefront at the CCD center is:

- $U_c = u_o e^{ik(l_1+l_c)}$, for source 1;
- $U_c = u_o e^{ik(l_2+l_c)}$, for source 2;

where the optical phase may be defined as $\varphi_1 = k(l_1 + l_c)$ and $\varphi_2 = k(l_2 + l_c)$, so that the phase difference is

$$\Delta\varphi = k\,(l_2 - l_1). \tag{10.23}$$

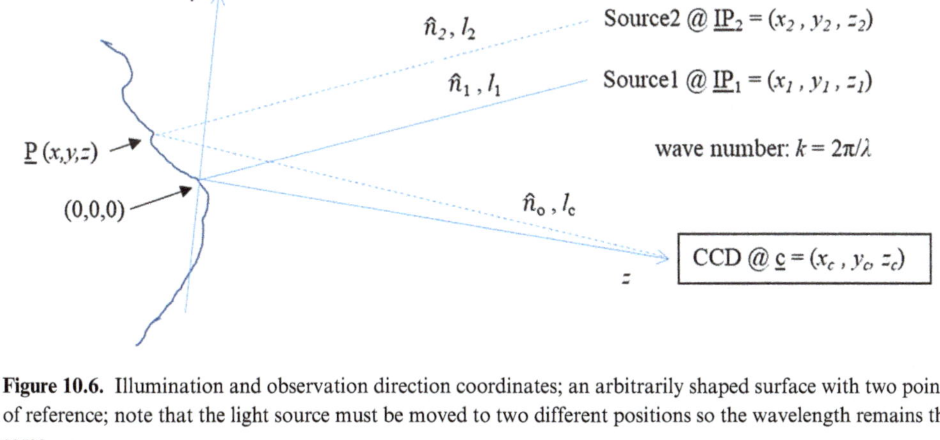

Figure 10.6. Illumination and observation direction coordinates; an arbitrarily shaped surface with two points of reference; note that the light source must be moved to two different positions so the wavelength remains the same.

This phase difference can be rewritten as

$$\frac{\Delta\varphi}{2\pi} = \frac{(l_2 - l_1)}{(\lambda)} \text{ or}$$

$$l = l_2 - l_1 = \frac{\Delta\varphi}{2\pi}\lambda. \tag{10.24}$$

Now,

$$l_2 - l_1 = \sqrt{(\underline{s}_2 - \underline{P})^2} - \sqrt{(\underline{s}_1 - \underline{P})^2}, \tag{10.25}$$

where

$$l_1 = \sqrt{(x - x_1)^2 + (y - y_1)^2 + (z - z_1)^2}, \tag{10.26}$$

and

$$l_2 = \sqrt{(x - x_2)^2 + (y - y_2)^2 + (z - z_2)^2}. \tag{10.27}$$

In order to compute z (the object height), using all the known quantities, namely all the coordinates (within the associated uncertainties while measuring these coordinates), the phase $\Delta\varphi$, and the wavelength λ, it is required to go through the following process.

Recall that $(l_2 - l_1) = l$, upon squaring it $l^2 + 2l_1l_2 - l_1^2 - l_2^2 = 0$, and squaring this leads to

$$l^4 + (l_1^2 - l_2^2)^2 - 2l^2(l_1^2 + l_2^2) = 0. \tag{10.28}$$

Using equations (10.26) and (10.27) to find the second and third terms of equation (10.28), and after some straightforward algebraic manipulation, it is found that

$$z = -\frac{B}{2} \pm \frac{1}{2}\sqrt{B^2 - D}, \tag{10.29}$$

where B is defined as

$$B = \frac{l^2(z_1 + z_2) + \left[((x - x_1)^2 + (y - y_1)^2 + z_1^2) - ((x - x_2)^2 + (y - y_2)^2 + z_2^2)\right](z_2 - z_1)}{(z_2 - z_1)^2 - l^2}.$$

Moreover, D is defined as

$D = \frac{l^4 - 2l^2\beta + \alpha^2}{(z_2 - z_1)^2 - l^2}$, with

$\alpha = ((x - x_1)^2 + (y - y_1)^2 + z_1^2) - ((x - x_2)^2 + (y - y_2)^2 + z_2^2)$,

$\beta = ((x - x_1)^2 + (y - y_1)^2 + z_1^2) + ((x - x_2)^2 + (y - y_2)^2 + z_2^2)$.

Thus equation (10.29) gives the object contour through equation (10.24) and the use of all coordinates and geometric arrangement in the optical set-up.

References

[1] Joenathan C, Pfister B and Tiziani H J 1990 Contouring by electronic speckle pattern interferometry employing dual beam illumination *Appl. Opt.* **29** 1905–11

[2] Quan C, Chen W and Tay C J 2008 Shape measurement by multi-illumination method in digital holographic interferometry *Opt. Commun.* **281** 3957–64

[3] Uribe López U, del Socorro Hernández-Montes M and Mendoza-Santoyo F 2016 Fully automated digital holographic interferometer for 360 deg contour and displacement measurements *Opt. Eng.* **55** 121719

[4] De Nicola S, Finizio A, Pierattini G, Alfieri D, Grilli S, Sansone L and Ferraro P 2005 Recovering correct phase information in multiwavelength digital holographic microscopy by compensation for chromatic aberrations *Opt. Lett.* **30** 2706–8

[5] Hansel T, Steinmeyer G, Grunwald R, Falldorf C, Bonitz J, Kaufmann C, Kebbel V and Griebner U 2009 Synthesized femtosecond laser pulse source for two-wavelength contouring with simultaneously recorded digital holograms *Opt. Express* **17** 2686–95

[6] Yamaguchi I, Ida T, Yokota M and Yamashita K 2006 Surface shape measurement by phase-shifting digital holography with a wavelength shift *Appl. Opt.* **45** 7610–6

[7] Wyant C 1971 Testing aspherics using two-wavelength holography *Appl. Opt.* **10** 2113–8

Fernando Mendoza-Santoyo, Manuel De la Torre-Ibarra, María del Socorro Hernández-Montes and Jorge Mauricio Flores Moreno

Chapter 11

Optical coherence tomography

Optical coherence tomography (OCT) is a non-invasive optical method that is able to inspect beneath the surface of low-scattering media [1, 2]. OCT retrieves high-resolution cross-sectional images employing light, making it possible to observe internal echoes corresponding to an object's structural information. It was first used for optical fiber inspection, but it quickly found use in the medical field, particularly in ophthalmology [3]. It is now applied in a broad spectrum of engineering and medical fields [4].

Its working principle is based on interferometry, where the light source is usually divided into two, the reference and object beams. Common path interferometric configurations only need one arm to retrieve the object's optical phase. In this case, the OCT signal—the backscattered light coming from the object (internal echoes)—interferes with the unaltered reference beam. The data coming from this interference pattern is directly associated with different depths within the object [5]. Figure 11.1 shows a general representation of an OCT system, where the object is a sample from a slice of lemon.

In OCT the illumination source is not monochromatic; it has a bandwidth ($\Delta\lambda$) centered at a wavelength (λ_c) that is used to specify the resolution and penetration of the optical system. It is worth mentioning that most medical OCT devices are centered at the near-infrared (NIR) band, so it is common to find λ_c at 840, 980, or 1300 nm. These wavelengths have been widely used as they approximately match the so-called therapeutic window between 650 and 1300 nm. The term window is used because most body constituents have no absorption of light in this wavelength band (some of them are melanin, collagen, hemoglobin, and water) [6]. These constituents are found in low-scattering tissue that is of interest in disciplines such as ophthalmology, cardiology, dermatology, and gastroenterology, among others. As a result, the OCT signal is then formed by the backscattering coming from other structural materials in the sample that create the tomographic view of the object.

doi:10.1088/978-0-7503-3027-5ch11

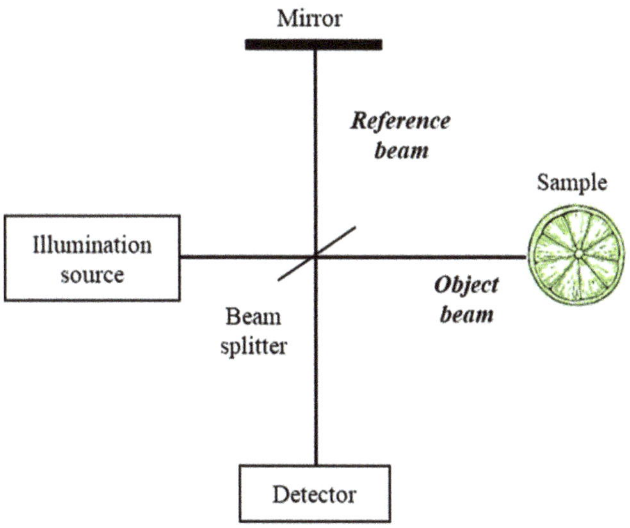

Figure 11.1. OCT schematic view based on the Michelson interferometer.

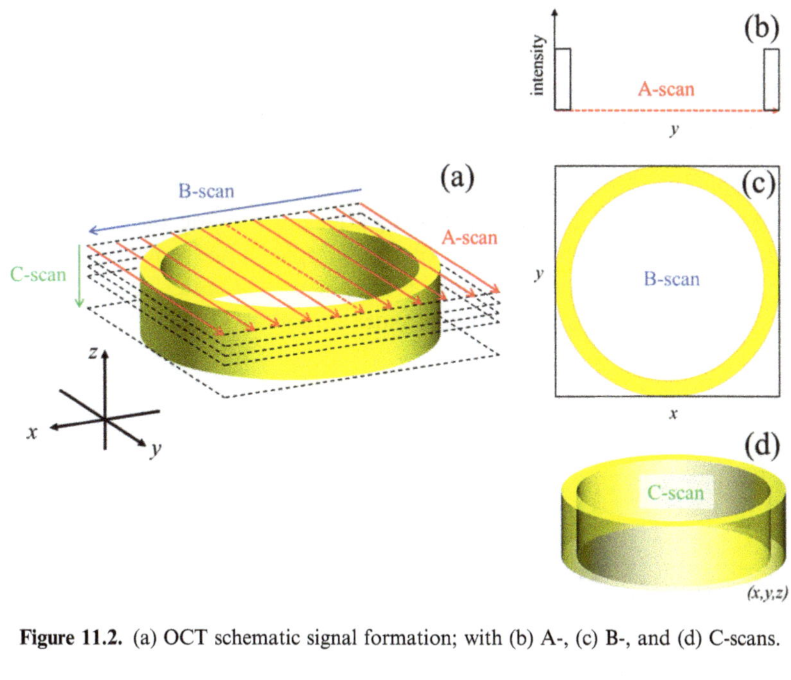

Figure 11.2. (a) OCT schematic signal formation; with (b) A-, (c) B-, and (d) C-scans.

The backscattering waves from the object re-enter the OCT interferometer; then they travel with the reference beam, where they overlap with the detection hardware. Each wavelength interference pattern is processed to create the depth profile or OCT tomographic signal. According to the OCT set-up, the interference signal can be identified as the A-, B-, or C-scan (see figure 11.2(a)).

For the first case, let us assume that a single point is illuminated on the sample's surface. This point creates in the OCT system a 1D interference pattern which is converted into a 1D tomographic signal referred to as the A-scan (figure 11.2(b)).

However, a B-scan is obtained when a line is drawn on the object's surface by continuous and adjacent A-scans. In this case, the tomographic image will show a 2D depth profile of the object (figure 11.2(c)). This B-scan is the most commonly displayed result in OCT systems as it offers a high-resolution internal slice of the sample. Scanning in the perpendicular direction of the B-scan will produce several adjacent B-scans that create a 3D set of information known as the C-scan (figure 11.2(d)).

As OCT is based on interferometry, the resulting interference could come from the illumination source emitting the entire $\Delta\lambda$, or just part of it at the time. For example, the first case typically uses a superluminescent diode (SLD), while the second is a tunable laser (swept-source OCT). In addition to this difference, the available tomographic information will depend on the configuration used to gather the interference signal. The most common arrangements are the temporal, spectral, and quantum OCT configurations.

11.1 OCT configurations

The temporal OCT configuration is a set-up that uses a Michelson interferometer associated with a single point sensor (typically, a photodetector) [7]. This system scans the sample utilizing a moveable device attached to the reference mirror (i.e. a piezo Pzt driver). Here, high-speed scanning devices are required for this purpose. Figure 11.3 shows a schematic representation of a time-domain OCT.

The reference beam scanning creates the effect of moving the reference plane of the interferometer below the surface of the sample. This movement modifies the signal on the photodetector, which retrieves a 1D interference pattern (see figure 11.4). Once recorded, another adjacent point is illuminated by moving the

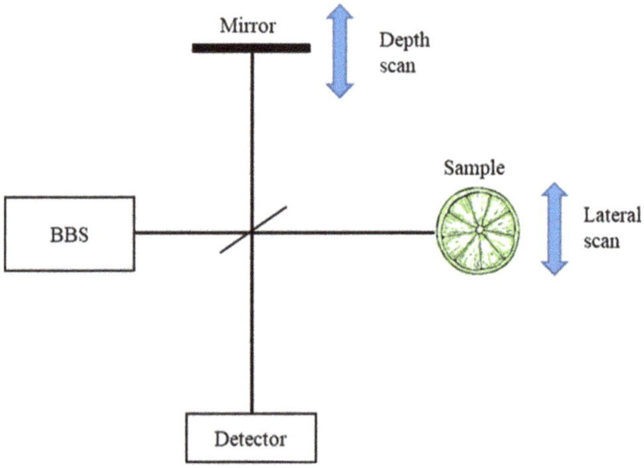

Figure 11.3. Schematic view of temporal OCT.

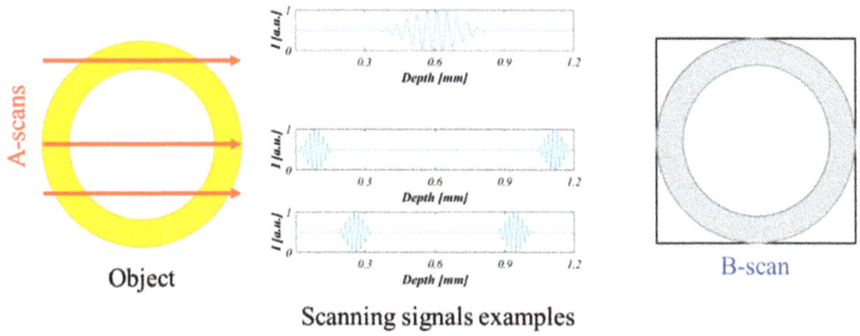

Figure 11.4. Temporal OCT signal retrieval. The vertical axis of the middle plots refers to the intensity (in arbitrary units (a.u.)).

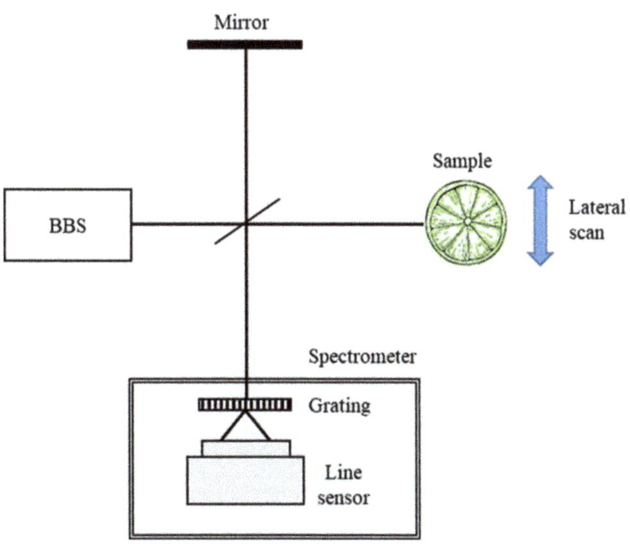

Figure 11.5. Spectral OCT set-up.

object or its illumination (see figure 11.3). In this configuration, the scanning process is not just required at the reference beam; the object also needs to be scanned.

The spectral version of the OCT uses a spectrometer to split all the $\Delta\lambda$ wavelengths into a line sensor (figure 11.5) [8]. There is no need to add a scanning device to the reference mirror as the spectrometer performs this task. The spectrometer uses a diffraction grating to separate and resolve the optical source bandwidth for a single point of the object. For this reason, a match between the grating's resolution and the size of the line sensor is required. From figure 11.5 it is seen that the only scanning device is now placed in the object. The interference signal is registered at once by the line camera, i.e. it is possible to obtain an A-scan in a single exposure (figure 11.6).

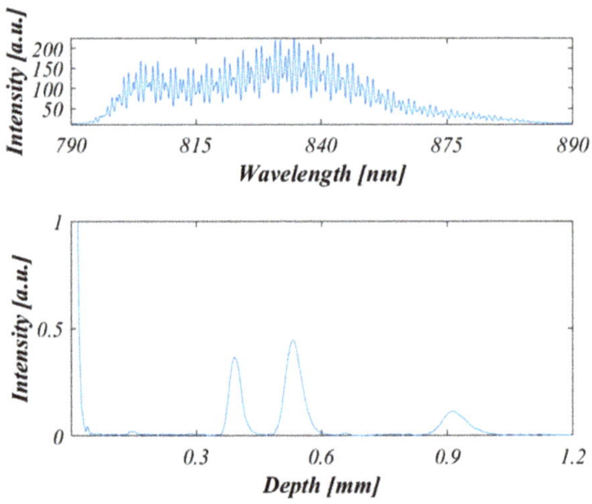

Figure 11.6. Spectral OCT A-scan signal retrieval from a glass coverslip. See the paragraph immediately after equation (11.4) for an explanation of both panels.

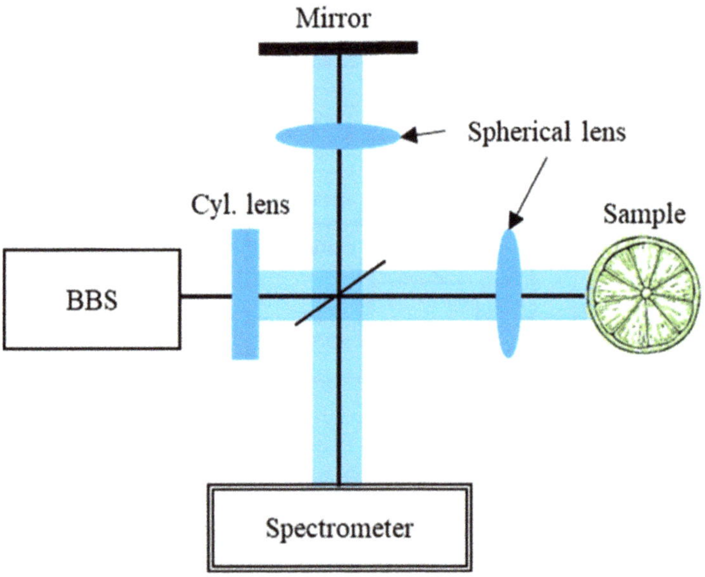

Figure 11.7. B-scan single-shot OCT system.

Taking this same configuration but modifying the optics and the sensor, it is possible to set up a B-scan single-shot system. This spectral OCT configuration will use a cylindrical lens that produces an illumination line over the object's surface [9, 10]. This line has the entire bandwidth of the illumination source, so it penetrates the sample in a 2D form (see figure 11.7).

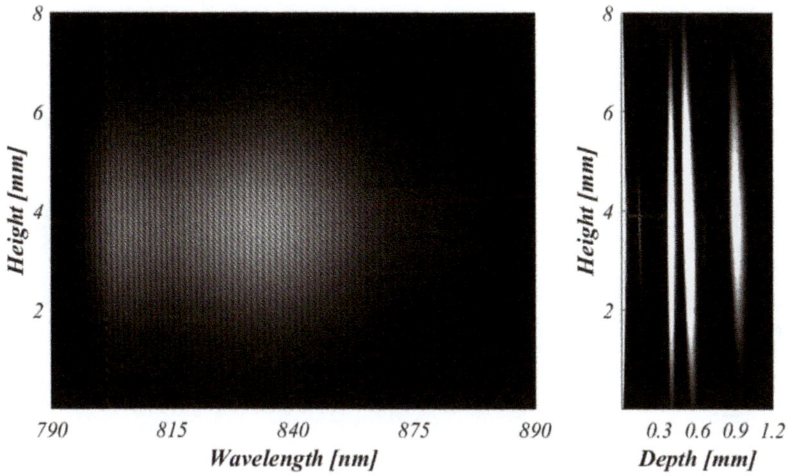

Figure 11.8. B-scan single-shot spectral OCT signal retrieval.

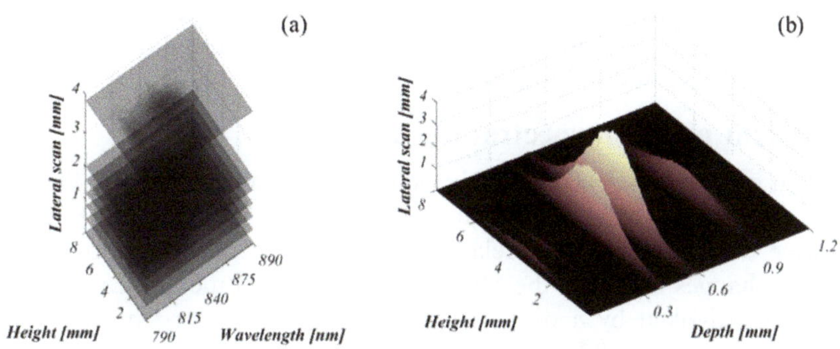

Figure 11.9. C-scan formation from a stack of B-scan single-shot images.

The diffraction grating splits a complete line over a 2D sensor for each wavelength. Here a B-scan is imaged directly, so no scanning devices are required (figure 11.8).

This set-up requires a 2D image sensor to retrieve the full B-scan in a single image. This feature helps to record high-speed events as the A-scan recording is simplified. Another advantage is the possibility of reconstructing a C-scan at high speed by moving the sample or using a tilting mirror in front of it. In this way, several B-scans are registered and stacked in a C-scan, as figure 11.9 shows.

OCT configurations are continuously evolving in speed, resolution, and detection. An example is the quantum OCT configuration (Q-OCT). It is based on the presence of a pair of entangled photons generated in a spontaneous parametric down-conversion process using a nonlinear crystal [11]. One photon propagates in the reference arm while the other propagates along the object arm. Then, they are reflected to a couple of detectors via a beam splitter (figure 11.10). The overlapped

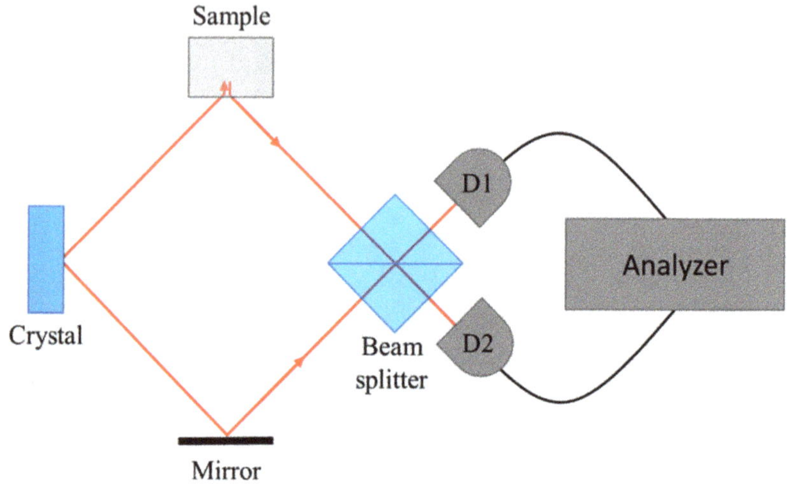

Figure 11.10. Q-OCT schematic set-up.

signals are processed and a tomographic image is generated. This configuration has the advantage of being insensitive to dispersion effects in multi-layered scattering media, allowing high-resolution images.

11.2 Optical phase in spectral OCT

A particular case of the spectral OCT is the Fourier domain OCT (FD-OCT), where it is possible to retrieve the optical phase between two tomographic images. Here, the Fourier transform uses the relative optical phase difference to detect inner mechanical changes affecting the object (here, the word mechanical refers to two object states provoked by a displacement of deformation or simply two distinct objects' inner positions). This phase gives the relation between the backscattered light from different inner layers (jth slices) and the reference beam plane (a zero-delay line):

$$\varphi_j(\lambda) = \varphi_{j0} + 4\pi\lambda^{-1}z_j. \tag{11.1}$$

As it was mentioned, OCT λ covers the entire $\Delta\lambda$, i.e. from $\lambda_c - \Delta\lambda/2$ to $\lambda_c + \Delta\lambda/2$. The term φ_{j0} is the phase of the jth slice, and z_j represents the optical path difference between the reference plane and the jth slice. Simplifying in terms of the wave-number k,

$$\varphi_j(k) = \varphi_{j0} + 2kz_j. \tag{11.2}$$

The latter helps sample each A-scan intensity uniformly along the k-axis instead of the λ-axis, implying non-uniform sampling due to the diffraction grating. The depth range (Δz) represents the maximum optical path difference of the OCT system and, in this case, it is limited by the spectrometer resolution as

$$\Delta z = N\lambda_c^2/4\Delta\lambda, \tag{11.3}$$

where N represents the number of pixels used to sample the A-scan in the camera's sensor, the signal retrieved on the camera (after it has gone through the spectrometer) can be expressed as

$$I(k) = I_0 + 2\sum_{j=1}^{M}\sqrt{I_R I_j}\cos\left(\varphi_j(k)\right) + 2\sum_{i=1}^{M}\sum_{j=i+1}^{M}\sqrt{I_i I_j}\cos\left(\varphi_i(k) - \varphi_j(k)\right). \tag{11.4}$$

The first term, I_0, is the background or dc term without phase. I_R and I_j are the intensities coming from the reference and the jth layer of the sample, respectively. Furthermore, the second term represents the consecutive interference of the reference beam and the backscattering coming from all the object slices within the Δz, and the third term with a double summation means the auto interference of the light coming from all the object slices. When expression (11.4) is Fourier transformed, these three terms are identified in the spectrum, as figure 11.6 shows. By modifying the distance between the object in relation to the zero-delay line (reference plane), the third and first terms are grouped within the range of low frequency. In this manner, the second term is filtered, and the optical phase can be extracted. However, it is worth mentioning that before the A-scans are Fourier transformed, a series of signal processing steps should be used to improve the tomographic reconstruction (this process is graphically shown in figure 11.11). The A-scan raw fringe pattern is shown in figure 11.11(a), while the Fourier spectrum is in figure 11.11(b). First, consider the even sampling of the interference signal from wavelength to wavenumber. For this purpose, a spectrometer's characterization is performed, where the grating dispersion is identified and set as a dechirp vector. An even sampling is obtained by multiplying this vector with each A-scan (figures 11.11(c) and (d)). This evenly spaced interference pattern is multiplied by an apodization/window vector that stretches the peaks' signals (figure 11.11(e)) in the tomographic view (figure 11.11(f)).

Once the interference signal is in the Fourier domain, it is possible to determine the relative phase difference between two object positions (or object displacement/deformation states) using the optical phase. Each point or layer inside the object has associated a frequency (f) depending on its depth

$$f_k = z_j/\pi. \tag{11.5}$$

This principle makes it possible to relate each f_k from different tomographic images to obtain a phase map as follows:

$$\Delta\varphi_j = \tan^{-1}\left[\frac{\mathrm{Re}\left(\tilde{I}_1(f_{kj})\right)\mathrm{Im}\left(\tilde{I}_2(f_{kj})\right) - \mathrm{Im}\left(\tilde{I}_1(f_{kj})\right)\mathrm{Re}\left(\tilde{I}_2(f_{kj})\right)}{\mathrm{Im}\left(\tilde{I}_1(f_{kj})\right)\mathrm{Im}\left(\tilde{I}_2(f_{kj})\right) + \mathrm{Re}\left(\tilde{I}_1(f_{kj})\right)\mathrm{Re}\left(\tilde{I}_2(f_{kj})\right)}\right]. \tag{11.6}$$

$\Delta\varphi_j$ is the phase difference up to the jth layer inside the object. $\tilde{I}_1(f_{kj})$ and $\tilde{I}_2(f_{kj})$ are the Fourier transforms of two processed A-scans at different times of the object deformation, and Re and Im are the real and imaginary parts of a complex number.

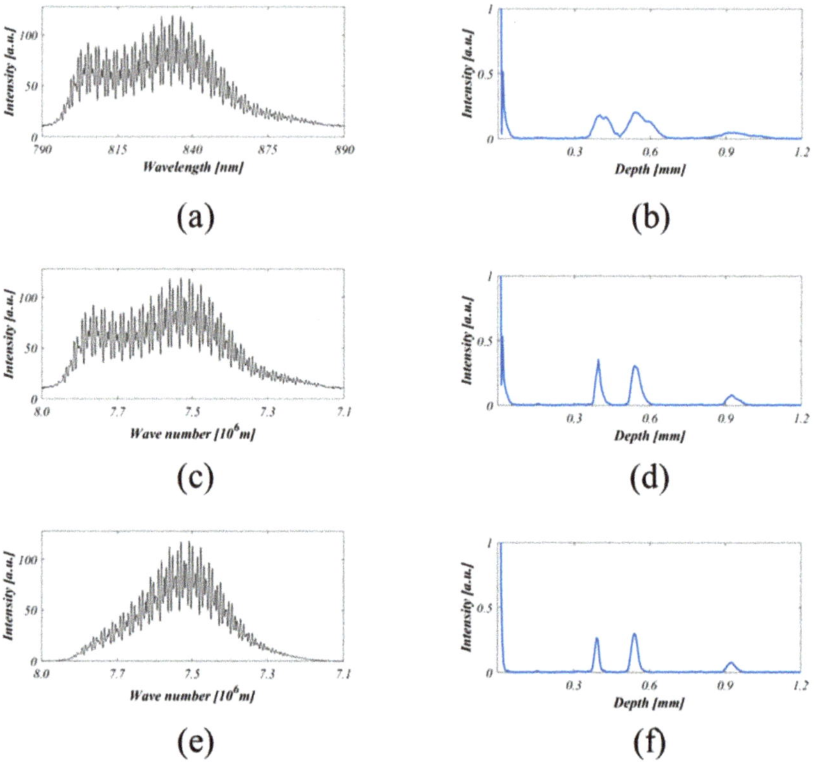

Figure 11.11. Example of signal processing and its tomographic result. (a) Raw interference signal and (b) its Fourier transform (see also figure 11.6); (c) dechirped for the same interference signal and its (d) Fourier transform; (e) dechirped and apodized the same interference signal and its (f) Fourier transform.

The wrapped phase map has values between $-\pi$ and π observed as dark and bright intensities, respectively. The unwrapping $\Delta\varphi_j$ could be related to a displacement (w_j) similar to those described in DHI; see, for example, chapter 10. However, as the FD-OCT illuminates the object in a perpendicular form, the relation is reduced as

$$w_j = \frac{\lambda_c}{4\pi}\Delta\varphi_j. \tag{11.7}$$

This phase sensitivity will detect deformations smaller than the illumination wavelength; for example, a 2π phase change will represent $\lambda_c/2$ (half of the central wavelength of the broadband light source). For this reason, a more significant pixel number (N) in the sensor will be able to detect smaller displacements with a smoother phase. Nevertheless, as each scattering point inside the sample has an associated frequency, the object deformation introduced needs to be small enough to avoid decorrelation in the phase. The latter requires a change of $<2\pi$ at every pixel of the wrapped phase to have a useful displacement map (one that can easily be interpreted).

11.3 Examples of depth-resolved phase measurement

11.3.1 Mechanical test: bending

One of the most significant advantages of OCT is its versatility in data processing to obtain several measurements coming from the tomographic image and the optical phase [12, 13]. One example is the mechanical testing of objects in optical nondestructive evaluation to predict their behavior during a controlled test remotely. One of the most common methods to deform a solid object to observe its flexure mechanical properties during an examination is bending. A simple and easy way to perform it is the three-point bending (TPB), where a well-controlled load is applied to the sample while it remains fixed at two points (to avoid any rigid body motion), as shown in figure 11.12.

Most mathematical theories applied to solids, e.g. metal bars, are mainly dedicated to finding their elasticity modulus, strain, stress, etc. The deformation is assumed to be primarily present in the center of the sample, and it is proportional to the applied load producing a displacement in one point of the composite's surface, affecting the entire object. If the applied load is exerted with a micrometric screw, it is possible to control it within a few microns providing a wide range of measurements during the test. It is possible to find the mechanical deformation, but it will be limited by the sensitivity and resolution of the optical system. For instance, for an out-of-plane sensitive system, i.e. one able to detect deformations along the normal to the object's surface, it will be possible to measure optical phase variations of 2π within ~200 nm in air.

If it is required to obtain phase maps during a lengthy deformation test, it becomes necessary to record a series of images, see Img in figure 11.13, and correlate them with a reference state. Nevertheless, this will involve a temporal algorithm due to the expected large deformation. To avoid this, the images are re-referenced with the previous ones, given the partial deformation state for that particular instant. In a lengthy deformation, the starting position of the sample will change with respect to its final one, so the re-referencing will help minimize the effect of this motion. This process results in a constant time separation between consecutive interference

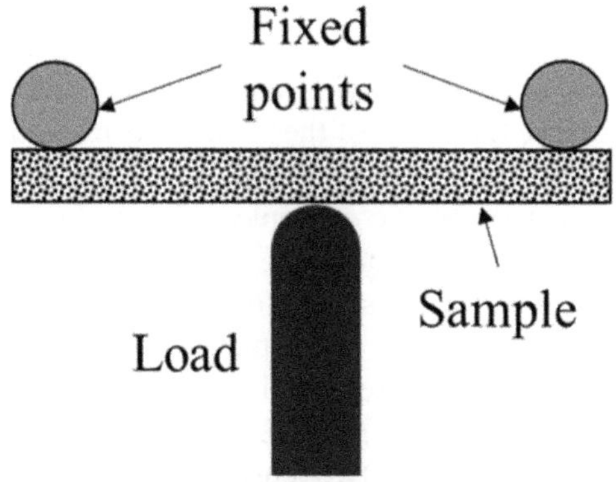

Figure 11.12. Schematic representation for the TPB test.

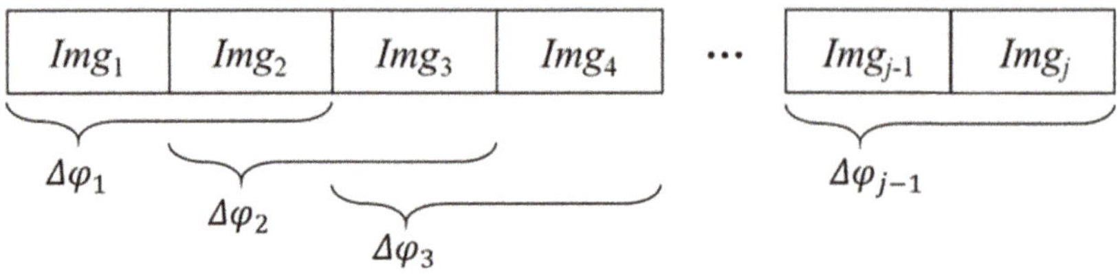

Figure 11.13. Relative phase states coming from a long series of OCT images.

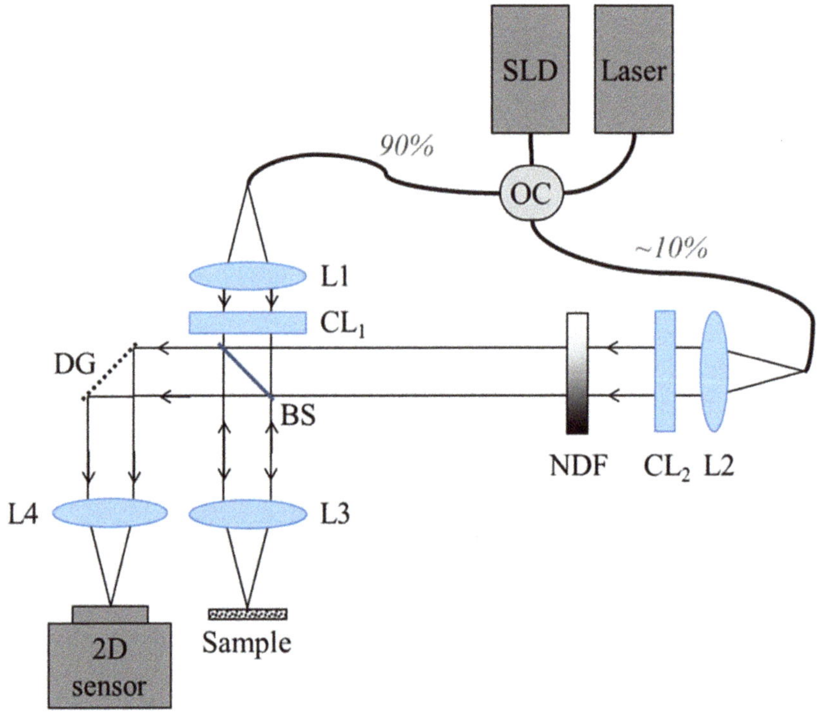

Figure 11.14. Schematic view of the experimental set-up for the three-point bending test.

images. Under these assumptions, having consecutive relative displacement maps makes it possible to observe the effect of the same deformation but at different times during the test.

Let us have a look at two specific examples. Figure 11.14 shows the schematic view of the single-shot Fourier domain OCT system that uses two illumination sources. The first one is a superluminescent diode (SLD) at 840 nm with a bandwidth of 50 nm and optical output power of 15 mW. The second is a He–Ne laser at 632.8 nm used for alignment purposes. Both are launched into the OCT interferometer using an optical coupler (OC) in a 2 × 2 configuration (transmission ratio 90:10). At the two outputs of the OC, two lenses are used to generate two collimated beams (L1 and L2). These are called the object (90% of total beam intensity) and the reference (10% of total beam intensity) beams. The reference beam is further attenuated using a neutral density filter (NDF). The reference beam has a cylindrical lens (CL2) with a focal length of 250 mm used to generate a slit (a line of light) which is diffracted due to the diffraction grating (DG) and conveyed to the camera's sensor via the spherical lens (L4) with a focal lens of 150 mm. The DG has 1200 l mm^{-1} and is blazed within the SLD bandwidth to maximize the first-order term. The object beam focuses a thin slit of light onto the sample using a combination of a cylindrical lens (CL1) and a spherical lens (L3) with a focal length of 160 mm and 100 mm, respectively. The backscattered light coming from the sample returns to the optical system via L3, which collimates it, and it is then reflected through a pellicle beam splitter (BS) optimized for the SLD bandwidth. It gets further diffracted employing the DG and focuses onto the camera using L4. Both reference and object paths are matched to have a zero optical path difference between them (an easy task thanks to the use of optical fibers of the OC).

The interference signal due to the reference and the object beams is captured by the camera's sensor and then processed to obtain, in the first instance, the tomographic view and the relative deformation map by comparing consecutive states of the deformation. In this process, two images are correlated at a time, with the result that a series of relative phase maps are obtained as wrapped phase maps showing variations between $-\pi$ and π (black and white). Each wrapped phase map has the deformation information codified, and it can be recovered using an unwrapping phase algorithm (see chapter 5 for this topic). In this form, it is possible to observe the micro-deformation magnitude inside the sample as it occurs during the test.

To apply the bending/load, a micrometric screw with a round end tip is used (it is painted in black to avoid any reflections coming from the tip end). A thin bar made with a composite seeded with tiny metal particles is used as the object (whose thickness is 1 mm), see figure 11.12. The metal particles embedded in the matrix are micro copper spheres (<5 μm), and they act as a cloud instead of independent and resolved elements due to the depth resolution (δz) of the OCT set-up, that could be expressed as

$$\delta z = \gamma \frac{\lambda_c^2}{\Delta\lambda}. \tag{11.8}$$

The depth resolution represents the minimum axial distance between two adjacent points inside the sample whose corresponding interference signals can be fully resolved in the frequency domain in OCT. The term γ is a constant depending on the apodization method and approximates the illumination source's coherence length as $2\ln 2/\pi$. If any apodization or window is applied $\gamma = 1$ (rectangular window), using different windows such as Hanning, Blackman, Gaussian, Kaiser, or Hamming, it will change this value and increase or decrease the δz.

The object is preloaded with a displacement of 100 μm, and the first image is recorded. After that, an additional load was applied to the composite layer, and a second image was recorded (deformed state). Following the process described for FD-OCT, the tomographic view of the sample before the deformation is shown in

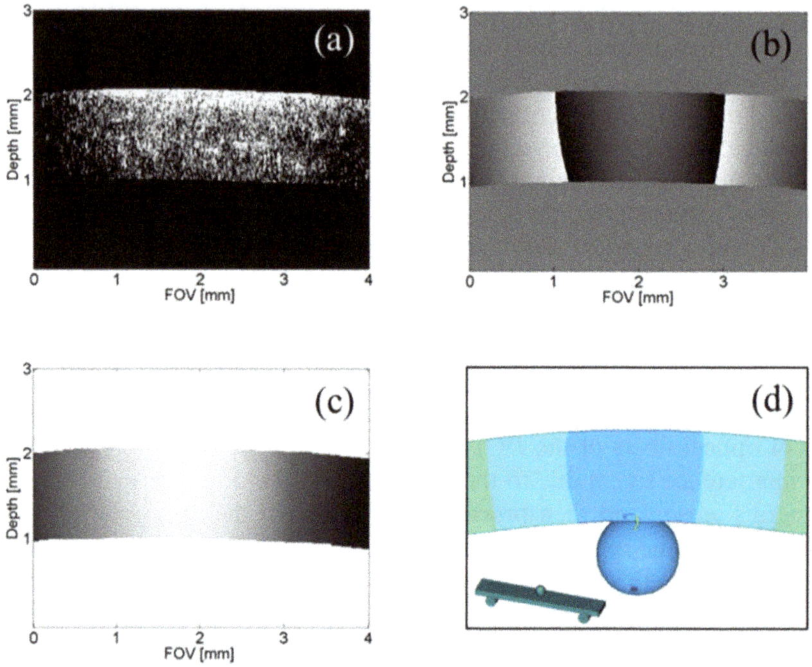

Figure 11.15. Three-point bending results. (a) Tomographic view; (b) wrapped phase map; (c) unwrapped phase map (displacement); and (d) the FEM prediction.

figure 11.15(a). The figure shows that the central part of the composite sample is brighter than at its edges due to the beam shape of the illumination source used. The wrapped phase map (containing the codified displacement map) is shown in figure 11.15(b), where phase variations from black to white ($-\pi$ to π, respectively) are observed. The unwrapped phase map (displacement map) is shown in figure 11.15(c), where a smooth deformation is observed, having a similar response as the finite element model (FEM) calculation shown in figure 11.15(d). These magnitude maps describe the same quantitative values between the experimental and the simulated theoretical results.

11.3.2 Mechanical test: a crack

Once it was proved that the three-point bending results are in good agreement with the expected response of the theoretical simulation, a second object is used to generate a crack. Once the object is firmly supported in the mechanical system, it is preloaded with 100 μm. A series of images are recorded with constant load changes, as indicated in figure 11.13. A depth-resolved full-field of view within the composite specimen will show the crack generation and propagation during the test.

Figure 11.16 shows some instants during the crack formation; each shows its OCT image and corresponding displacement map. The first set of images shows the elastic range of the composite where just a bending appears, figure 11.16(a). As the

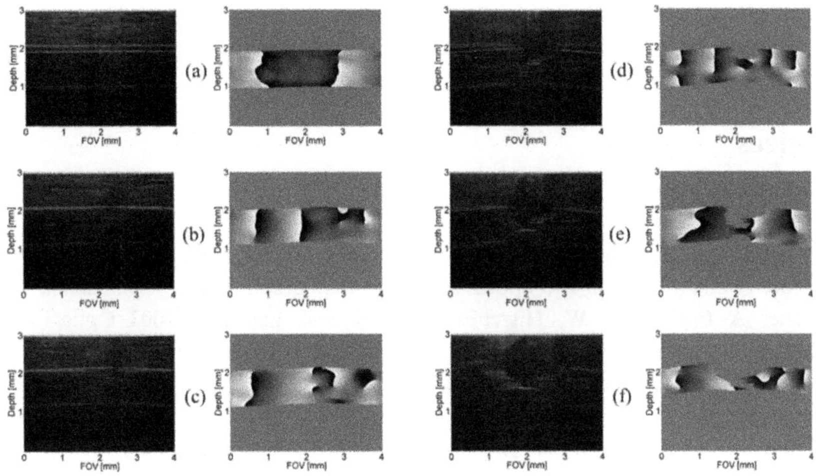

Figure 11.16. Crack test results: (left images) tomographic views; (right images) deformation maps at different instants of the deformation.

screw continues its movement, an irregular deformation starts to appear near the outer layer of the sample (figure 11.16(b)). Then, an indentation effect starts to appear at the deeper layer of the composite, and a fissure on the external surface appears (figure 11.16(c)). At this point, less than half a millimeter of deformation has been applied behind the sample. The appearance of a notorious fissure on the object's external surface, starting a conical-like behavior moving deeper into it, is also observed. The composite layer begins to have a crack moving along the internal layers, as figure 11.16(d) shows. The damage expands twice its width in less than 20 μm of displacement (figure 11.16(e)). After this point, the sample breaks completely, and both sides look like independent plaques where the deformation is no longer radial. At this point, the round tip of the micrometric screw is observed due to the absence of the composite in front of it (figure 11.16(f)). A non-optical surface technique will struggle to follow the crack until it makes an indentation on the surface (figure 11.16(c)). The tomographic images clearly show how the structure cracks. It is possible to detect a tomographic image variation only when the load is tens of micrometers. However, it is possible to see differences of tens of nanometers between two consecutive images with the phase maps.

The optical phase coming from an OCT system can observe the crack creation and propagation under particular conditions in a composite. The high resolution and high sensitivity of the spectral version are crucial to achieve this. The above also showed that tomographic images offer a high level of detail but limited discrimination in minor variations. Under those limits, phase detection has shown to be a powerful tool to detect micro deformations otherwise impossible to see. The displacement data can be used to know the sample's mechanical properties at different depths. However, the signal–noise ratio limits the threshold of sensitivity where the system can detect small variations. For more significant deformations, a temporal unwrapping technique will be necessary to deal with the motion introduced by the object. This motion and the OCT image processing is a matter of continuous research in this area.

References

[1] Fercher A F, Mengedoht K and Werner W 1988 Eye-length measurement by interferometry with partially coherent light *Opt. Lett.* **13** 186–8

[2] Huang D *et al* 1991 Optical coherence tomography *Science* **254** 1178–81

[3] Fercher A F *et al* 1993 *In vivo* optical coherence tomography *Am. J. Ophthalmol.* **116** 113–4

[4] Fercher A F, Drexler W, Hitzenberger C K and Lasser T 2003 Optical coherence tomography—principles and applications *Rep. Prog. Phys.* **66** 239–303

[5] Bouma B E and Tearney G J 2002 *Handbook of Optical Coherence Tomography* (New York: Marcel Dekker)

[6] Boudoux C 2016 *Fundamentals of Biomedical Optics* (Montréal: Pollux)

[7] Leitgeb R, Hitzenberger C and Fercher A 2003 Performance of Fourier domain vs time domain optical coherence tomography *Opt. Express* **11** 889–94

[8] de Boer J F, Cense B, Hyle Park B, Pierce M C, Tearney G J and Bouma B E 2003 Improved signal-to-noise ratio in spectral-domain compared with time-domain optical coherence tomography *Opt. Lett.* **28** 2067–9

[9] Teramura Y, Suekuni M and Kannari F 2000 Two-dimensional optical coherence tomography using spectral domain interferometry *J. Opt.* A **2** 21–6

[10] De la Torre-Ibarra M H, Ruiz P D and Huntley J M 2006 Double-shot depth-resolved displacement field measurement using phase-contrast spectral optical coherence tomography *Opt. Express* **14** 9643–56

[11] Hee M R, Izatt J A, Jacobson J M, Fujimoto J G and Swanson E A 1993 Femtosecond transillumination optical coherence tomography *Opt. Lett.* **18** 950–2

[12] Briones-R M de J, De la Torre-Ibarra M H, Mendoza-Santoyo F and Pedroza-G J 2015 Simultaneous birefringence imaging and depth phase resolved measurement using a Fourier domain OCT system *Opt. Lasers Eng.* **68** 104–10

[13] De la Torre Ibarra M H, Ruiz P D and Huntley J M 2009 Simultaneous measurement of in-plane and out-of-plane displacement fields in scattering media using phase-contrast spectral optical coherence tomography *Opt. Lett.* **34** 806–8

Fernando Mendoza-Santoyo, Manuel De la Torre-Ibarra, María del Socorro Hernández-Montes and Jorge Mauricio Flores Moreno

Chapter 12

Digital image correlation

Contributing author: Bing Pang

12.1 A brief introduction to digital image correlation

Surface deformation measurement of materials and structural components subjected to various external loads (e.g. mechanical or thermal loads) is fundamental in experimental solid mechanics. The measured surface displacements and strains reveal the test material's mechanical behavior, deformation mechanism, and structure. Then, various mechanical parameters of a material, e.g. the Young's modulus, Poisson's ratio, coefficient of thermal expansion, and stress intensity factor, can be determined readily. Important enough is that these obtained kinematic fields help verify the correctness of theoretical predictions and FEM analyses.

Aside from the widely used pointwise strain gauge technique, which only provides averaged strain readings within the gauge length, various non-contact full-field optical measurement techniques have been developed in experimental mechanics to measure deformations in various objects and materials. These techniques are interferometric techniques such as holographic interferometry [1, 2], electronic speckle pattern interferometry [1], and moiré interferometry [3]. Also, non-interferometric techniques such as the grid method [4] and digital image correlation (DIC) are used [5–7]. Interferometric techniques require a coherent light source, and the measurements typically should be conducted in a vibration-isolated optical table in the laboratory. Interferometric techniques measure the deformation by recording the phase difference of the scattered, reflected, or diffracted light wave from the test object surface before and after deformation. The measurement results are often presented in fringe patterns; thus, further fringe processing and phase analysis techniques are required. In contrast, non-interferometric techniques determine the surface deformation by directly comparing the grayscale changes in the images of the object surface before and after deformation and generally have less stringent requirements on experimental equipment and conditions.

doi:10.1088/978-0-7503-3027-5ch12

As a representative non-interferometric and non-contact optical technique, the DIC method has been widely accepted and commonly used as a powerful and flexible tool for surface deformation measurement in experimental solid mechanics and various scientific research and engineering fields. Since the 1980s plenty of research and development activities have been dedicated to this easy-to-use but effective optical measurement technique. Also, the applications of DIC have expanded from regular metal or polymeric materials to special composite and biological materials, from the macroscopic to the microscopic scale, from static or quasi-static loading to high-speed dynamic loading, and from common laboratory conditions to extreme high-temperature environments.

DIC is a non-contact, full-field yet easy-to-implement optical metrology based on digital image recording and digital image processing. Researchers first developed it at the University of South Carolina in the 1980s [8–10], when digital image processing and numerical computing were still in their infancy. During the past four decades, the DIC method has been extensively investigated and significantly improved to reduce computation complexity, enhance measurement accuracy, and expand its application range. For example, the two-dimensional (2D) DIC method using a single fixed camera is limited to in-plane deformation measurements of nominal flat objects. In order to acquire accurate measurements, some requirements on the specimen deformation, loading device, and measuring system must be strictly met [11, 12]. A drawback is that if the test object is of a curved surface or a three-dimensional (3D) deformation occurs after loading, the 2D-DIC method is no longer applicable. The disadvantages of 2D-DIC are overcome with the 3D-DIC (or, more precisely, stereo-DIC) based on the principle of binocular stereovision [13, 14], which was developed for the 3D shape and deformation measurement of both flat and curved surfaces. Also, the digital volume correlation (DVC) method, as a direct 3D extension of the 2D-DIC method, was proposed by Bay and Smith *et al* [15, 16] to perform full-field internal deformation measurements of solid objects: it works by tracking the movement of the volume unit within 3D digital images recorded before and after loading. The DVC method emerges as a novel and effective tool for quantifying the internal deformation of opaque cellular solid objects (e.g. trabecular bone, wood, rock, and cellular foam) and certain semi-transparent biological tissues such as cells (this method is outside the scope of this chapter). In addition, combined with the projection of computer-generated speckle patterns with a digital projector and the 2D-DIC method, the speckle projection method [17, 18] was also developed for out-of-plane displacement and shape measurements. Table 12.1 lists the three main correlation-based DIC techniques and their applicability.

Compared with the high-sensitivity interferometric optical techniques mentioned above, DIC methods offer the following unique and attractive advantages:

Simple specimen preparation and experimental set-up. Only one fixed camera (2D-DIC) or two cameras (3D-DIC) are required to record the digital images of the test specimen surface before and after deformation. Specimen preparation is unnecessary if the natural texture of the specimen surface has a random gray intensity distribution or can be achieved simply by spraying paint onto the specimen surface.

Low requirements for the measurement environment. DIC does not require a laser source. A white light source, natural light, or monochromatic light can be used for illumination during loading. Thus, it is particularly suitable for both laboratory and field conditions.

A wide range of sensitivity and resolution in measurements. Since the DIC method deals with digital images, these can be recorded by various high-spatial-resolution acquisition devices and thus can be directly processed by the DIC method. For example, 2D-DIC can be coupled with various microscopes [19–29], such as optical microscopy, laser scanning confocal microscope (LSCM), scanning electron microscopy (SEM), and atomic force microscopy (AFM) to obtain micro-scale to nanoscale deformation measurements. 3D-DIC can be used with light stereo microscopes. Similarly, the instantaneous 2D (or 3D) deformation measurement can be realized by analyzing the dynamic sequence of digital images recorded with high-speed digital image recording equipment using the 2D- (or 3D-)DIC method [30–32].

Table 12.1. Three correlation-based DIC techniques and their applicability.

DIC technique	Imaging devices used	Applicability
2D-DIC	Optical imaging systems, SEM, AFM, etc	In-plane deformation measurement of nominal flat surface objects
3D-DIC	Stereovision system, light stereo microscope, etc	3D deformation or shape measurement of flat or curved surface objects
DVC	x-CT, LSCM, micro-MRI, etc	3D internal deformation measurement of solid objects or biological tissues

DIC is combined with x-ray computed tomography (x-CT) and micro magnetic resonance imaging (MRI) to acquire images of the test object.

More importantly, with the constant emergence of high-spatial-resolution and high-temporal-resolution imaging equipment and continuous improvements in correlation algorithms, the DIC method can be applied easily in diverse academic and engineering areas.

Hence it is fair to say that DIC is one of the most active optical measurement technologies currently used in the field of experimental mechanics and demonstrates increasingly broad applications.

12.2 Two-dimensional digital image correlation

2D-DIC using a fixed camera is limited to in-plane deformation measurements of nominally flat objects. In general, the implementation of this method comprises the following three consecutive steps: (i) specimen and experimental preparation; (ii) recording the digital images of the flat object surface before and after loading;

and (iii) processing the acquired images using computer algorithms to obtain the desired displacement and strain information. In what follows, issues in object preparation and image capture are introduced, followed by a description of the basic principles and concepts of 2D-DIC. Next, the most commonly used displacement and strain estimation algorithms are outlined. Finally, a typical but straightforward application of 2D-DIC for the determination of the tensile strains and elastic properties of an aluminum specimen is demonstrated.

12.2.1 Object preparation and image capture

Figure 12.1 shows a typical experimental set-up using an optical imaging device for the 2D-DIC method. The test object surface must have a random gray intensity distribution, so upon illumination, it scatters light as a random speckle pattern, which deforms together with the object surface as a carrier of deformation information.

As pointed out by various researchers, in general, a good speckle pattern should fulfill the following empirical qualitative requirements [33–43]: (i) *high contrast*—with varying grayscale intensities and relatively large intensity gradients; (ii) *randomness*—non-periodic and non-repetitive pattern to facilitate full-field displacement mapping; (iii) *isotropy*—non-obvious directionality in the pattern; and (iv) *stability*—tightly adhere to the object surface and deform together with its surface without evident changes in geometric and grayscale characteristics. The

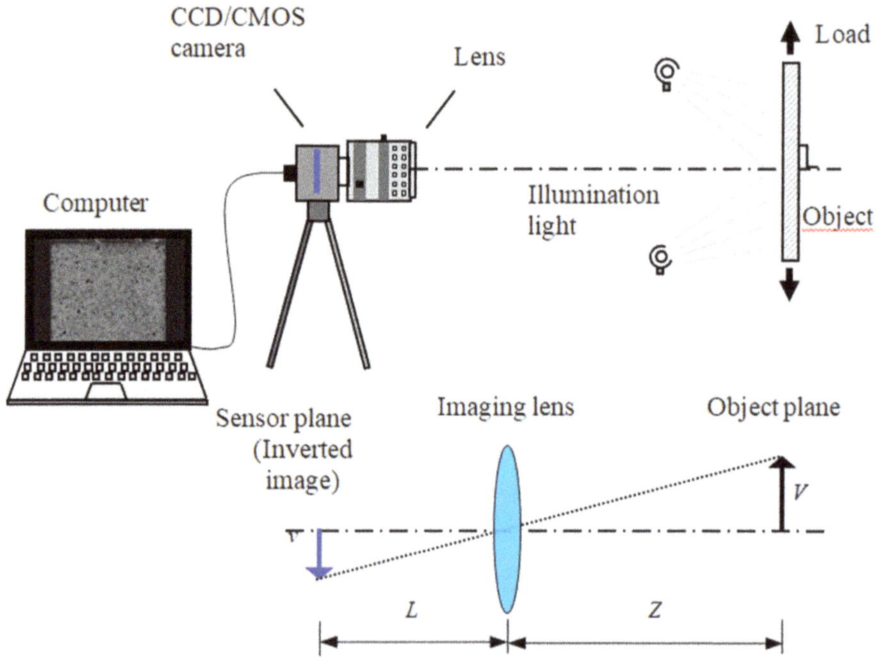

Figure 12.1. Typical optical image acquisition system for the 2D-DIC method and the corresponding pinhole camera model.

speckle pattern can be produced by the natural texture of the specimen surface or be artificially made by using either additive (an airbrush or mark pen for macro speckles) or subtractive approaches (high focused ion beam technique, spin-coating, compressed air technique, e-beam lithography or vapor-assisted remodeling for micro/nano speckles) [44].

The camera is placed with its optical axis perpendicular to the nominal object's flat surface and acquires images of different loading states on its sensor. A simple ideal pinhole camera model describes the mathematical relationship between the coordinates of an object point and its corresponding image point on the sensor plane [12], shown at the bottom of figure 12.1. The camera sensor plane displacements (u, v) are linearly proportional to the object plane displacements (U, V) through the simple linear relationship of $u = MU$, $v = MV$, where M denotes the time-invariant constant lens magnification. However, in the practical application of 2D-DIC, the magnification factor M cannot be seen as a constant across the entire image plane unless the following requirements are met (from geometrical optics considerations):

(i) The imaging system is perfect and does not suffer from geometric aberrations (image distortions). Hence, the magnification factor of M can be assumed constant across the image plane.
(ii) The object surface (or a small part of the object) must be flat and placed parallel to the camera sensor plane. Both the object and image surfaces should remain on the same plane without any motion and rotation during the experiment. This assumption implies that the object distance Z and image distance L will not change after loading. Thus, the magnification factor M can be regarded as a time-invariant constant.

As mentioned above, regrettably, it has been convincingly shown [12, 19] that the ideal pinhole imaging model assumed for an ordinary 2D-DIC system using a typical imaging lens is neither perfect nor stable. The magnification factor M varies at different pixel locations and changes slightly with time. As one might have thought, the reasons are as follows (see figure 12.2). (i) The pinhole imaging model is not perfect because any actual camera lens, particularly a low-cost one, unavoidably contains a certain amount of distortions due to the inherent aberrations made during their fabrication, namely machining errors and assembly errors [45, 46]. In other high-resolution imaging systems, including SEM, LSCM, and AFM, geometric distortions are more or less present, impairing the ideal linear correspondence between the physical point and its image, thus producing additional displacements, a source of potential errors. Due to the non-uniform characteristics of these lens distortions, the small in-plane motion on the object surface will cause different amounts of image motions, as schematically shown in figure 12.2(d). The theoretical analysis and experimental results carried out in [12] convincingly verified that the magnification factor M depends on several factors and is not a constant across the image. (ii) The pinhole imaging model is not stable because after a test object was loaded, its distance in the 2D-DIC system may change due to the out-of-plane

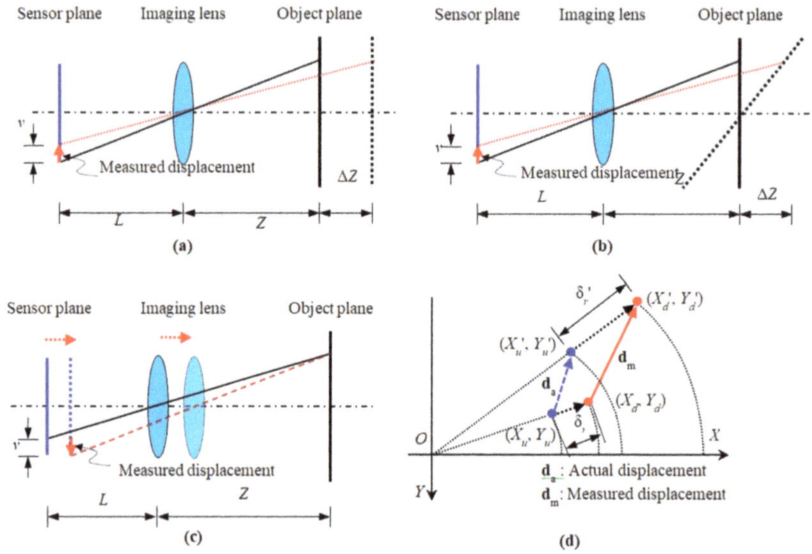

Figure 12.2. (a) Object out-of-plane translation along the X-direction and (b) out-of-plane rotation of the test object—both alter the object distance Z of the imaging system; (c) self-heating of the camera changes the imaging lens distance L and/or the object distance Z; and (d) radial lens distortion alters the displacement vector measured by 2D-DIC.

motion of the surface [11], as shown in figure 12.2(a) and (b). Also, the image and/or object distances change due to a camera lens's self-heating and/or temperature variation [47], as indicated in figure 12.2(c). The object changes and/or image distances alter the magnification factor M of the imaging system. In some instances, the imperfectness and instability of a low-cost 2D-DIC system may give rise to hundreds or even thousands of microstrain errors.

Consequently, in order to obtain accurate 2D-DIC measurements, a near-perfect (with negligible lens aberrations) and very stable optical system (insensitive to out-of-plane motions in object surface and imaging plane) using a well-design bilateral telecentric lens is highly recommended [12], see figure 12.3(a). The system can be further improved by introducing a blue light source and bandpass filters mounted before the imaging lens. Furthermore, to better suppress the thermal radiation emitted from hot objects at a high range of temperatures, ultraviolet-DIC (UV-DIC) systems have also been developed [48, 49]. These are based on the combined use of ultraviolet (UV) light, a UV light transmission filter, and a UV camera. In addition, cross-polarization using orthogonal linear polarizers at the light source and camera has also been introduced to DIC systems for higher robustness against direct specular reflections from the object surface [50]. In order to compensate for errors associated with imperfect and unstable imaging and obtain high-accuracy deformation measurements using a standard lens (i.e. one that may contain a tolerable amount of aberrations, particularly sphericity that causes severe object image distortions). A reference sample compensation (RSC) method was developed by attaching a thin, hollow rectangular compensation specimen onto the surface of the testing object [51], figure 12.3(b).

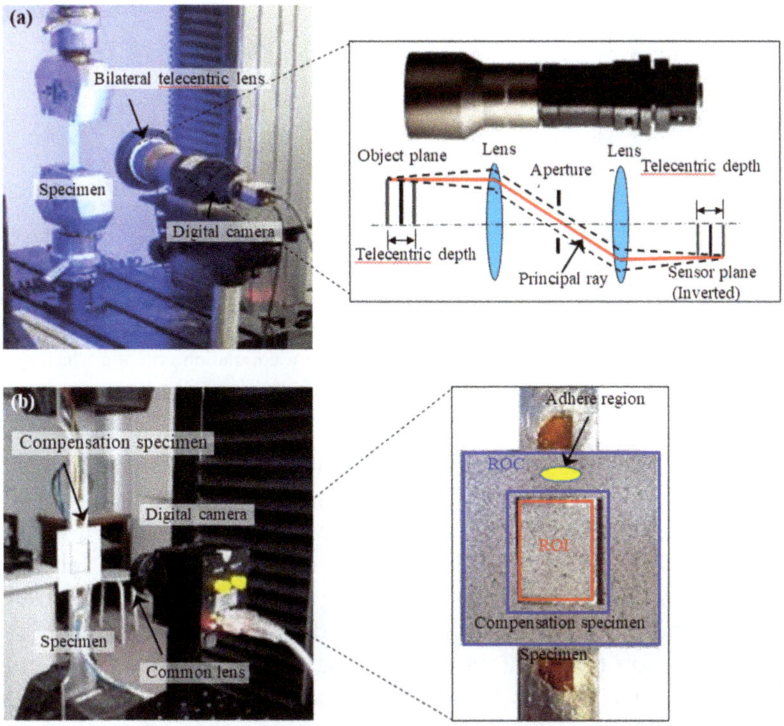

Figure 12.3. (a) 'Near-perfect and ultra-stable' 2D-DIC system using a high-quality bilateral telecentric lens with active imaging; (b) high-accuracy 2D-DIC measurement based on the reference sample compensation (RSC) method.

12.2.2 Basic principles and concepts

After recording the digital images of the object surface before and after deformation, the DIC computes the motion of each point of interest by comparing the digital images of the test object surface at different states. The image displacements (in pixels) can be further differentiated using an appropriate numerical algorithm to estimate full-field strains. In the following, the basic principles and concepts involved in 2D-DIC are introduced.

12.2.2.1 Basic principles

In the routine implementation of 2D-DIC, a region of interest (ROI) in the reference image should be specified first and further divided into evenly spaced virtual grids, as shown in the left part of figure 12.4. The displacements are computed at each point of the virtual grids to obtain full-field deformation. The basic principle of 2D-DIC is the tracking (or matching) of the same pixel points between the two images recorded before and after deformation, as schematically illustrated in figure 12.4. So, in order

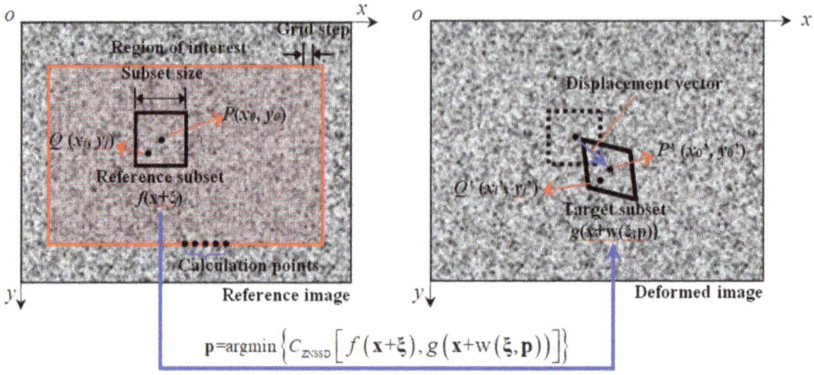

Figure 12.4. Left: reference image; the imposed black square is the subset used for tracking the motion of its center point, and the intersection points of the red grid denote the points to be calculated. Right: the displacement vector of the point P is computed by optimizing a correlation criterion which quantifies the similarity degree between the reference square subset and target (or deformed) subset.

to compute the displacements of point P, a square reference subset of $(2M + 1) \times (2M + 1)$ pixels centered at a point $P(x_0, y_0)$ from the reference image is chosen and used to track its corresponding location in the deformed image. A square subset rather than an individual pixel is selected for matching because the subset comprising a broader variation in gray levels will distinguish itself from other subsets and be more uniquely and accurately identified in the deformed image.

A cross-correlation (CC) criterion or sum-squared difference (SSD) correlation criterion must be predefined as an objective function to quantitatively evaluate the similarity degree between the reference subset and the deformed subset. The matching procedure is completed by searching the peak position of the distribution of the correlation coefficient. Once the correlation coefficient maximum is detected, the position of the deformed subset is determined. The differences in the positions of the reference subset center and the target subset center give the in-plane displacement vector at point **p**, as illustrated in figure 12.4.

12.2.2.2 Shape function/displacement mapping function

We can say that without loss of generality, one can assume that both the position and shape of the reference subset are changed in the deformed image, as figure 12.4 shows. However, based on the assumption of the deformation continuity of a deformed solid object, a set of neighboring points in a reference subset should remain as neighboring points in the target subset. Thus, the coordinates of the point $Q(x_i, y_j)$ around the subset center $P(x_0, y_0)$ in the reference subset can be mapped to the point $Q'(x_i', y_j')$ in the target subset according to the so-called 'shape function' [52] or 'displacement mapping function' [53]

$$\begin{cases} x'_i = x_i + \xi(x_i, y_j) \\ y'_j = y_j + \eta(x_i, y_j) \end{cases} (i, j = -M{:}\ M). \tag{12.1}$$

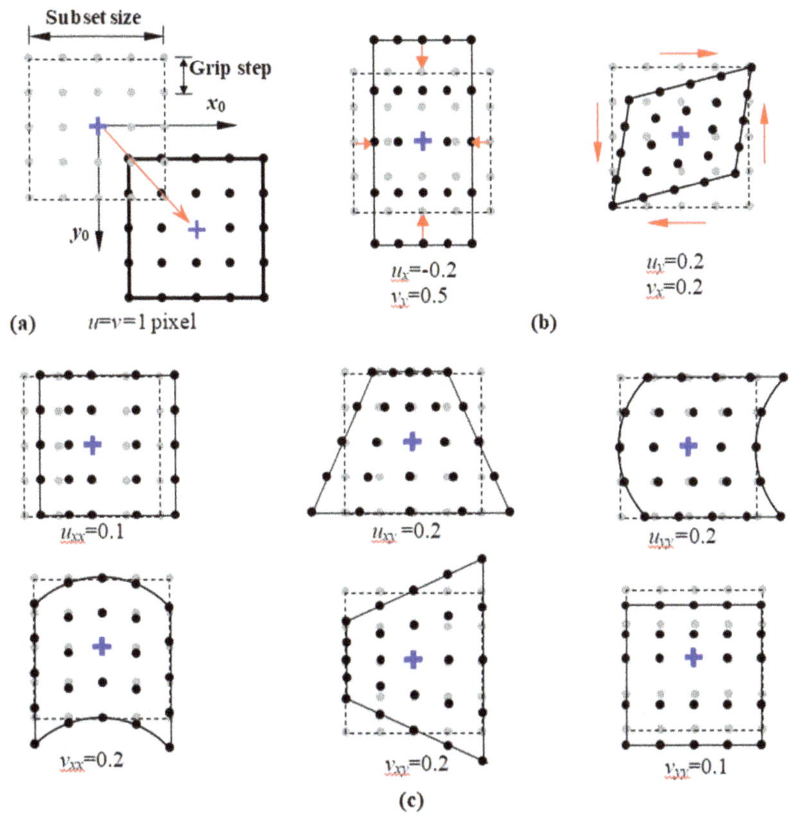

Figure 12.5. Examples showing the effects of different displacement parameters on the shape of the target subset: (a) zeroth-order, (b) first-order, and (c) second-order shape function.

If only rigid body translation exists in the reference subset and deformed subset, in other words, the displacements of each point in the subset are the same (see figure 12.5(a)), then the zeroth-order shape function can be used

$$\begin{cases} \xi_0(x_i, y_j) = u \\ \eta_0(x_i, y_j) = v \end{cases}. \tag{12.2}$$

The zeroth-order shape function is not sufficient to depict the shape change of the deformed subset. Thus, the first-order shape function, shown in figure 12.5(b), that allows translation, rotation, shear, normal strains, and their combinations of the subset, is most commonly used

$$\begin{cases} \xi_1(x_i, y_j) = u + u_x \Delta x + u_y \Delta y \\ \eta_1(x_i, y_j) = v + v_x \Delta x + v_y \Delta y \end{cases}. \tag{12.3}$$

In addition, the second-order shape functions, figure 12.5(c), proposed by Lu *et al* [53], can be used to depict more complicated deformation states of the deformed subset:

$$\begin{cases} \xi_2(x_i, y_j) = u + u_x\Delta x + u_y\Delta y + \dfrac{1}{2}u_{xx}\Delta x^2 + \dfrac{1}{2}u_{yy}\Delta y^2 + u_{xy}\Delta x\Delta y \\ \eta_2(x_i, y_j) = v + v_x\Delta x + v_y\Delta y + \dfrac{1}{2}v_{xx}\Delta x^2 + \dfrac{1}{2}v_{yy}\Delta y^2 + v_{xy}\Delta x\Delta y \end{cases}. \quad (12.4)$$

In equations (12.2), (12.3), and (12.4) $\Delta x = x_i - x_0$, $\Delta y = y_j - y_0$, u and v are the x- and y-directional displacement components of the reference subset center $P(x_0, y_0)$, u_x, u_y, v_x, and v_y are the first-order displacement gradients of the reference subset, and u_{xx}, u_{xy}, u_{yy}, v_{xx}, v_{xy}, and v_{yy} are the second-order displacement gradients of the reference subset. The influences of various deformation parameters on the final shape of the target subset are illustrated in figure 12.5.

Since the actual deformation in the deformed subsets cannot be known *a priori* in most practical measurements, mismatched (under-matched or overmatched) problems inevitably arise, leading to additional errors in the measured displacements. For clarity, the theoretically derived systematic and random errors due to mismatched shape functions are summarized in figure 12.6. The under-matched systematic error in displacement is related to subset size and second-order displacement gradient. In contrast, the overmatched random error in displacement depends on the noise level of the speckled image and the sum of the square of subset intensity

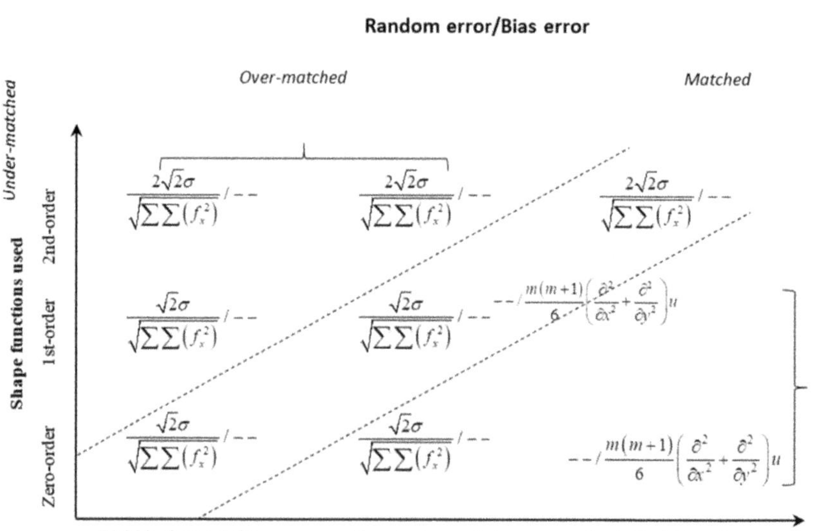

Figure 12.6. Random errors and under-matched bias errors in the case of various deformation modes (the symbol '—' indicates that the corresponding random error or bias error can be neglected). The denominator $\left(\sqrt{\Sigma\Sigma(f_x^2)}\right)$ denotes the SSSIG in the x-direction and σ is the standard deviation.

gradients (SSSIG) of the interrogated subsets [54, 55]. Thus, it can be explicitly claimed that second-order shape functions, capable of depicting more complex local deformation, should be used for practical DIC applications as the default [56].

In addition to the shape functions with varying orders of Taylor's expansions (e.g. zero-, first- and second-order), more practical displacement shape functions have been developed for specific applications. To characterize displacement discontinuity caused by cracks, crevices, dislocations, or shear bands, Poissant and Barthelat proposed a 'subset splitting' procedure into conventional DIC [57], which allows splitting subsets in two sections when a discontinuity is detected. Valle *et al* and Hassan *et al*, respectively, introduced the Heaviside function [58, 59], and the Burgers vectors [60] to modify subset functions with additional variables that can model the discontinuous displacement. Han *et al* developed an improved shape function by describing a displacement jump and strain variation in subsets near discontinuity that enhance displacement and strain accuracy [61]. In addition to these spatial descriptions of displacement, temporal components in the subset shape function have been considered in several challenging scenarios, including dynamic testing, object tracking, and high-temperature environments. Broggiato *et al* [62] employed linear regression in the post-processing stage to obtain the temporal components in displacements in consecutive images. Wang *et al* [63] introduced the first-order temporal shape function where the temporal component of velocity is considered under the assumption of constant velocity within a time. Chi *et al* [64] presented a framework of a spatial-temporal DIC algorithm by considering first- and second-order temporal shape functions to describe the displacement and strain rates.

12.2.2.3 Correlation criterion

As noted earlier, to evaluate the similarity (or difference) degree between the reference subset and its deformed counterpart, a correlation criterion must be defined, which is of fundamental importance in DIC. The commonly used correlation criteria can be classified into three categories according to their mathematical definitions [65], namely cross-correlation (CC), the sum of squared differences (SSD), and the parametric sum of squared differences (PSSD), as listed in table 12.2.

In table 12.2 $\overline{f} = \frac{1}{n}\sum_{i=1}^{n} f_i$, $\overline{g} = \frac{1}{n}\sum_{i=1}^{n} g_i$, $\overline{f_i} = f_i - \overline{f}$, and $\overline{g_i} = g_i - \overline{g}$, with n denoting the total number of pixels in a subset. $\overline{f}$, $\overline{g}$ are the mean intensity values of the reference ($f(x, y)$) and deformed ($g(x, y)$) subsets. Also in table 12.2:

- C_{CC} is the cross-correlation (CC) coefficient.
- C_{ZNSSD} is the coefficient of the robust zero-mean normalized sum of squared difference (ZNSSD) criterion.
- C_{ZNCC} is the zero-mean normalized cross-correlation (ZNCC) coefficient.
- C_{PSSDab} is the coefficient of a parametric sum of squared difference (PSSD*ab*) criterion.
- C_{NCC} is the normalized cross-correlation (NCC) coefficient.
- C_{ZSSD} is the coefficient of the normalized sum of squared difference (NSSD) criterion.

Table 12.2. Four types of correlation criteria are commonly used in DIC for evaluating the similarity (or difference) degree between the reference subset and the deformed subset. Reproduced with permission from [7]. Copyright 2011 Springer.

Performance	CC criterion	SSD criterion	PSSD criterion
Sensitive to all changes of the deformed subset intensity	$C_{CC} = \sum f_i g_i$	$C_{SSD} = \sum (f_i - g_i)^2$	
Insensitive to offset changes of the deformed subset intensity	$C_{ZCC} = \sum (\bar{f}_i \bar{g}_i)$	$C_{ZSSD} = \sum [\bar{f}_i - \bar{g}_i]^2$	$C_{PSSDb} = \sum (f_i + b - g_i)^2$
Insensitive to scale changes of the deformed subset intensity	$C_{NCC} = \frac{\sum f_i g_i}{\sqrt{\sum f_i^2 \sum g_i^2}}$	$C_{NSSD} = \sum \left(\frac{f_i}{\sqrt{\sum f_i^2}} - \frac{g_i}{\sqrt{\sum g_i^2}} \right)^2$	$C_{PSSDa} = \sum (af_i - g_i)^2$
Insensitive to the scale and offset changes of the deformed subset intensity	$C_{ZNCC} = \frac{\sum \bar{f}_i \bar{g}_i}{\sqrt{\sum \bar{f}_i^2 \sum \bar{g}_i^2}}$	$C_{ZNSSD} = \sum \left(\frac{\bar{f}_i}{\sqrt{\sum \bar{f}_i^2}} - \frac{\bar{g}_i}{\sqrt{\sum \bar{g}_i^2}} \right)^2$	$C_{PSSDab} = \sum (af_i + b - g_i)^2$

In practical measurement, the local intensity within the target image may differ significantly from that of the reference image. Various reasons may induce the intensity changes between images recorded at different configurations. Thus, a robust correlation criterion should be used to accommodate the variations in the intensity of the deformed image; otherwise, significant displacement measurement errors may occur due to the mismatch of the intensity change model [65–67]. Based on this consideration, the three robust correlation criteria listed in the bottom row of table 12.2, the C_{ZNCC}, C_{ZNSSD}, and $C_{\mathrm{PSSD}ab}$ criteria with two additional unknown parameters, have been highly recommended for practical use, as they are invariant against linear intensity changes of the target subset. At first glance, the mathematical expressions of these three robust correlation criteria are entirely different, although these three correlation criteria have the same performance against the offset and scale changes in the deformed subset intensity. Recently, rigorous theoretical analysis performed by Pan [65] convincingly proved that these three correlation criteria are mathematically equivalent. Specifically, the C_{ZNCC} criterion is related to the C_{ZNSSD} criterion according to

$$C_{\mathrm{ZNSSD}} = \sum\left(\frac{\bar{f}^{2}_{i}}{\sum\bar{f}_{i}^{2}} - 2\frac{\bar{f}_{i}^{2}\,\bar{g}^{2}_{i}}{\sqrt{\sum\bar{f}_{i}^{2}}\sqrt{\sum\bar{g}_{i}^{2}}} + \frac{\bar{g}^{2}_{i}}{\sum\bar{g}_{i}^{2}}\right) = 2-2\frac{\sum\bar{f}_{i}^{2}\,\bar{g}^{2}_{i}}{\sqrt{\sum\bar{f}_{i}^{2}}\sqrt{\sum\bar{g}_{i}^{2}}} \qquad (12.5)$$

$$= 2[1 - C_{\mathrm{ZNCC}}].$$

Similarly, it was also proved that the $C_{\mathrm{PSSD}ab}$ criterion is also equivalent to the C_{ZNNC} criterion, with the mathematical relationship between written as

$$C_{\mathrm{PSSD}ab} = \sum\bar{g}_{i}^{2}(1 - C^{2}_{\mathrm{ZNCC}}). \qquad (12.6)$$

However, the C_{ZNSSD} criterion is highly recommended in practical applications because it can be efficiently computed using the sum-table approach [68] and optimized using nonlinear numerical algorithms. In addition to the correlation criteria listed in table 12.2, a $C_{\mathrm{PSSD}abc}$ criterion (a, b, c denoting the offset, scale, and second-order intensity changes, respectively) and an intensity gradient-based SSD criterion have been developed to deal with possible nonlinear intensity changes in target subsets during experiments. These two last correlation criteria can be used in exceptional cases where noticeable nonlinear intensity changes occurred between the reference and target images due to non-uniform illumination or other reasons.

12.2.2.4 Interpolation scheme

As can be seen from equation (12.1), the coordinates of a point (x_i', y_j') in the deformed subset may locate between pixels (i.e. a sub-pixel location). The intensity of these points with sub-pixel locations must be provided before evaluating the similarity between the reference and deformed subsets using the correlation criteria defined in table 12.2. Thus, a particular sub-pixel interpolation scheme should be used. In the literature, various sub-pixel interpolation schemes have been used:

bilinear interpolation, bicubic interpolation, bicubic B-spline interpolation, bi-quintic B-spline interpolation, and bicubic spline. The detailed algorithms of these interpolation schemes can be found in numerical computing books [69], so we do not deal with them here. However, high-order interpolation schemes, such as the bicubic spline interpolation or the bi-quintic spline interpolation, are highly recommended [70] since they provide a higher registration accuracy and better convergence character of the algorithm than the simple interpolation schemes.

Nevertheless, bicubic interpolation has been regularly used due to its efficiency and accuracy. In using the generalized bicubic interpolation scheme [71], the gray values at sub-pixel locations can be estimated as

$$g(\Delta x, \Delta y) = \sum_{m=0}^{3} \sum_{n=0}^{3} a_{\mathrm{mn}} (\Delta x)^m (\Delta y)^n. \tag{12.7}$$

In this bicubic interpolation, the unknown 16 coefficients, i.e. a_{00}, a_{01}, …, a_{33}, in equation (12.7), can be determined by the gray intensity of the neighboring 4 × 4 pixels centered at the sub-pixel location, as illustrated in figure 12.7. It should be emphasized here that the 16 coefficients remain the same for each interpolation block, regardless of the specific sub-pixel locations within it. For this reason, an efficient interpolation coefficient look-up table approach can be used to speed the computational efficiency of DIC significantly.

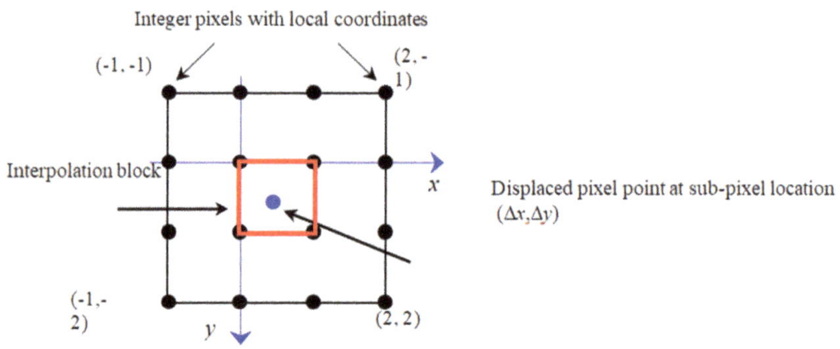

Figure 12.7. A schematic shows bicubic interpolation at the sub-pixel location using the surrounding 16 integer pixels.

Now let us work with the bicubic B-spline interpolation, divided into two steps [72]: (i) a 2D direct B-spline filter for the deformed image and (ii) discrete convolution to reconstruct the intensity at each sub-pixel position. In practice, a 2D direct B-spline filter can be decomposed into two 1D filters. Suppose that the grayscale array of each row or column within the deformed image is expressed as $G(k)$, then the filtered arrays $I(k)$ can be calculated as

$$I(k) = (B^n)^{-1} {}_{*} G(k), \qquad \text{with } k = 0,1,\ldots, N-1, \tag{12.8}$$

where N is the height or width of the image, and $(B^n)^{-1}$ is the nth-order direct B-spline filter. The operator of the three-order direct B-spline filter is denoted as

$$(B^3)^{-1} \xrightarrow{z} \frac{6}{z + 4 + z^{-1}} = 6\left(\frac{1}{1 - z_1 z^{-1}}\right)\left(\frac{-z_1}{1 - z_1 z}\right), \quad z_1 = -2 + \sqrt{3}. \tag{12.9}$$

The filtered intensity matrix $\boldsymbol{I}$ with the same size as the image $g(x, y)$ can be determined by applying the direct B-spline filter in two directions. Then, a discrete convolution between matrix $\boldsymbol{I}$ and the kernel function is done as follows:

$$g(\Delta x, \Delta y) = \sum_{j=-1}^{2} \sum_{i=-1}^{2} I(i, j)\beta^3(\Delta x - i)\beta^3(\Delta y - j), \tag{12.10}$$

where $(\Delta x, \Delta y)$ is the sub-pixel position in the local coordinated system, as shown in figure 12.7, i, j are the coordinates of the integer-pixel point in the local coordinated system, and $\beta^3(x)$ is the kernel function of cubic B-spline interpolation written as

$$\beta^3[(\Delta x + 1)\ \Delta x\ (1-\Delta x)\ (2-\Delta x)] = \frac{1}{6}[\Delta x^3\ \Delta x^2\ \Delta x\ 1]\begin{bmatrix} -1 & 3 & -3 & 1 \\ 3 & -6 & 3 & 0 \\ -3 & 0 & 3 & 0 \\ 1 & 4 & 1 & 0 \end{bmatrix}. \tag{12.11}$$

In order to enhance the accuracy of the 2D-DIC analysis, an optimized B-spline-based recursive filter [73, 74] and 4-, 6-, and 8-tap B-spline interpolation methods were introduced for higher sub-pixel interpolation precision. Although high-accuracy interpolation schemes ensure reduced errors in measured displacements, the errors due to sub-pixel interpolation schemes still prevent DIC from a better and more accurate deformation measurement [75, 76]. Therefore, diverse image pre-filtering techniques [77–79], random subset offset strategies, and self-correction schemes [80] have been made to reduce and eliminate the interpolation-induced errors in DIC.

12.2.3 Displacement field measurement using DIC

Strains are typically estimated by differentiating the displacement fields, so measuring the latter is generally considered a critical step in DIC. Various displacement tracking algorithms have been developed in the literature, including the subset-based local approach [81–83] and the element-based global approach [84–86]. However, the Newton–Raphson algorithm, initially proposed by Bruck *et al* [81] and refined by other researchers [82], has been considered a classic and standard algorithm. A detailed comparison between the NR algorithm and other sub-pixel registration algorithms (e.g. the peak-finding and gradient-based algorithms) reveals that the NR offers the most accurate and stable measurements [83]. To further eliminate the redundant computations involved in the classic NR algorithm without sacrificing its sub-pixel registration accuracy, Pan *et al* [87] proposed a more efficient inverse-compositional Gauss–Newton (IC-GN) algorithm for fast, robust, and accurate full-field displacement measurement. Since its formal advent in the DIC community, IC-GN has been a research must-do and thus has been continually refined by several researchers [88–90].

12.2.3.1 Forward additive Newton–Raphson algorithm

If we consider the relative deformation (i.e. shape change and rotation) between the reference and target subsets, the correlation function becomes a nonlinear function with respect to the desired vector mapping parameters. For example, if the first-order shape function is used, the desired vector mapping parameters are $\boldsymbol{p} = (u, u_x, u_y, v, v_x, v_y)^T$. These desired deformation parameters can be computed using certain optimization algorithms. In what follows, a forward additive matching strategy and the widely accepted NR algorithm are described. In the Newton–Raphson (NR) algorithm, the target subset is deformed by $\boldsymbol{W}(\boldsymbol{\xi}; \boldsymbol{p})$ with $\boldsymbol{\xi} = (\Delta x, \Delta y, 1)^T$ as the local coordinates of the pixel points in each subset $\boldsymbol{p}$ denotes an initial guess of deformation. Then, an incremental parameter vector $\boldsymbol{\Delta p}$ is further applied to the deformed subset and compared with the original reference subset to solve for $\boldsymbol{\Delta p}$, as schematically shown in figure 12.8.

Specifically, to quantitatively evaluate the similarity between the reference and target subsets and determine the desired deformation parameters, the most robust C_{ZNSSD} criterion, combined with an affine transform warping function, is used:

$$C_{\mathrm{ZNSSD}}(\boldsymbol{\Delta p}) = \sum_{\boldsymbol{\xi}} \left\{ \frac{[f(\boldsymbol{x} + \boldsymbol{\xi}) - \overline{f}\,]}{\Delta f} - \frac{[g(\boldsymbol{x} + \boldsymbol{W}(\boldsymbol{\xi}; \boldsymbol{p} + \boldsymbol{\Delta p})) - \overline{g}\,]}{\Delta g} \right\}^2, \qquad (12.12)$$

where $f(\boldsymbol{x})$ and $g(\boldsymbol{x})$ denote the grayscale levels at $\boldsymbol{x} = (x, y, 1)^T$ of the reference image and deformed image, $\overline{f}$ and $\overline{g}$ are the mean intensity value of the two subsets, and N denotes the total number of points within the subset; $\Delta f = \sqrt{\sum_{\xi}[f(\boldsymbol{x} + \boldsymbol{\xi}) - \overline{f}\,]^2}$ and $\Delta g = \sqrt{\sum_{\xi}[g(\boldsymbol{x} + \boldsymbol{W}(\boldsymbol{\xi}; \boldsymbol{p} + \boldsymbol{\Delta p})) - \overline{g}]^2}$; and $\boldsymbol{\xi} = (\Delta x, \Delta y, 1)^T$ is the local coordinate

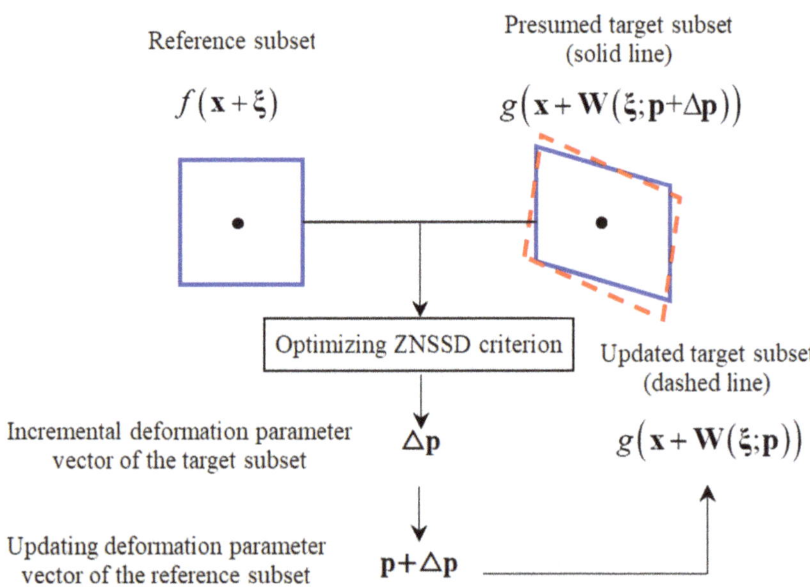

Figure 12.8. Schematic of the forward matching strategy for the classic NR algorithm.

of the pixel point in each subset. For brevity of notation, all image coordinates are expressed in the form of homogenous coordinates. $\boldsymbol{W}(\boldsymbol{\xi}; \boldsymbol{p})$ is the warp function, also known as the displacement mapping function, depicting the position and shape of the target subset relative to the reference subset and $\Delta\boldsymbol{p}$ is the desired incremental deformation vector exerted on the target subset deformed by an initial guess of deformation $\boldsymbol{p}$.

In general, a practical affine warp function (the linear displacement mapping function) allowing rigid body translation, rotation, normal strain, shear strain, and their combinations of the target subsets is used

$$\boldsymbol{W}(\boldsymbol{\xi}; \boldsymbol{p}) = \begin{bmatrix} 1 + u_x & u_y & u \\ v_x & 1 + v_y & v \\ 0 & 0 & 1 \end{bmatrix} \begin{pmatrix} \Delta x \\ \Delta y \\ 1 \end{pmatrix}, \tag{12.13}$$

where $\boldsymbol{p}$ is the pre-computed deformation vector parameter exerted on the target subset.

It should be noted that although only first-order displacement mapping is used, the following derivation can be easily extended to the second-order shape function. In order to solve for $\Delta\boldsymbol{p}$, we first perform a first-order Taylor expansion of equation (12.12) with respect to $\Delta\boldsymbol{p}$, which yields

$$C_{\text{ZNSSD}}(\Delta\boldsymbol{p}) = \sum_{\xi} \left[\frac{f(\boldsymbol{x} + \boldsymbol{\xi}) - \bar{f}}{\Delta f} - \frac{g(\boldsymbol{x} + \boldsymbol{W}(\boldsymbol{\xi}; \boldsymbol{p})) + \nabla g \frac{\partial \boldsymbol{W}}{\partial \boldsymbol{p}} \Delta\boldsymbol{p} - \bar{g}}{\Delta g} \right]^2, \tag{12.14}$$

where $g\,(\boldsymbol{x} + \boldsymbol{W}(\boldsymbol{\xi}; \boldsymbol{p}))$ represents the intensities of the target subset. Note that these values are located in sub-pixel positions; thus, a sub-pixel intensity interpolation algorithm, e.g. the bicubic interpolation scheme described before, should be used. $\partial \boldsymbol{W}/\partial \boldsymbol{p}$ is the Jacobian of the warp function (i.e. linear displacement mapping function). Minimization of the $C_{\text{ZNSSD}}(\Delta\boldsymbol{p})$ with respect to $\Delta\boldsymbol{p}$ gives the least-squares solution of $\Delta\boldsymbol{p}$:

$$\Delta\boldsymbol{p} = -\boldsymbol{H}_{6\times6}^{-1} \times \sum_{\xi} \left\{ \left(\nabla g \frac{\partial \boldsymbol{W}}{\partial \boldsymbol{p}} \right)^T_{6\times1} \times \left[\frac{\Delta g}{\Delta f} (f(\boldsymbol{x} + \boldsymbol{\xi}) - \bar{f}\,) - (g(\boldsymbol{x} + \boldsymbol{W}(\boldsymbol{\xi}; \boldsymbol{p})) - \bar{g}) \right] \right\}, \tag{12.15}$$

where $\boldsymbol{H}$ is the 6×6 Hessian matrix given as

$$\boldsymbol{H} = \sum_{\xi} \left[\left(\nabla f \frac{\partial \boldsymbol{W}}{\partial \boldsymbol{p}} \right)^T \times \left(\nabla f \frac{\partial \boldsymbol{W}}{\partial \boldsymbol{p}} \right) \right]. \tag{12.16}$$

Based on the obtained incremental parameter vector $\Delta\boldsymbol{p}$, the updated deformation vector parameter of the target subset can be computed as $\boldsymbol{p} + \Delta\boldsymbol{p}$. Equation (12.15) is repeated until a predefined convergence condition is satisfied. According to results in [91], the convergence conditions ensure that variations in the norm of the incremental displacements are equal to or less than 0.001, i.e. $\|\Delta \boldsymbol{p}\| \leqslant 0.001$, or a maximum iteration has been achieved.

12.2.3.2 Inverse-compositional Gauss–Newton algorithm

Figure 12.9 schematically shows the principle and flowchart of the IC-GN algorithm. Based on a pre-estimated initial guess of deformation, the IC-GN algorithm aims to optimize the following zero-mean normalized sum of squared difference (ZNSSD) criterion to determine the deformation parameters of the target subset:

$$C_{\mathrm{ZNSSD}}(\Delta \boldsymbol{p}) = \sum_{\xi}\left\{\frac{[f(\boldsymbol{x} + \boldsymbol{W}(\boldsymbol{\xi};\Delta \boldsymbol{p})) - \bar{f}\,]}{\Delta f} - \frac{[g(\boldsymbol{x} + \boldsymbol{W}(\boldsymbol{\xi};\boldsymbol{p})) - \bar{g}\,]}{\Delta g}\right\}^2, \quad (12.17)$$

where $f(\boldsymbol{x})$ and $g(\boldsymbol{x})$ denote the grayscale levels at $\boldsymbol{x} = (x, y, 1)^T$ of reference and deformed images, $\boldsymbol{\xi} = (\Delta x, \Delta y, 1)^T$ represent the local coordinates of the pixel point in each subset (again, note that all image coordinates are expressed in the form of homogeneous coordinates). $\bar{f} = \frac{1}{N}\sum_{\xi} f(\boldsymbol{x} + \boldsymbol{W}(\boldsymbol{\xi};\Delta \boldsymbol{p}))$, $\bar{g} = \frac{1}{N}\sum_{\xi} g(\boldsymbol{x} + \boldsymbol{W}(\boldsymbol{\xi};\boldsymbol{p}))$ are the mean intensity value of the two subsets, and $\Delta f = \sqrt{\sum_{\xi}[f(\boldsymbol{x} + \boldsymbol{W}(\boldsymbol{\xi};\Delta \boldsymbol{p})) - \bar{f}\,]^2}$ and $\Delta g = \sqrt{\sum_{\xi}[g(\boldsymbol{x} + \boldsymbol{W}(\boldsymbol{\xi};\boldsymbol{p})) - \bar{g}]^2}$. Recall that the practical first-order shape functions are considered, thus $\boldsymbol{p} = (u, u_x, u_y, v, v_x, v_y)^T$. $\boldsymbol{W}(\boldsymbol{\xi}; \boldsymbol{p})$ is the warp function,

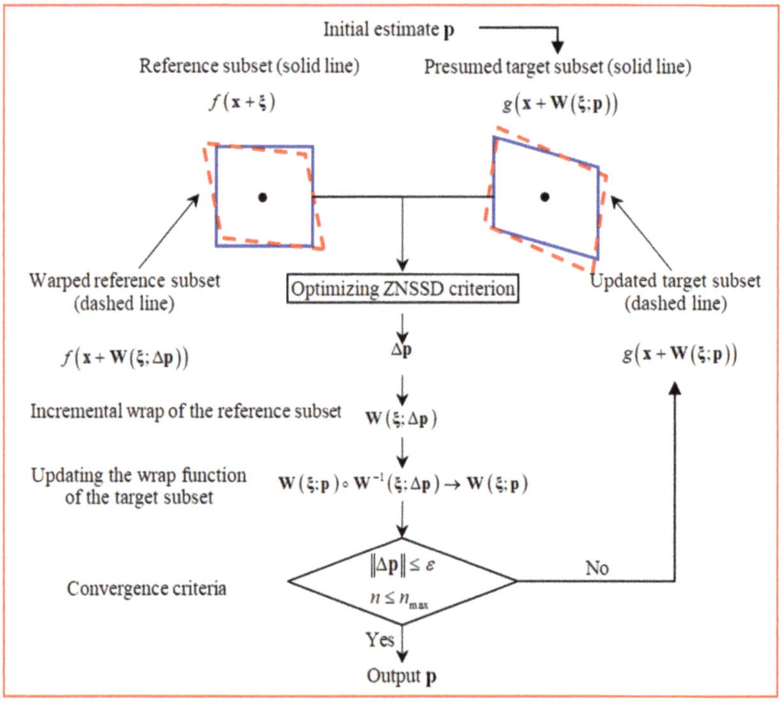

Figure 12.9. Schematic illustration of the principle of inverse-compositional Gaussian–Newton algorithm.

also known as the displacement mapping function in DIC, depicting the position and shape of the target subset relative to the reference subset. In addition, in the inverse-compositional algorithm, $\boldsymbol{W}(\boldsymbol{\xi};\, \Delta\boldsymbol{p})$ is the incremental affine warp function exerted on the reference subset, where $\Delta\boldsymbol{p}$ denotes the incremental vector parameter to be determined.

The nonlinear ZNSSD correlation criterion given in equation (12.17) can be minimized with respect to $\Delta\boldsymbol{p}$ using the Gaussian–Newton algorithm. Specifically, the incremental deformation vector $\Delta\boldsymbol{p}$ of the reference subset can be iteratively solved as

$$\Delta\boldsymbol{p} = -\boldsymbol{H}_{6\times 6}{}^{-1} \times \sum_{\boldsymbol{\xi}} \left\{ \left(\nabla f \frac{\partial \boldsymbol{W}}{\partial \boldsymbol{p}} \right)^{T}_{6\times 1} \times \left[(f(\boldsymbol{x} + \boldsymbol{\xi}) - \bar{f}\,) - \frac{\Delta f}{\Delta g}(g(\boldsymbol{x} + \boldsymbol{W}(\boldsymbol{\xi};\, \boldsymbol{p})) - \bar{g}) \right] \right\}, \quad (12.18)$$

where $\boldsymbol{H} = \sum_{\boldsymbol{\xi}} \left[\left(\nabla f \times \frac{\partial \boldsymbol{W}}{\partial \boldsymbol{p}} \right)^{T} \times \left(\nabla f \times \frac{\partial \boldsymbol{W}}{\partial \boldsymbol{p}} \right) \right]$ is the Hessian matrix with $\nabla f = \left(\frac{\partial f(\boldsymbol{x} + \boldsymbol{\xi})}{\partial x}, \frac{\partial f(\boldsymbol{x} + \boldsymbol{\xi})}{\partial y} \right)$ being the gradients of the reference subset. Because the Hessian matrix relies only on the gray value derivatives of the reference subset, it is therefore constant and can be pre-computed. This is the primary efficiency advantage of the IC-GN algorithm over the classic FA-NR algorithm.

After computing the vector parameter $\Delta\boldsymbol{p}$, the incremental warp $\boldsymbol{W}(\boldsymbol{\xi};\, \Delta\boldsymbol{p})$ of the reference subset, can be determined. Subsequently, the incremental warp is inverted and composed with, rather than simply added to, the current estimate $\boldsymbol{W}(\boldsymbol{\xi};\, \boldsymbol{p})$ to generate the updated warp function of the target subset:

$$\boldsymbol{W}(\boldsymbol{\xi};\, \boldsymbol{p}) \leftarrow \boldsymbol{W}(\boldsymbol{\xi};\, \boldsymbol{p}) \circ \boldsymbol{W}^{-1}(\boldsymbol{\xi};\, \Delta\boldsymbol{p}) = \begin{bmatrix} 1 + u_x & u_y & u \\ v_x & 1 + v_y & v \\ 0 & 0 & 1 \end{bmatrix} \begin{bmatrix} 1 + \Delta u_x & \Delta u_y & \Delta u \\ \Delta v_x & 1 + \Delta v_y & \Delta v \\ 0 & 0 & 1 \end{bmatrix}^{-1}, \quad (12.19)$$

where $\boldsymbol{W}^{-1}(\boldsymbol{\xi};\, \Delta\boldsymbol{p})$ denotes the inverted incremental warp function, and the operator $\circ$ signifies the composition of two warps. It is worth pointing out that the IC-GN algorithm is free from re-evaluating the Hessian matrix and reconstructing sub-pixel intensity gradients in each iteration; it generates similar results but is about 3–5 times faster than the FA-GN algorithm [87]. Furthermore, the IC-GN algorithm has proven to have better noise robustness (eliminating noise-induced bias error) than the FA-GN algorithm and gives rise, on average, to smaller random errors in displacement estimation in the cases of high noise levels or using small subsets [90]. Because of the enhanced sub-pixel registration accuracy, better noise-proof performance, and higher computational efficiency, the IC-GN algorithm has been strongly recommended as a standard sub-pixel registration algorithm for practical DIC applications.

12.2.3.3 Initialization of the NR algorithm and the calculation path

Recall that the regularly spaced pixel points within the ROI are considered points to be correlated, and the NR algorithm should compute the displacements of all these calculation points. However, the NR algorithm requires an initial guess of deformation close to the true value for converging rapidly and correctly as a local optimization algorithm. Otherwise, the algorithm may converge to incorrect values, or it does not converge at all. Although we can estimate the initial guess separately for each point, this approach is impractical because it is exceptionally time-consuming or impossible if large deformation and/or rotation is present in the deformed image. Therefore, the calculation path is crucial in DIC for full-field displacement measurement. Path-independent analysis, which has been explicitly used since DIC's inception, deals with each calculation point independently without an initial guess predicted from neighboring calculated points. It generally consists of integer-pixel displacement searching using a cross-correlation algorithm either implemented in the spatial domain [68] or the frequency domain employing the FFT [92]. Of course, path-independent DIC analysis can be speeded by high-performance parallel computing with a powerful GPU [93]. However, it cannot deal with images with inter-frame large rotation and considerable deformation.

Now, to increase the robustness and efficiency of conventional path-dependent DIC analysis, a robust reliability-guided displacement tracking (RGDT) algorithm has been proposed [94]. In the method, the ZNCC coefficient is used to identify the reliability of a computed point. The correlation calculation begins with a seed point guided by the calculated ZNCC coefficient. The neighbors of the point with the highest ZNCC coefficient in a queue for computed points will first be processed using the IC-GN algorithm. Meanwhile, the computed deformation parameters of this point will be transferred to its neighbors according to the deformation continuity assumption. The advantages of the reliability-guided displacement tracking algorithm are three-fold. First, the calculation path is always along the most reliable direction, and possible error propagation of the conventional DIC method can be entirely avoided. Second, because a very accurate initial guess can be predicted for the points to be processed, the NR algorithm performed on these points generally converges rapidly after minimal iterations. Third, it is universally applicable to the deformation measurement of images with the area and/or deformation discontinuities.

However, the path-dependent RGDT strategy can only be implemented point by point using serial computing; thus, it limits the computational efficiency of DIC calculation and wastes the powerful parallel computation capability of computers with multicores. In order to remedy this weakness, an improved RGDT strategy was proposed [95]. This method maintains the robustness of the existing RGDT strategy and can be implemented using multithread parallel computation to dramatically enhance the computational efficiency of the DIC analysis. Two empty queues should be built for the calculated points to facilitate the parallel implementation of the improved RGDT strategy using multi-threads. A high-reliability queue Q_1 contains calculated points whose ZNCC coefficient is greater than or equal to a threshold T_c (0.7). In Q_1, all points are inserted at the tail of the queue without ordering. A low-reliability queue Q_2 is established for computed points whose ZNCC coefficient is smaller than the preset threshold. In Q_2 points are ordered according to their ZNCC coefficients. Also, to realize multithread computation, the number of child threads

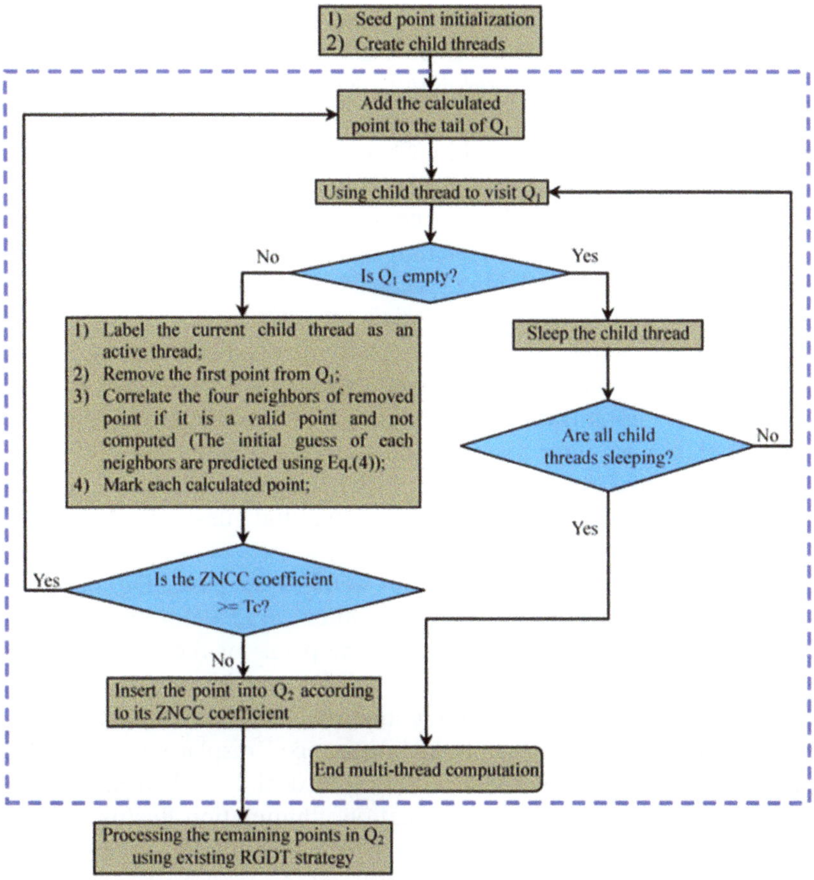

Figure 12.10. Flowchart of the parallel DIC method using the improved reliability-guided displacement tracking (RGDT) strategy and multithread computation.

should be defined *a priori* to correlation analysis. Then the improved RGDT algorithm can be implemented using multithread computing according to the following steps: (i) seed point initialization; (ii) multithread computation for high-reliability points in Q_1 (see the flowchart illustrated in figure 12.10); and (iii) reliability-guided displacement tracking for the remaining low-reliability points in Q_2.

12.2.3.3.1 The initial guess for the seed point

As for the seed point, if the relative deformation or rotation between the reference subset and the deformed subset is relatively small, it is easy to obtain an accurate estimation of the initial displacements of the seed point by a simple searching scheme implemented either in the spatial domain or in the frequency domain. In the spatial domain, the accurate locations of the target subset can be determined by a search routine, pixel by pixel, performed within the specified range in the deformed image.

Some methodologies such as the coarse-to-fine, the nested searching scheme [97], and the sum-table approach [68] can be used to speed up the calculation. These search methods yield a one-pixel resolution. Alternatively, the correlation between the reference subset and the deformed subset can also be implemented in the Fourier domain as conducted and advocated by Chen *et al* [98]. In the Fourier domain, the correlation between two subsets is calculated as the complex multiplication of the first subset's Fourier spectrum by the complex conjugate of the second subset's spectrum. The Fourier domain correlation method is also extremely fast since the FFT can be implemented at very high speeds. However, the Fourier domain method implicitly assumes the in-plane translation, so minor strains and/or rotations occurring between the two subsets will lead to significant errors. Indeed, the accuracy of the estimated initial guess directly affects the convergence characteristics of these nonlinear optimization algorithms. Thus, some novel algorithms are commonly used in computer vision, such as genetic algorithms [99], the scale-invariant feature transform (SIFT) [100, 101], and the Fourier–Mellin transform (FMT) [102], have been introduced to provide an initial guess for FA-NR or IC-GN algorithms.

The technique mentioned above performs well in most cases. Difficulties occur in some instances when significant rotation and/or large deformation are present between the reference and target subsets. Some pixels of the reference subset run out of an area of the assumed subset within the deformed image. Consequently, the similarity between the reference subset and the assumed deformed subset will decrease substantially and thus fails in integer-pixel displacement searching. An incremental DIC method was developed [103] to deal with deformed images with serious decorrelation due to large deformation, illumination fluctuations, or other reasons. In this method, an automatic reference image updating scheme was used to update the reference images based on the estimated correlation coefficient of seed points. For deformed images with serious decorrelation, incremental displacements extracted by comparing the current deformed image and the updated reference image are accumulated to determine the overall deformation.

12.2.3.3.2 Initial guess for the rest calculation points using the reliability-guided transfer scheme

To provide an accurate initial guess for the rest calculation points, a simple but effective initial guess transfer scheme is used, guided by the ZNCC correlation coefficients of computed points. Precisely, remove the first point, which has the maximum ZNCC coefficient, from the top of the queue. Then, analyze each of its four (or eight) neighboring points. If it is a valid point and has not been computed, it will then be computed using the NR algorithm to refine its correlation coefficient and deformation parameters. Note that the initial guess of the neighboring points, denoted by $\boldsymbol{p}_{\mathrm{NP}}^{0} = \left(u_{\mathrm{NP}}^{0}, u_{x\_\mathrm{NP}}^{0}, u_{y\_\mathrm{NP}}^{0}, v_{\mathrm{NP}}^{0}, v_{x\_\mathrm{NP}}^{0}, v_{y\_\mathrm{NP}}^{0}\right)$, is computed from the computed deformation parameters of the removed point, denoted as $\boldsymbol{p}_{\mathrm{RP}} = (u_{\mathrm{RP}}, u_{x\_\mathrm{RP}}, u_{y\_\mathrm{RP}}, u_{y}, v_{\mathrm{RP}}, v_{x\_\mathrm{RP}}, v_{y\_\mathrm{RP}})$, through the following formula

$$\begin{cases} u_{\mathrm{NP}}^{0} = u_{\mathrm{RP}} + u_{x\mathrm{RP}} \times \Delta x + u_{y\mathrm{RP}} \times \Delta y \\ u_{x\mathrm{NP}}^{0} = u_{x\mathrm{RP}} \\ u_{y\mathrm{NP}}^{0} = u_{y\mathrm{RP}} \\ v_{\mathrm{NP}}^{0} = v_{\mathrm{RP}} + v_{x\mathrm{RP}} \times \Delta x + v_{y\mathrm{RP}} \times \Delta y \\ v_{x\mathrm{NP}}^{0} = v_{x\mathrm{RP}} \\ v_{y\mathrm{NP}}^{0} = v_{y\mathrm{RP}} \end{cases}, \tag{12.20}$$

where Δx, Δy are the distances from the neighboring point to the removed point. We should note that the initial guess will provide a very accurate initial guess estimation by considering the local displacement gradients. When the deformed images involve large relative deformation or rotation, the NR algorithm, whose initial guess is provided by equation (12.20), generally exhibits better converge characteristics, resulting in a higher average ZNCC coefficient (indicating higher registration accuracy) and a less than average iteration number (denoting higher computational efficiency) [104]. Afterward, each just calculated point is labeled as calculated points and inserted into the queue according to its ZNCC coefficient value. The initial guess described above is repeated until the queue is empty, which means that all the valid points defined in the ROI have been computed, and the DIC analysis is completed.

12.2.3.4 An example of super-fast DIC with the IC-GN algorithm and the multi-path RGDT strategy

Two images of a chromium–nickel austenite stainless steel specimen were recorded to validate the super-fast DIC method's robustness, accuracy, and efficiency with the IC-GN algorithm and multi-path RGDT strategy during high-temperature testing [95]. The reference and deformed images shown in figure 12.11, with a resolution of 1280 × 1024 pixels at 8-bit gray levels, were recorded at 28 °C and 1200 °C, respectively. It is worth noting that these images were captured using an active imaging system composed of an actively illuminated monochromatic light and a digital CMOS camera outfitted with an optical bandpass filter.

Figure 12.11 shows u- and v-displacement fields and the ZNCC coefficient distributions computed by this DIC method using a 41 × 41 pixels subset. As expected, the evenly spaced contour lines in u- and v-displacement fields clearly show the uniform thermal expansion of the alloy. It is also seen that the ZNCC coefficients of most calculated points are more significant than 0.96, with an average ZNCC value of 0.979. The high ZNCC coefficients confirm the correctness of the DIC analysis. However, due to local surface oxidation that occurred during heating the specimen, some local regions on the image are matched with very low ZNCC coefficients.

In figure 12.12 the computational efficiency of the proposed DIC technique using three different subset sizes is plotted as a function of the number of employed child threads. It is apparent that the computational efficiency first increases almost

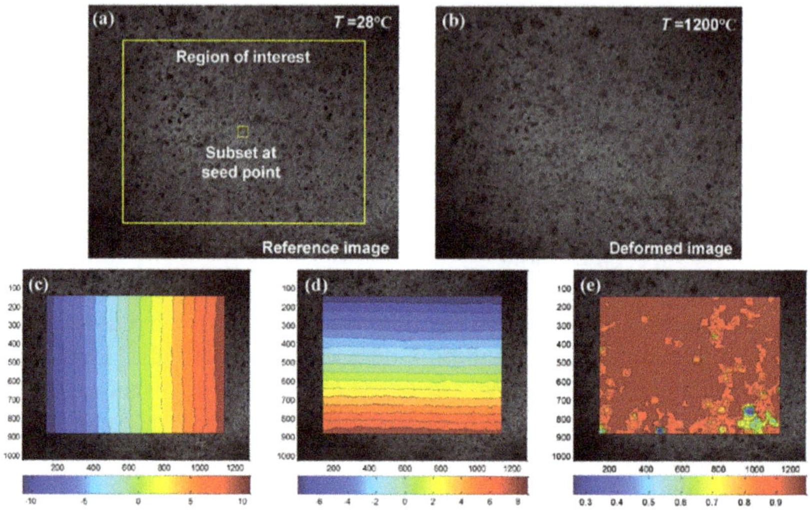

Figure 12.11. Experimental images and the calculated results: (a) reference image, (b) deformed image, (c) *u*-displacement field, (d) *v*-displacement field, and (e) zero-mean normalized cross-correlation (ZNCC) coefficient distributions.

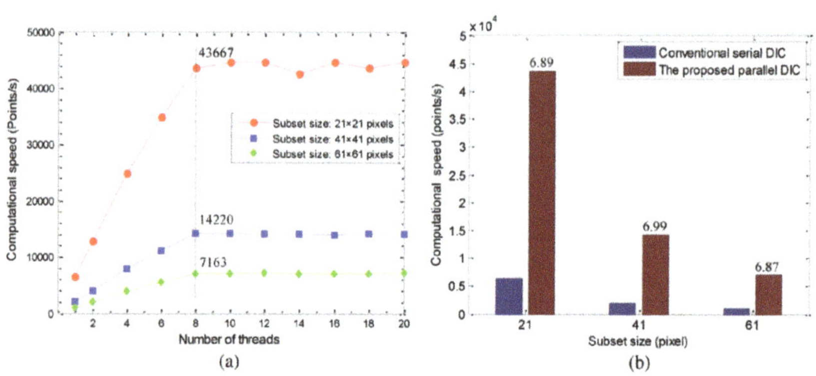

Figure 12.12. (a) The computational speed of the proposed super-fast DIC using various threads and different subset sizes and (b) comparison of the computational speed of the existing DIC using the RGDT strategy and the proposed super-fast DIC using an improved RGDT strategy with various subset sizes (the numbers on each bar denote the computational speed ratios of the proposed super-fast DIC to existing DIC).

linearly with the increase of child threads. However, when more than eight child threads (which are equal to the number of logical cores on the system) were used, the calculation speed no longer continued to increase but remained stable. Compared with the existing DIC method implemented on a single thread, the parallel DIC method using eight child threads achieves a speed-up of nearly seven times. The same ROI defined in figure 12.11 was analyzed using various subset sizes to exemplify the computational efficiency of the improved RGDT strategy executed with parallel computing.

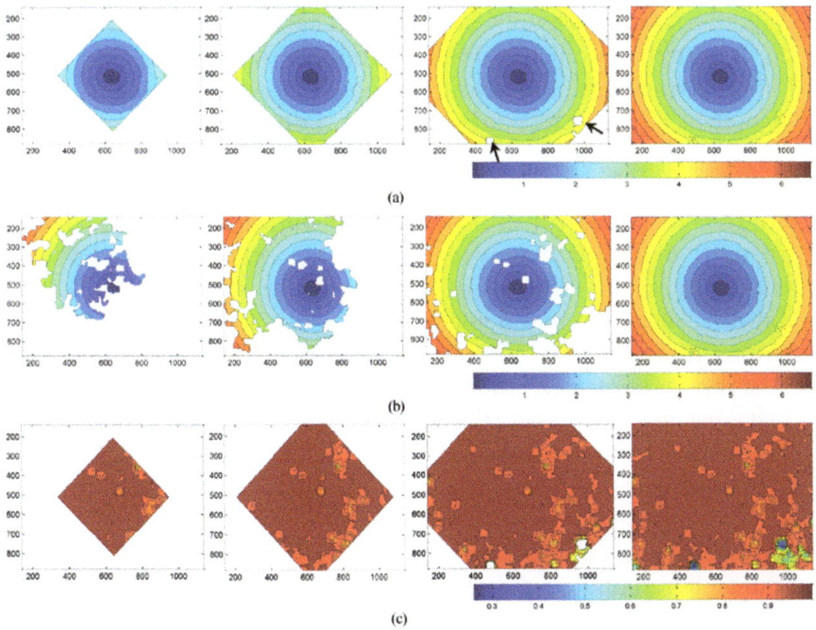

Figure 12.13. Three intermediate stages and final results of the radial displacement field are computed using (a) the improved RGDT strategy and (b) the existing RGDT strategy. (c) The ZNCC coefficient distribution.

Figure 12.12 shows the computational speed of the proposed super-fast DIC using the first-order shape function at different subset sizes. For a relatively small subset size of 21×21 pixels, the computational speed of the super-fast DIC is 43 667 points/second. With the increase of subset size, the computational efficiency reduces accordingly. However, for the three different subset sizes, the speed-up ratios of the parallel DIC to the existing serial DIC are all close to 7 times.

To more clearly indicate the calculation procedure of the improved RGDT strategy, the three intermediate stages of the computed radial displacement field and the ZNCC coefficient distribution are plotted in figures 12.13(a) and (c). For comparison, the corresponding results obtained using the existing RGDT strategy are given in figure 12.13(b). In the improved RGDT strategy, all the points in the center of ROI, whose ZNCC coefficients are greater than the preset threshold, are given the same priority to propagate through the DIC analysis. Thus, the region of calculated points grows uniformly along the left, top, right, and bottom directions and forms a diamond-shaped region. However, in particular, in small regions (e.g. the white hole indicated by the arrows), where the local oxidation occurs, and the resulting ZNCC coefficients are lower than the threshold; the process of these points is postponed, and the correlation analysis is executed using the existing RGDT strategy.

12.2.4 Strain field estimation with a pointwise least-squares algorithm

Using the displacement mapping algorithm previously described, now we can obtain the full-field displacements to sub-pixel accuracy. However, in many tasks of experimental solid mechanics such as mechanical testing of materials and structure stress analysis, object full-field strain distributions are more important and desirable because strains are closely related to many mechanical and physical quantities such as stress state, elastic modulus, Poisson's ratio, and coefficient of thermal expansion. Although the displacement gradients (i.e. strains) can be directly calculated using the displacement estimation algorithms (e.g. the FA-NR algorithm), the resulting error in the estimated strains limits its use only to local strains greater than approximately 0.010 [67]. In addition, although the relationship between the strain and displacement can be described as a numerical differentiation process in mathematical theory, unfortunately the numerical differentiation is considered an unstable and risky operation because it can amplify the noise contained in the computed displacement. Therefore, the resultant strains are untrustworthy if they are calculated by directly differentiating the estimated noisy displacements. An error to that extent will probably hide the tested object's underlying strain information, an unbearable issue in most cases.

Therefore, it is believed that the accuracy of strain estimation will be improved by smoothing the computed displacement fields first (to eliminate most of the noise) and subsequently differentiating them to calculate strains. Based on these considerations, several researchers [106–108] proposed a technique that involves smoothing the computed displacement fields with a finite element method first and subsequently differentiating them to calculate strains. In addition, the thin plate spline smoothing technique and the other three smoothing algorithms were also introduced [109–112], respectively, to remove the noise contained in displacement fields. Because the noise level in the displacement field significantly decreases after smoothing operations, these techniques substantially increase the precision of the resulting strain estimation. However, these smoothing techniques offer two shortcomings: it is challenging to conduct an optimal smoothing, i.e. the resulting displacement field tends to be over-smoothed or not sufficiently smoothing, and the numerical differentiation can also amplify the tiny fluctuations in the smoothed displacement fields.

The more practical technique for strain field estimation is the pointwise local least-squares fitting technique used and advocated by Wattrisse *et al* [113] and Pan *et al* [67]. The local least square method can estimate the strains from the discrete and noisy displacement fields computed by the DIC method, even for the image boundary points, holes, cracks, and other discontinuity regions.

The implementation of the local least-squares fitting technique for strain estimation can be explained as follows; see figure 12.14. Suppose we want to compute the strains of a calculation point (x, y); we first select a square window (also known as the strain calculation window) containing $(2m + 1) \times (2m + 1)$ discrete points around it. Note that the real size of the strain calculation window is $(2m \times \Delta L)^2$ pixels, with ΔL standing for the grid step between adjacent points. If the strain calculation window is small enough, the local displacement distributions can be approximated as a bilinear function. Thus, we have

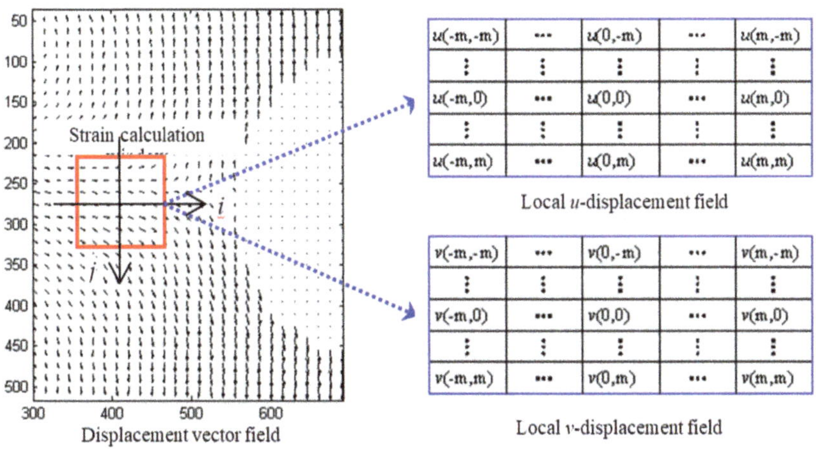

Figure 12.14. Local strain calculation window containing $(2m + 1) \times (2m + 1)$ discrete displacement data used for strain estimation.

$$\begin{cases} u(x+i,\, y+j) \cong a_0 + a_1 i + a_2 j + a_3 ij \\ v(x+i,\, y+j) \cong b_0 + b_1 i + b_2 j + b_3 ij \end{cases}, \tag{12.21}$$

where $i, j = -m \times \Delta L : m \times \Delta L$ are the local coordinates within the strain calculation window, $u(x + i, y + j)$ and $v(x + i, y + j)$ are the original displacements obtained by DIC, and $a_{i\,=\,0,1,2,3}$, $b_{i\,=\,0,1,2,3}$ are the unknown polynomial coefficients to be determined. These unknown parameters can be estimated by minimizing the following two objective functions:

$$\begin{cases} \chi^2(a_0, a_1, a_2, a_3) = \sum\limits_{i,j \in W_S} [u(x+i,\, y+j) - (a_0 + a_1 i + a_2 j + a_3 ij)]^2 \\ \chi^2(b_0, b_1, b_2, b_3) = \sum\limits_{i,j \in W_S} [v(x+i,\, y+j) - (b_0 + b_1 i + b_2 j + b_3 ij)]^2 \end{cases}. \tag{12.22}$$

It is important to note that, for the points located at the image boundary or the vicinity of a discontinuity region, the strain calculation window may contain fewer than $(2m + 1) \times (2m + 1)$ points. However, we can still compute strain components using least-squares fitting by neglecting these invalid points within the local strain calculation window from equation (12.22) [67].

Once these coefficients are obtained, the desired Cauchy strains or Green strains at the center point of the local sub-region can be computed. For example, the Cauchy strains can be calculated as

$$\varepsilon_x = \frac{\partial u}{\partial x} = a_1, \quad \varepsilon_y = \frac{\partial v}{\partial y} = b_2, \quad \gamma_{xy} = \frac{\partial u}{\partial y} + \frac{\partial v}{\partial x} = a_2 + b_1. \tag{12.23}$$

Since the random noise can be removed mainly in local fitting, the accuracy of the calculated strains is greatly improved. However, it is necessary to notice that, in order to obtain a reasonable and accurate full-field strain estimation, the following three aspects are essential and should be considered: (i) the accuracy of the displacement fields obtained by DIC; (ii) the size of the local strain calculation window; and (iii) the order of the fitting polynomials for strain estimation. Apparently, larger strain windows lead to a better noise suppression effect with reduced random errors in the estimated strain fields. In contrast, a significant strain calculation window may lead to an unreasonable linear approximation of the deformation within the strain calculation window. Since, in general, the actual local deformation of a test sample is complicated and cannot be known prior, the mismatch between the adopted polynomials and the underlying local deformation gives rise to either a bias error (due to under-matched polynomials) or random errors (because of overmatched polynomials) [124]. Strain information is often more valuable than displacement data in many mechanical applications. Accurate estimation of strain distributions from noisy displacement fields for samples subjected to non-homogeneous deformation or strain concentration is undoubtedly challenging. In this case, the proper size of the strain calculation window and proper order of the fitting polynomials should be selected to balance strain accuracy and smoothness.

12.2.5 Application of 2D-DIC for strain measurement

Benefitting from these scientific and technological advances, various high-accuracy video extensometers that can deliver accurate, real-time strain measurements have been proposed. Specifically, by combining active imaging, a bilateral telecentric lens, and state-of-the-art DIC algorithms, Pan *et al* [114] proposed an advanced video extensometer for real-time, high-accuracy strain measurements in mechanical testing. A schematic illustration of this video extensometer for non-contact, real-time, high-accuracy strain measurement is shown in figure 12.15. The system consists of a high-speed area scan monochrome camera (Genie HM1024, Teledyne DALSA, Ontario, Canada), a high-quality bilateral telecentric lens (Xenoplan 1:5, Schneider Optics, Inc., Germany), an optical bandpass filter with a center wavelength of 450 ± 2 nm, a monochromatic source emitting at 450–455 nm and a laptop (ThinkPad T440p, Lengend, Intel(R) Core(TM) i7-4700MQ CPU, 2.40 GHz of frequency and 8G RAM). The camera allows for a maximum image capture rate of 117 fps with 1024 × 768 pixel resolution and 8-bit quantization and is connected to the laptop using a Gigabit Ethernet standard LAN wire.

To capture high-fidelity digital images of a test object and lay the basis for subsequent high-accuracy strain measurement, a 'near-perfect and ultra-stable' imaging system is used—it combines active imaging with a high-quality bilateral telecentric lens. Compared with conventional optical imaging systems that use a regular lens and white light illumination, the present imaging system offers the following three distinct advantages: (i) better performance in minimizing the virtual displacements/strains caused by unavoidable out-of-plane motions occurred to the

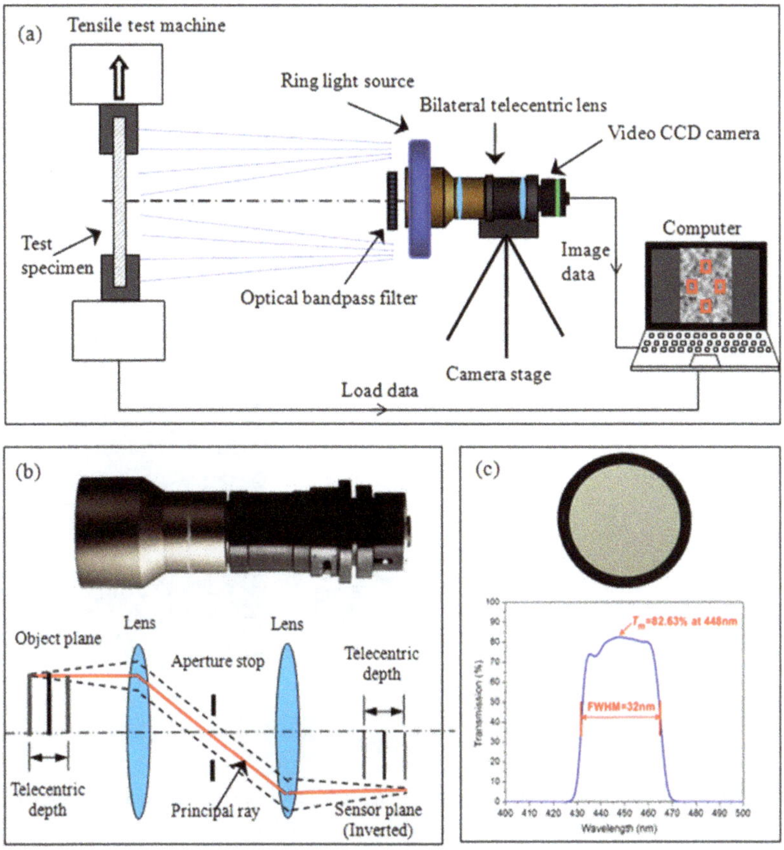

Figure 12.15. (a) Schematic diagram of the established advanced video extensometer. (b) The bilateral telecentric lens and (c) the transmission spectrum of the optical bandpass filter.

specimen surface; (ii) negligible errors in displacement and strain fields due to in-plane motions because of very low lens aberrations; and (iii) capability of acquiring stable video images with almost constant brightness and contrast.

As a typical application example, this video extensometer was used to characterize the mechanical properties of a carbon fiber multifilament sample. According to the Chinese national standard GB/T 3362-2005 (Test methods for tensile properties of carbon fiber multifilament), carbon fiber tensile properties involve three important indicators, namely tensile strength and tensile modulus, and elongation at break. The carbon fiber multifilament specimen was bought from Toray carbon fiber (Type: T300-3000) with a total length of 250 mm. The testing machine was clamped at both ends of the carbon fiber cardboard stick, where the distance between two cardboards was about 150 mm (see figure 12.16). White and black points were premade on the carbon fiber filament. The video extensometer tracked the axial strain of the sample. The tensile stress–strain relationship and the elongation at the breaking point of the sample were measured, from which the tensile modulus and

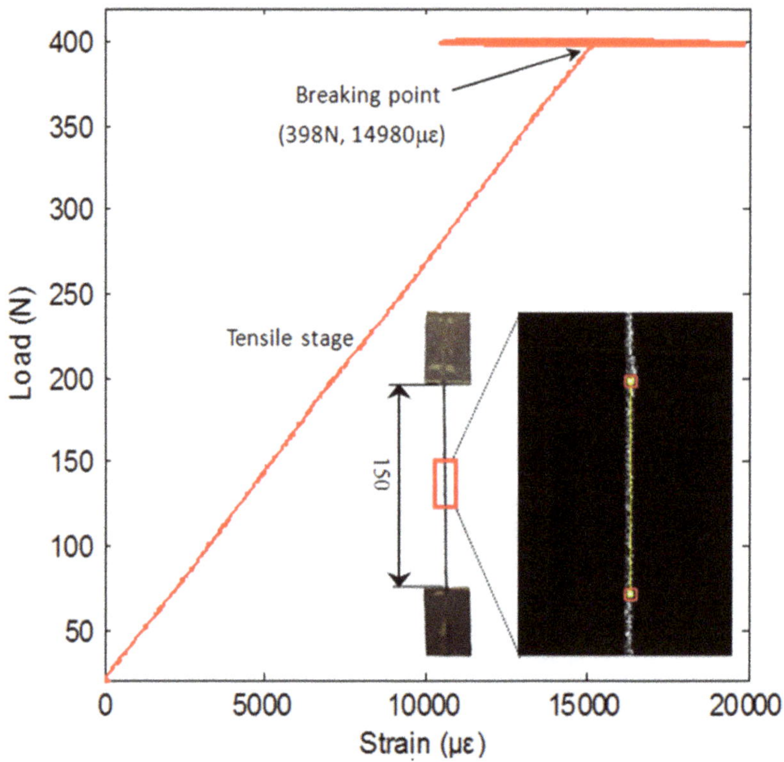

Figure 12.16. The video extensometer measured real-time stress–strain curves of the carbon fiber multifilament sample. The two photos inserted show an image of the specimen and its reference image with specified subsets.

tensile strength can be estimated. Also, to demonstrate the accuracy of the video extensometer, these experimental values were compared with the reference values provided by the manufacturer.

Before starting the tensile test, a tensile load of 20 N was applied to the specimen as the initial load. Then the machine was loaded at a speed of 20 N s^{-1} to the maximum tensile loading of 1000 N. Meanwhile, the advanced video extensometer was started to measure the strain until the sample broke. Two points with a distance of 450 pixels ($\approx$ 14.021 mm) were used to extract the longitudinal strains for the sample subjected to this tensile loading.

The strain data appeared abnormal when the specimen was broken. From figure 12.16, it is observed that on the breaking moment, the strain was 14 980 $\mu\varepsilon$, and the corresponding loading force was 398 N. The cross-sectional area of the specimen can be estimated based on the provided carbon fiber volume density of 1.76 kg m^{-3} and the line density of 198 g km^{-1}. Thus, the tensile strength can be estimated as 3.54 GPa. Considering the linear segment's stress–strain data from 5 s to 60 s, an elastic modulus of 231 GPa was calculated with least-squares fitting.

Table 12.3 summarizes the sample's elastic modulus and maximum elongation measured by the advanced video extensometer. In contrast to the corresponding

Table 12.3. Comparison between the experimental obtained mechanical properties and the reference values. Reproduced with permission from [114]. Copyright 2016 Optical Society of America.

	Measured results	Reference	Relative error (%)
Tensile modulus (GPa)	231 GPa	235 GPa	1.7%
Tensile strength (GPa)	3.54 GPa	3.53 GPa	0.3%
Elongation at break	1.498%	1.5%	0.1%

reference values provided by the manufacturer, the relative percentage errors are found to be 1.7% and 0.1%, respectively. In addition to its high-accuracy measurements, this video extensometer also demonstrates another apparent advantage by determining all the three mechanical properties in single testing.

High-accuracy video extensometers based on modified or simplified RSC methods were also developed [115, 116]. With reference samples attached to the test object, virtual strains caused by out-of-plane motions and lens distortion can be effectively compensated in real-time. In addition, several other techniques were also introduced into 2D-DIC-based video extensometers to minimize or compensate for the virtual strains due to out-of-plane motions [117–119]. For instance, Hoult *et al* [117] presented a correction scheme by using two cameras to view both sides of the loaded specimen to improve the strain measurement accuracy of the 2D-DIC method. Inspired by Hoult's idea, Zhu *et al* [118–120] proposed an easy-to-implement high-accuracy optical extensometer using a single-camera and a dual-reflector imaging technique. This technique can be considered as using two virtual cameras to record the front and rear surfaces of a specimen. By averaging the two strain results of the front and rear optical extensometers, the virtual strains due to out-of-plane motion, non-perpendicularity between the camera optical axis and specimen surface, and the initial bending of the specimen can all be automatically eliminated, as demonstrated in these reports.

12.3 Three-dimensional digital image correlation

The 3D-DIC method using two cameras was developed for shape and deformation measurements of both flat and curved objects. Although the subset-based matching algorithm can be directly used in 3D-DIC to determine disparities and image motions, stereovision calibration techniques are needed to estimate the intrinsic and extrinsic parameters of the two cameras, which are critical to the final measurements. Next, we will briefly describe the basic principles and implementation procedures of 3D-DIC and finally present two typical applications for shape and deformation measurements.

12.3.1 Stereo-DIC imaging system

Recall that the 2D-DIC method described previously can only be used for in-plane deformation of nominal flat objects. So, for deformation measurements of macroscopic objects such as structural components and industrial products with curved surfaces, the advanced 3D-DIC method is more practical and effective. It can be used for 3D profiles (the words shape or contour are synonyms) and deformation measurements of both flat and curved surfaces; also, it is capable of detecting, rather than affecting, the out-of-plane displacements. Furthermore, to meet the requirement of measuring the shape and deformation of small-scale objects, the 3D-DIC method has been implemented on light stereo microscopes [121] as well as fluorescent stereo microscopy systems, which enables shape and deformation measurements of small objects with sizes varying from several millimeters to several centimeters. It is also interesting to note that 3D-DIC measurements can be realized by combining a single camera with mirrors [123, 130], a bi-prism lens [124], as well as a diffraction grating [125, 126]. The main advantages of the single-camera 3D-DIC system lie in its relatively low cost and the convenience for time-resolved measurements, and in this manner, the need for hardware or software synchronization of two expensive high-speed cameras is avoided.

12.3.1.1 Conventional binocular stereo-DIC system

Existing 3D-DIC imaging systems generally use a white light source or natural light to illuminate the test object surface, and then the reflected light is collected by two CCD or CMOS cameras with ordinary imaging lenses (figure 12.17). This optical

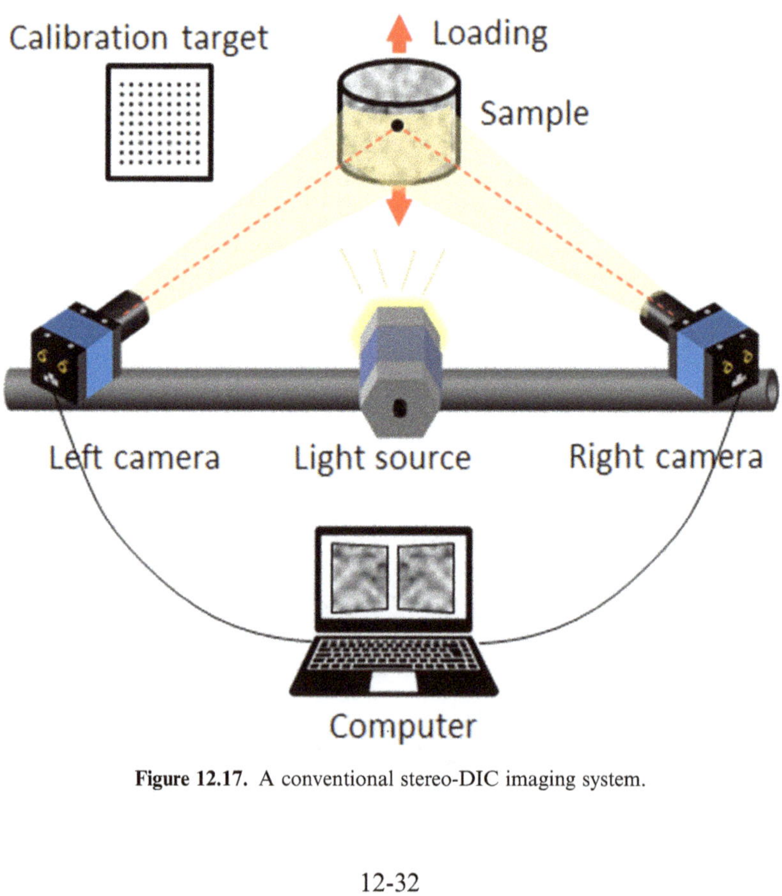

Figure 12.17. A conventional stereo-DIC imaging system.

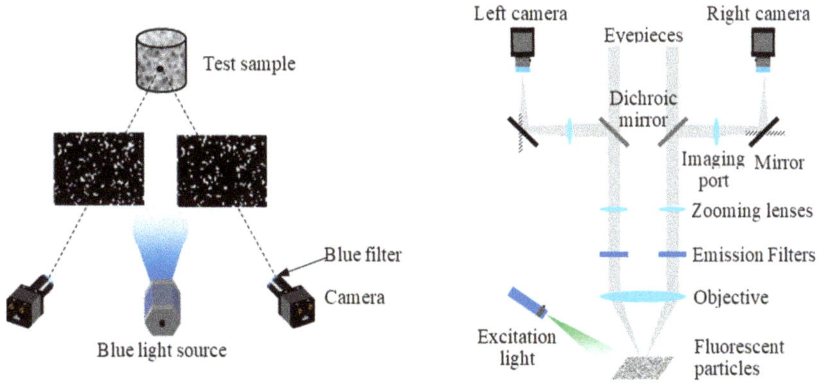

Figure 12.18. (a) An optimized stereo-DIC system combining monochromatic blue illumination with a bandpass filter; (b) microscopic stereo-DIC system based on a fluorescent stereo microscope.

imaging system can receive all light within the sensitive wavelength range of the camera. Being an image-based optical measuring technique, the contrast of the recorded speckle pattern has an important influence on the 3D-DIC measurements. Therefore, to guarantee an accurate measurement, the illumination intensity and ambient light should remain stable during the experiment.

Several optical components have been used to enhance the robustness of conventional stereo-DIC systems to extend the scope of stereo-DIC to more challenging environments (outdoor or high-temperature). By employing monochromatic blue illumination with a bandpass filter, Pan *et al* [127] developed an optimized stereo-DIC system (figure 12.18(a)) to suppress ambient variations in an outdoor environment and thermal radiation of a hot object in extremely high-temperature environments. Polarized light and polarized filters were introduced to suppress specular reflection in a conventional stereo-DIC system to realize the shape measurement of a coin [128]. Instead of using standard digital cameras and regular optical lenses, new imaging devices have also been introduced for various purposes. Nguyen *et al* [129] employed the infrared (IR) cameras in two Kinect sensors and one infrared-radiation projector for accurate 3D motion and vibration measurements of texture-less objects. Chen *et al* [130] developed a telecentric stereo micro-vision system for shape and deformation measurements within limited fields of view. To realize deformation and profile measurements at the micro-scale, Hu *et al* [131, 132] established a microscopic stereo-DIC system using a fluorescent stereo microscope (figure 12.18(b)) and Zhu *et al* [133] developed quantitative stereovision in an SEM. Furthermore, for high-speed 3D deformation measurements, a high-speed stereo-DIC imaging system using two synchronized high-speed or ultra-high-speed cameras was developed and applied in a large variety of practical situations [134–141].

12.3.1.2 Single-camera pseudo-stereo-DIC system

Due to the requirement of two synchronized cameras, constructing a regular binocular stereo-DIC system is both expensive and complicated, particularly in

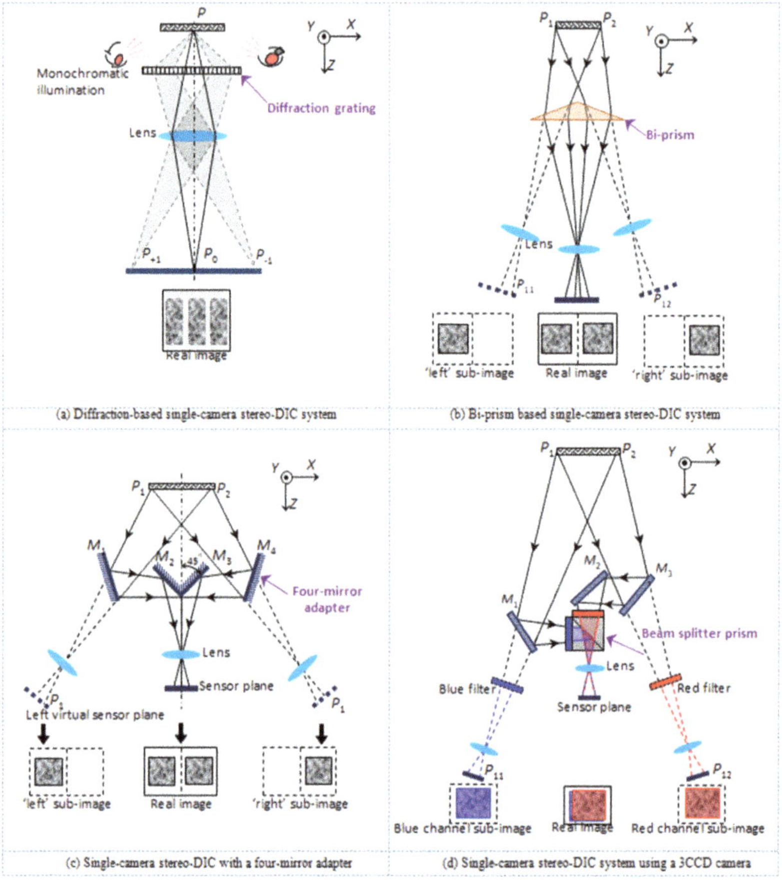

Figure 12.19. Optical arrangements of four different single-camera stereo-DIC systems.

high-speed deformation measurements. Therefore, single-camera stereo-DIC techniques have attracted significant attention in recent years [124, 126, 142–149] and were comprehensively overviewed by Pan *et al* [150]. These techniques can dramatically reduce the hardware investment and eliminate the need for precise camera synchronization by placing additional light splitting elements (e.g. a diffraction grating, a bi-prism, or a four-mirror adapter) in front of an existing imaging lens (figure 12.19). However, as two or more object views are separately imaged on non-overlapped regions of the sensor, the effective area of the region of interest (ROI) is reduced, resulting in the spatial resolution being halved. To take full advantage of the spatial resolution of the camera sensor, Yu and Pan proposed a novel single-camera stereo-DIC method using a single 3CCD [148] or CMOS color camera [149] and a skillfully-designed color separation device, shown in figure 12.19(d). In this method, the lights reflected by the object's surface pass through two optical paths with two different optical bandpass filters and are then imaged on the blue and red channels of the sensors to record an overlapped two-channel image of the object. Then, the object surface's full-field shape and 3D deformation information can be readily retrieved by comparing the blue and red channel sub-images using a regular stereo-DIC algorithm.

12.3.2 Basic principles and concepts

The 3D-DIC method is based on DIC and binocular stereovision. It retains the advantages of the non-contact, full-field measurement of 2D-DIC but overcomes its inherent disadvantages. Compared with 2D-DIC, the advantages of 3D-DIC are embodied in the following two aspects: (i) 3D-DIC is suitable for the shape and deformation measurements of both flat and curved objects and (ii) 3D-DIC can measure all three components of displacement (x, y, z) of the test object. Thus, small amounts of out-of-plane motion that occurred during loading can be accurately detected without affecting the accuracy of the other two in-plane displacement components [11].

Figure 12.20 schematically illustrates the basic principle of the 3D-DIC method. In figure 12.20 O_L and O_R are optical centers of the left and right cameras, respectively. Observe that a physical point P is imaged as point P_L in the image plane of the left camera and as point P_R in the image plane of the right camera. The 3D-DIC aims to accurately recover the 3D coordinates of point P with respect to a world coordinate system from image points P_L and P_R. The world coordinate system can be created by camera calibration techniques widely used in computer vision. Generally, a flat calibration target using evenly spaced features (circular dots or corners) is used. The two cameras first simultaneously capture images of the calibration target from different views. Then, the positions of these feature points on the calibration target images are detected using an image processing technique and inputted to the calibration model to optimize the desired intrinsic parameters (including effective focal length, principal point coordinates, and lens distortion coefficients) and extrinsic parameters (including the 3D position and orientation of the camera relative to the world coordinate system) of each camera. Together with the disparity data of each calculation point, these calibration parameters will be used

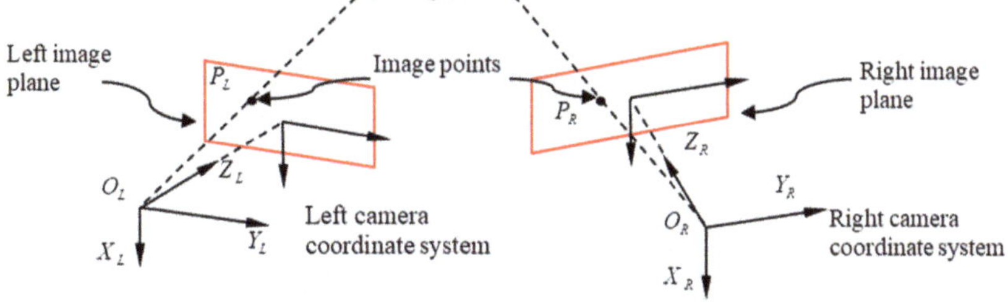

Figure 12.20. Schematic drawing of the binocular stereovision principle used in a 3D-DIC system.

to calculate the 3D position according to the triangulation principle. The differences between the 3D coordinates reconstructed before and after deformation give the 3D displacement vector of a measurement point. From this procedure, it can be concluded that there are two crucial steps in the implementation of the 3D-DIC method: (i) camera calibration (i.e. stereovision sensor calibration) and (ii) stereo matching [14].

12.3.2.1 Imaging model in stereovision system

Camera calibration is the procedure to determine the intrinsic parameters (e.g. effective focal length, the principle point, and lens distortion coefficient) and extrinsic parameters (including the 3D position and orientation of the camera relative to the world coordinate system) of a camera. For 3D-DIC, the calibration also involves determining the relative 3D position and orientation between the two cameras. As the accuracy of the final measurement dramatically depends on the obtained calibration parameters, camera calibration plays a crucial role in the 3D-DIC measurement.

As shown in figure 12.20, the projection of a 3D point P (x_w, y_w, z_w) in units of millimeters onto the image plane (u, v) in units of pixels can be represented by an ideal pinhole model. The relationship between a 3D point P and its image points P_L and P_R are

$$s\begin{bmatrix} u \\ v \\ 1 \end{bmatrix} = \mathbf{A}\begin{bmatrix} \mathbf{R} & \mathbf{t} \\ \mathbf{0} & \mathbf{1} \end{bmatrix}\begin{pmatrix} x_{\mathrm{w}} \\ y_{\mathrm{w}} \\ z_{\mathrm{w}} \\ 1 \end{pmatrix},\ \mathbf{A} = \begin{bmatrix} \alpha & \gamma & u_0 & 0 \\ 0 & \beta & v_0 & 0 \\ 0 & 0 & 1 & 0 \end{bmatrix}, \tag{12.24}$$

where s is an arbitrary scale factor, $\mathbf{A}$ are intrinsic parameters, and $\mathbf{R}$, $\mathbf{t}$ are extrinsic parameters (three for rotation and three for translation of each camera). For the camera, intrinsic matrix $\mathbf{A}$, (u_0; v_0) are the principal point coordinates, α and β are the scale factors in the image u- and v-axes, and γ describes the skewness of the two images axes [151].

Nonlinear optical distortions exist in practical applications, so the above equation is insufficient for accurate camera calibration. As reported in the literature, it is likely that the distortion function is dominated by the radial components and mainly dominated by the first term. It has also been found that any more elaborated modeling would not help provide higher accuracy and cause numerical instability. For this reason, the radial distortion considering only the first two terms is generally used.

Thus, let (u, v) be the distortion-free pixel image coordinates, and ($\tilde{u}$, $\tilde{v}$) the corresponding real observed image coordinates. If radial distortion is considered, we have

$$\begin{bmatrix} \tilde{u} \\ \tilde{v} \end{bmatrix} = \begin{bmatrix} u \\ v \end{bmatrix} + \begin{bmatrix} (u - u_0)[k_1 r^2 + k_2 r^4] \\ (v - v_0)[k_1 r^2 + k_2 r^4] \end{bmatrix}, \tag{12.25}$$

where k_1 and k_2 are the coefficients of the radial distortion, and r is the distance between an image point (u, v) to the principal point.

Although calibration of a single camera can be conducted using widely accepted techniques, e.g. Zhang's method [151], the calibration of the binocular stereovision system needs more additional work [152]. In a stereovision system, aside from the intrinsic parameters of each camera, the relative position and orientation of the two cameras should be determined precisely. As for a spatial point P, assume its 3D world coordinate and its image coordinate in the left and right images are denoted by x_w, x_l, and x_r, respectively, then

$$\begin{aligned} x_l &= R_l x_w + t_l \\ x_r &= R_r x_w + t_r. \end{aligned} \tag{12.26}$$

Elimination of x_w from these two formulas gives

$$x_l = R_l R_r^{-1} x_r + t_l - R_l R_r^{-1} t_r. \tag{12.27}$$

Let

$$\begin{aligned} R_{r2l} &= R_l R_r^{-1} \\ t_{r2l} &= t_1 - R_l R_r^{-1} t_r. \end{aligned} \tag{12.28}$$

Thus, we obtain

$$x_l = R_{r2l} x_r + t_{r2l}. \tag{12.29}$$

In summary, A_l, A_r, R_{r2l}, t_{r2l}, $k_{1,\,l}$, $k_{2,\,l}$, $k_{1,\,r}$, $k_{2,\,r}$ are parameters that need to be calibrated for a binocular stereovision system.

12.3.2.2 Stereovision system calibration

To achieve high-quality stereo-DIC measurements, accurate calibration of the two-camera unit is the first and essential step. Camera calibration provides the intrinsic (center point, lens distortion coefficients, focal length of each camera) and extrinsic parameters (translation and rotation between the two cameras) required for a correct triangulation. In computer vision, procedures for camera calibration can be roughly classified into two categories: photogrammetric calibration and self-calibration [151]. The former uses particular calibration objects with precise geometry, while only images of a static scene from different perspectives are required in self-calibration [153–155]. Numerous advances have been made in this field, among which the early works like the direct linear transformation method [156] and the two-stage method [157] established the foundation for camera calibration. Due to the accuracy and easy implementation, the landmark technique using a 2D planar pattern proposed by Zhang [151] has become the *de facto* standard method for calibration.

In Zhang's technique, camera calibration needs a 3D stereo or 2D plane calibration target [151], in which the characteristic points with known 3D coordinates can be accurately and automatically extracted using a digital image processing algorithm. Figure 12.21 shows several typical calibration templates used for 3D-DIC.

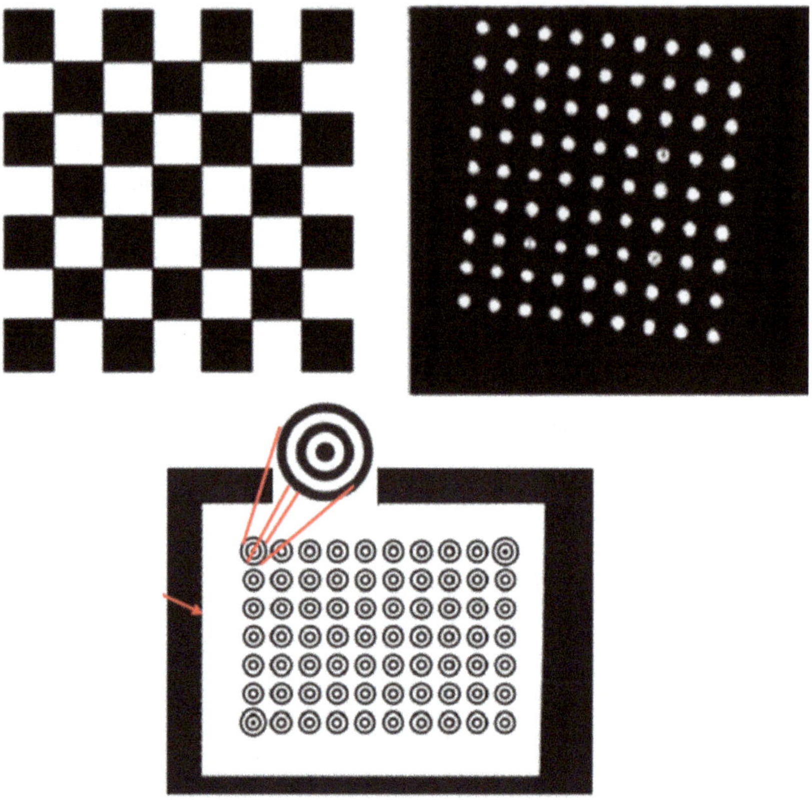

Figure 12.21. (Clockwise) A chessboard calibration target, a circular dot calibration target, and a ring template.

Using Zhang's technique [151], the intrinsic and extrinsic parameters of the left and right camera (i.e. A_l, $R_{l,\,i}$, $T_{l,\,i}$ and A_r, $R_{r,\,i}$, $T_{r,\,i}$) can be determined for each position of the calibration target. During the target image acquisition, the relative position and orientation of the camera are fixed. Thus, in theory, these parameters should be all the same. However, due to various noise and calculation errors, these parameters obtained using each calibration image are not equal. Therefore, a nonlinear least-squares optimization of the intrinsic parameters of the two cameras and their relative positions and orientations can be optimized globally using the following objective function [158]:

$$\begin{aligned}&\sum_{i=1}^{n}\sum_{j=1}^{m_{l,\,i}}\left\| m_{l,\,ij} - \tilde{m}\left(A_l,\, k_{1,\,l},\, k_{2,\,l},\, R_{1,\,i},\, t_{1,\,i},\, X_{1j}\right)\right\|^2 + \\ &\sum_{i=1}^{n}\sum_{j=1}^{m_{r,\,i}}\left\| m_{r,\,ij} - \tilde{m}\left(A_r,\, k_{1,\,r},\, k_{2,\,r},\, R_{1,\,i},\, t_{1,\,i},\, R_{r21},\, t_{r21},\, X_{1j}\right)\right\|^2,\end{aligned} \tag{12.30}$$

where $\tilde{m}(A_l, k_{1,l}, k_{2,l}, R_{1,i}, t_{1,i}, X_{1j})$ and $\tilde{m}(A_r, k_{1,r}, k_{2,r}, R_{1,i}, t_{1,i}, R_{r21}, t_{r21}, X_{1j})$ are the image coordinates of the jth feature point in the left and right image, as predicted using the nonlinear imaging model; $m_{l,ij}$ and $m_{r,ij}$ are the corresponding real image coordinates obtained using the feature extraction technique. The cost function can be optimized using the Levenberg–Marquart algorithm to determine the calibration parameters.

On the one hand, a slight improvement in calibration accuracy is appreciable to achieve higher stereo-DIC measurement accuracy. By adopting circular control points instead of square patterns in a checkboard calibration target and using a first-order elliptic approximation to consider the perspective effect, Reu [159] improved the camera calibration results with the retro-projection root-mean-squared-error (RMSE) by around 0.05 pixels. In addition, Vo *et al* [160] proposed an advanced geometric camera calibration technique by employing a frontal image concept in conjunction with a hyper-precise control point detection scheme based on DIC and further reduced the retro-projection RMSE to below 0.01 pixels. On the other hand, to realize deformation measurements of big engineering components or structures, Shao *et al* [161] adopted a self-calibration method by combining speckle analysis with a classical relative orientation algorithm. In this method, the intrinsic parameters of both cameras are calibrated in the laboratory beforehand, and then the corresponding points in image pairs matched using DIC are used for rotation and translation estimation between the cameras. In addition, on the grounds of the automatic system calibration using the speckled reference image of the specimen, the speckle-based self-calibration method also allows for real-time strain measurement using a single-lens 3D extensometer.

12.3.3 Stereo matching

In stereo-DIC, the correspondence of the projection of physical points in the left and right images must be established using correlation algorithms. Together with the calibrated parameters of the stereo-rig, the 3D coordinates of measurement points can be recovered with the classic triangulation method. Further, by subtracting the 3D coordinates of the same physical points at an initial state from that of deformed states, three displacement components caused by an external loading can be computed. Correlation analysis in stereo-DIC can be separated into two steps: stereo matching and temporal matching (or tracking). The goal of the stereo matching is to precisely register the same physical points in the two images captured by the left and right cameras; while temporal matching is to track the same physical points in the consecutive images recorded by the same (left or right) camera at different times or states. Temporal matching can be directly implemented using well-established 2D-DIC algorithms. Stereo matching is more challenging due to the significant distortion between the images captured by different cameras, and this task is commonly considered the most challenging part of stereovision measurement.

Three different matching strategies (figure 12.22) can be adopted for stereo-DIC. The first strategy is shown in figure 12.22(a) matches the left reference image

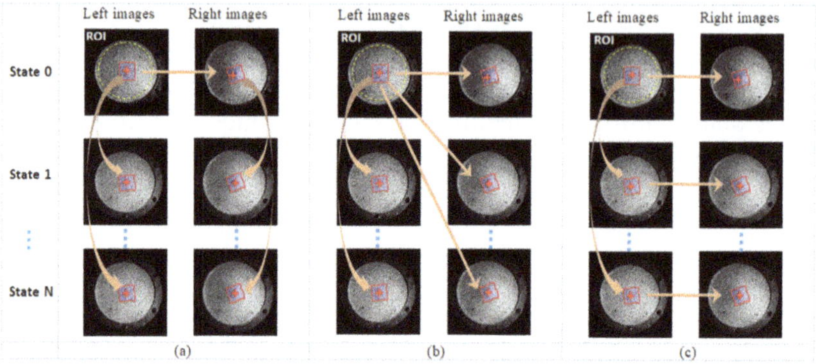

Figure 12.22. Three major strategies for the stereo-correlations in stereo-DIC: (a) stereo matching for initial left and right images and temporal matching for subsequent left and right image series; (b) temporal matching for left image series and stereo matching for all the right images using the original left image as the reference image; and (c) temporal matching for the left images and stereo matching for all the right images using the current left image as the reference image.

to the right reference image (i.e. stereo matching only for the initial state), and all the left and right deformed images are matched to their initial states (temporal matching). Figure 12.22(b) shows that all the deformed left and right images are correlated to the left reference image. In figure 12.22(c) all left images after deformation are compared to the left reference image using temporal matching; then, each corresponding right image is compared to the current left image with stereo matching. The detailed procedure for the first matching strategy is described as follows.

To reconstruct the profile of a sample surface of an object, a region of interest (ROI) should be first specified in the left image, within which all the calculation points are defined. Then, these calculation points with known image coordinates are searched in the right image using a subset-based correlation algorithm to determine its corresponding image coordinates, as schematically shown in the top part of figure 12.23. Afterward, these mapped image coordinates can be used to reconstruct the profile of the ROI. This figure also indicates the detailed procedure for 3D displacement measurements. After defining an ROI in the reference pair's left image, each measurement point within the ROI is then searched in three other different images (i.e. the right image in the reference state, left and right images in the deformed state) using the same subset-based matching method. Based on the disparity information and the calibration parameters of the two cameras, the world coordinate of the interrogated point P before and after deformation can be reconstructed as $P(x_{w0}, y_{w0}, z_{w0})$ and $P'(x_{w1}, y_{w1}, z_{w1})$, respectively. The differences in the two coordinates yield the three displacement components caused by external loading.

Finally, the detailed matching procedures are summarized as follows. (i) The robust ZNSSD criterion combined with a twelve-parameter displacement mapping function can be defined as the objective function. Note that, in stereovision

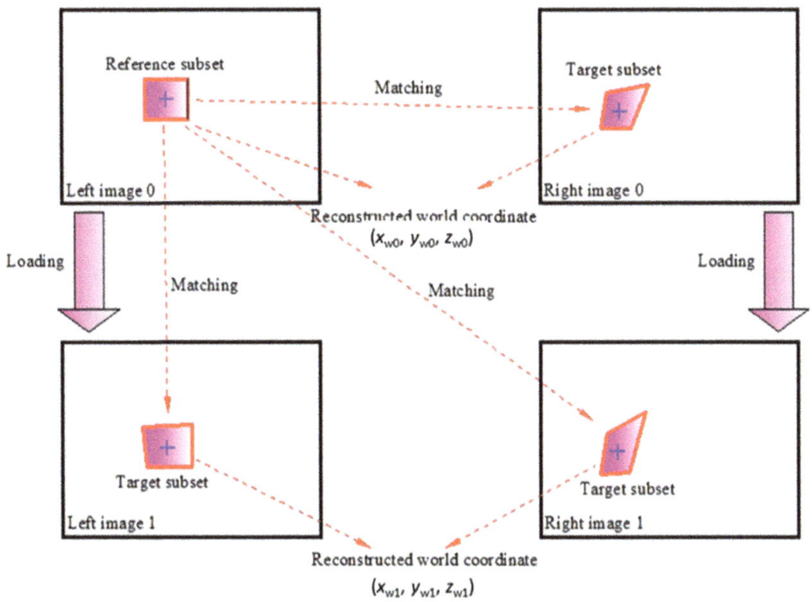

Figure 12.23. Subset-based 2D-DIC method is used to track the positions of each calculation point, defined in left image 0, in the rest of the images.

matching, the second-order shape function generally offers better results than the simple first-order shape function. It can realistically approximate a more complicated deformation of the target subset image by different cameras. (ii) The nonlinear ZNSSD criterion is subsequently optimized using the classic (NR) algorithm and a bicubic interpolation scheme to determine the desired parameters. It is also worth noting that a robust but straightforward reliability-guided displacement tracking scheme [127] is used to guide the calculation path of the NR algorithm, which can significantly improve the computational efficiency of the NR algorithm by ensuring an accurate initial guess transfer between consecutive measurement points. The same procedure is repeated to track the corresponding positions of the considered points of interest in the two images recorded after deformation.

12.3.3.1 3D profile reconstruction

Based on the obtained calibration parameters of each camera and the calculated disparities of points in the images, the 3D world coordinates of the points in the regions of interest on the specimen surface can be easily determined using the triangulation method

$$z_{c1}\begin{bmatrix} u_1 \\ v_1 \\ 1 \end{bmatrix} = \begin{bmatrix} m_{11}^1 & m_{12}^1 & m_{13}^1 & m_{14}^1 \\ m_{21}^1 & m_{22}^1 & m_{23}^1 & m_{24}^1 \\ m_{31}^1 & m_{32}^1 & m_{33}^1 & m_{34}^1 \end{bmatrix}\begin{pmatrix} x_w \\ y_w \\ z_w \\ 1 \end{pmatrix}, \tag{12.31}$$

$$z_{c2}\begin{bmatrix} u_2 \\ v_2 \\ 1 \end{bmatrix} = \begin{bmatrix} m_{11}^2 & m_{12}^2 & m_{13}^2 & m_{14}^2 \\ m_{21}^2 & m_{22}^2 & m_{23}^2 & m_{24}^2 \\ m_{31}^2 & m_{32}^2 & m_{33}^2 & m_{34}^2 \end{bmatrix}\begin{pmatrix} x_w \\ y_w \\ z_w \\ 1 \end{pmatrix}. \tag{12.32}$$

By eliminating the z_{c1} and z_{c2} from the above equations, four linear equations as a function of the desired world coordinates (x_w, y_w, z_w) can be obtained:

$$\begin{aligned} (u_1m_{31}^1 - m_{11}^1)x_w + (u_1m_{32}^1 - m_{12}^1)y_w + (u_1m_{33}^1 - m_{13}^1)z_w &= m_{14}^1 - u_1m_{34}^1 \\ (v_1m_{31}^1 - m_{21}^1)x_w + (v_1m_{32}^1 - m_{22}^1)y_w + (v_1m_{33}^1 - m_{23}^1)z_w &= m_{14}^1 - v_1m_{34}^1 \\ (u_2m_{31}^2 - m_{11}^2)x_w + (u_2m_{32}^2 - m_{12}^2)y_w + (u_2m_{33}^2 - m_{13}^2)z_w &= m_{14}^2 - u_2m_{34}^2 \\ (v_2m_{31}^2 - m_{21}^2)x_w + (v_2m_{32}^2 - m_{22}^2)y_w + (v_2m_{33}^2 - m_{23}^2)z_w &= m_{14}^2 - v_1m_{34}^2. \end{aligned} \tag{12.33}$$

Equations (12.31) and (12.32) have a geometrical interpretation, and that is that the spatial lines passing thorough $o_{c1}p$ and $o_{c2}p$, and thus the intersection point of these two lines gives the exact world coordinates of point P. However, it is clear that equation (12.33) is an over-determined function, and hence the world coordinates (x_w, y_w, z_w) are generally solved using linear least squares. By repeating the same procedure to other points of interest, the 3D shape of the ROI can therefore be reconstructed.

12.3.3.2 3D displacement and strain estimation

Tracking the coordinates of the same physical points before and after deformation, the 3D motions of each point can be determined. Similarly, the same technique proposed can be utilized for the strain estimation based on the obtained displacement fields. Matching the left reference image and a right deformed image can increase the computed accuracy instead of matching the left reference image and the corresponding right image. Moreover, the three displacement components caused by an external loading can be calculated by subtracting the world coordinates after and before deformation.

Strain estimation uses pointwise local least squares to deal with noisy discrete three displacement components $u(x_w, y_w, z_w)$, $v(x_w, y_w, z_w)$, and $w(x_w, y_w, z_w)$, with x_w, y_w, and z_w are in millimeter units. The method is written as

$$\begin{aligned} u(x_w, y_w, z_w) &= a_0 + a_1x_w + a_2y_w + a_3x_wy_w \\ v(x_w, y_w, z_w) &= b_0 + b_1x_w + b_2y_w + b_3x_wy_w \\ w(x_w, y_w, z_w) &= c_0 + c_1x_w + c_2y_w + c_3x_wy_w. \end{aligned} \tag{12.34}$$

The linear least squares can be used to calculate these fitting coefficients, based on which the six strain components $\frac{\partial u}{\partial x}, \frac{\partial u}{\partial y}, \frac{\partial v}{\partial x}, \frac{\partial v}{\partial y}, \frac{\partial w}{\partial x}, \frac{\partial w}{\partial y}$ can be estimated.

In practical applications, to warrant accurate and efficient surface 3D deformation measurements, some issues, such as the matching strategy, correlation algorithm, shape function, and initial guess, must be handled with care in stereo matching. On the one hand, due to the nonlinear nature of the perspective projection, the use of affine shape functions in stereo matching generally leads to substantial errors for large subset sizes and large stereo angles. For this reason, second-order shape functions should always be used in stereo matching. To realize high-accuracy stereo matching with high efficiency, Gao *et al* [88] developed an improved IC-GN algorithm using second-order shape functions (IC-GN2). On the other hand, to enable fast and accurate stereo matching when the images from different cameras undergo a large deformation due to a large stereo angle, feature-based matching using the famous scale-invariant feature transformation (SIFT) algorithm has been introduced to estimate the initial guess for stereo matching algorithms. The SIFT algorithm is robust and can cope well with the scaling, translation, rotation, distortion (affine transformation), occlusion, and illumination-variation issues encountered in real image pairs. In addition, Tang *et al* [162] proposed a large deformation measurement scheme for stereo-DIC by combining an improved coarse search method with a reshaped reference subset and a reference image updating scheme. With this and to eliminate the additional systematic error due to intensity interpolation during the reference image update, Zhou *et al* [163] proposed an adaptive subset offset scheme, in which the reference subsets in the updated reference images are translated to nearest integer positions, rather than being interpolated at sub-pixel locations.

12.3.4 Applications of 3D-DIC for SHAPE and deformation measurement

The optical arrangement of the color stereo-DIC system is schematically illustrated in figure 12.24. It consists of a single 3CCD color camera, a beam splitter prism placed before the camera lens, two optical bandpass filters with specific center

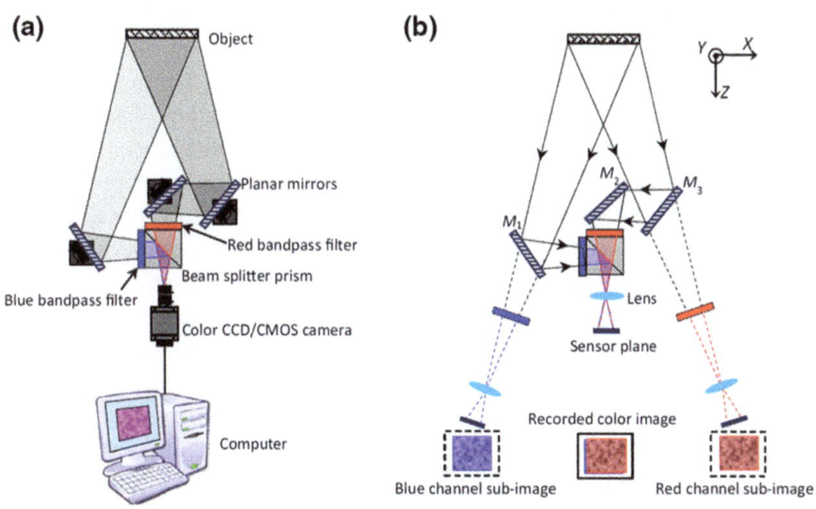

Figure 12.24. (a) Optical arrangements of the established single-camera color stereo-DIC system using three flat mirrors and a beam splitter prism; (b) the corresponding schematic of the optical paths from the object to the camera sensor.

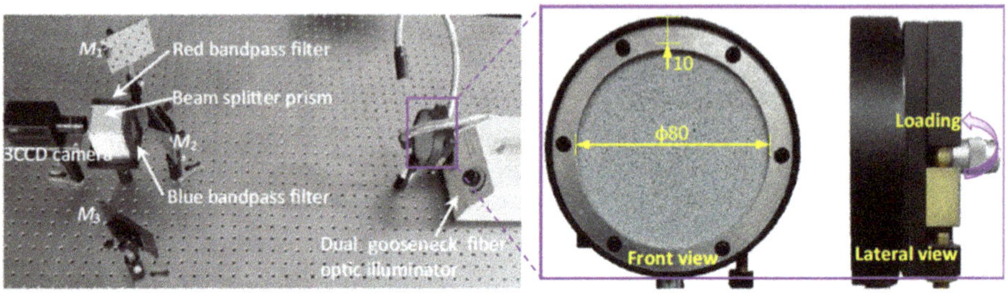

Figure 12.25. Photograph of the established single-camera color stereo-DIC system for 3D deformation measurements of a centrally loaded circular plastic plate.

wavelengths attached on the left and front of the beam splitter prism, and three flat (planar) mirrors (denoted as M_1, M_2, and M_3) arranged on the two optical paths or arms of the optical layout [148]. By adjusting the posing angles of the flat mirrors and the lens of the camera, two views of the object surface can pass through the optical bandpass filters via two different optical paths and then enter into the sensor target of the 3CCD camera with the aid of the beam splitter. As a result, an overlapped color image composed of one blue color image and one red color image is captured.

More details are shown in figure 12.25;: the imaging system is composed of a beam splitter prism, a blue (center wavelength: 450 ± 2 nm, full-width at half-maximum value: 30 nm), and a red (center wavelength: 630 ± 2 nm, full-width at half-maximum value: 30 nm) optical bandpass filters; three flat (planar) mirrors (denoted as M_1, M_2, and M_3); a 3CCD camera (AT-140CL, JAI Ltd, Japan) with a 1392 × 1040 pixel spatial resolution and a lens with a fixed focal length of 25 mm (LM25NC3, Kowa Ltd, Japan). Note that the two optical bandpass filters were attached on the left and front of the beam splitter prism. For the validation experiments, we picked two objects for the shape and deformation measurements, respectively: a cylinder with a nominal diameter of 65.80 mm (measured by a Vernier caliper with a resolution of 0.02 mm) and a circular plastic plate with a nominal diameter of 100 mm and a thickness of 1.96 mm (this object is shown on the right-hand side of figure 12.25). Random speckle patterns were produced on the three object surfaces by spraying white and black paints before the experiments.

Figure 12.26(a) shows the recorded color image of the specimen surface, which was then separated into the blue and red channel sub-images as indicated in figures 12.26(b) and (c). A rectangular ROI was first specified in the blue channel sub-image, and regularly spaced calculation points within the ROI were tracked in the red channel sub-image to determine their disparity data. By matching the blue and red channel sub-images using the stereo-DIC algorithm with a subset size of 41 × 41 pixels and a grid step of 5 pixels, the 3D coordinates of 47 × 80 points were measured based on the calibrated extrinsic and intrinsic parameters and the triangulation principle. The profile within the specified ROI was reconstructed as shown in figure 12.26(d), and the profile along the line AB was extracted as shown in figure 12.26(e). It can be observed that the reconstructed 3D coordinates within the

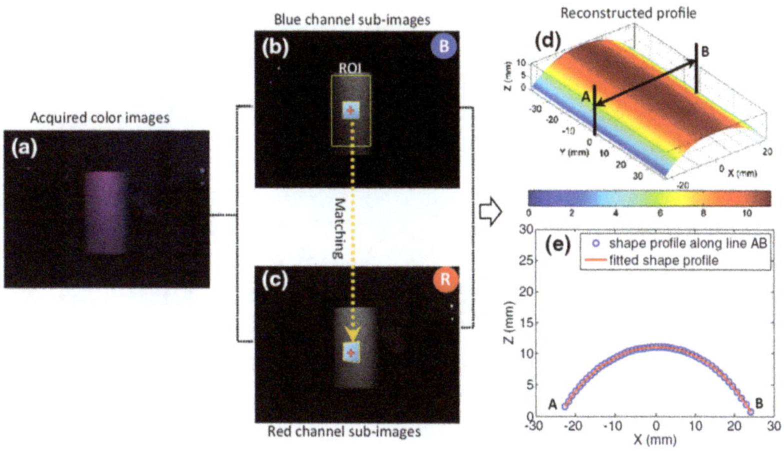

Figure 12.26. (a) The captured color image of the cylinder surface with the blue and red channel images extracted in (b) and (c), respectively; and reconstructed profile of the cylinder surface (d) corresponding to the profile specified by the ROI; (e) profile along with line AB.

ROI indicate a perfectly cylindrical surface, in which the fitted profile along with line AB closely agrees with the measured one. In addition, the radius of the cylinder surface was computed by fitting the obtained 3D coordinate data to obtain 65.34 mm. When compared with the physical size (65.80 mm) determined by the Vernier caliper, the relative error is less than 0.7%, confirming the accuracy of the single-camera color stereo-DIC system.

For the 3D deformation measurements of a centrally loaded circular plastic plate, a circular area located at the center of the blue channel sub-image of the initial state was adopted as the ROI. Comparing the blue and red channel sub-images of the initial and deformed states, we computed the 3D displacement components within the ROI with the approximate diameter of the ROI evaluated as 60 mm (as a reference, see the right-hand side of figure 12.26). Figure 12.27(a)–(c) show the determined 3D full-field u, v, and w components of the surface deformation of the centrally loaded circular plastic plate, with x- and y-axes lying in the plate plane and z-axis coming out of the plane of the page (outward). As shown in figure 12.27(a) and (b), the u displacement field presents an antisymmetric distribution with respect to the y-axis, while the v displacement field shows an antisymmetric distribution with respect to the x-axis. Meanwhile, the u and v displacement maps indicate zero displacements along the vertical and horizontal central axes.

Furthermore, the w displacement distribution presents a perfect concentric pattern with its center located at the point of the applied load. The maximum w displacement is measured as 6.56 mm, close to the applied displacement of 6.50 mm, obtained with the micrometer indenter. In short, these measured characteristic displacement distributions can be theoretically explained by the force boundary conditions and displacement boundary conditions, further verifying the effectiveness

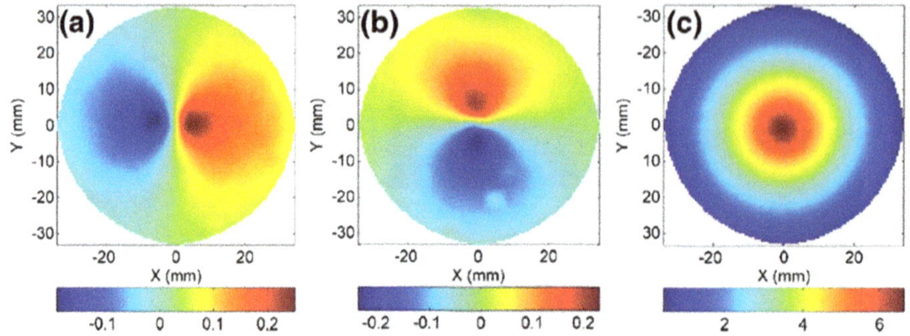

Figure 12.27. 3D deformation measurements of a circular plastic plate subjected to a central load normal to the specimen surface (at its back): contour plots of the displacement components (a) u, (b) v, and (c) w.

of the proposed single-camera color stereo-DIC technique for 3D deformation measurements.

The basic principles and implementation procedures described the 2D and 3D digital image correlation techniques for surface deformation and shape measurements. Representative applications of these techniques were also presented to demonstrate their effectiveness and practicality. With the development and popularity of the high-accuracy and easy-to-implement stereovision calibration technique, it is clear that 3D-DIC will attract many more applications for deformation and shape measurements of macroscopic objects and structures. However, 2D-DIC remains a simple, cost-effective yet practical tool for in-plane deformation measurement of nominally flat objects at various temporal and spatial scales. Notably, for observing and quantitatively measuring micro-scale deformations at reduced size, the 2D-DIC combined with a high-spatial-resolution imaging device (in particular SEM and AFM) will still be the most effective and irreplaceable tool.

Over the past thirty years, the DIC method has been in continuous improvement by many researchers around the globe and gradually evolved to become the most famous and popular tool in the field of experimental mechanics. Commercial DIC systems developed by various companies with friendly interfaces, powerful functions, and high efficiency are available. However, the ultimate objective of developing an intelligent DIC method for fully automatic, real-time, high-accuracy measurements remains, no doubt, a job for the young generations of scientists and engineers. Some critical challenges to this end involve, for example, automatic, adaptive, and optimal selection of various calculation parameters (in particular, a proper subset size and a displacement mapping function for displacement tracking, and an optimal strain calculation window for strain field estimation), and full-automatic, high-accuracy calibration of a stereovision system.

References

[1] Rastogi P K 2000 *PhotomechanicsTopics in Applied Physics* (Berlin: Springer)

[2] Schnars U and Jueptner W 2005 *Digital Holography* (Berlin: Springer)

[3] Post D, Ifju P and Han B T 1994 *High Sensitivity Moiré* (New York: Springer)

[4] Sirkis J S and Lim T J 1991 Displacement and strain-measurement with automated grid methods *Exp. Mech.* **31** 382–8
[5] Sutton M A, Orteu J J and Schreier H W 2009 *Image Correlation for Shape, Motion and Deformation Measurements* (Berlin: Springer)
[6] Pan B, Qian K M, Xie H M and Asundi A 2009 Two-dimensional digital image correlation for in-plane displacement and strain measurement: a review *Meas. Sci. Technol.* **20** 062001
[7] Pan B 2011 Recent progress in digital image correlation *Exp. Mech.* **51** 1223–35
[8] Peters W H and Ranson W F 1981 Digital imaging techniques in experimental stress analysis *Opt. Eng.* **21** 427–31
[9] Peters W H *et al* 1983 Application of digital correlation methods to rigid body mechancis *Opt. Eng.* **22** 738–42
[10] Chu T C *et al* 1985 Applications of digital-image-correlation techniques to experimental mechanics *Exp. Mech.* **25** 232–44
[11] Sutton M A, Yan J H, Tiwari V, Schreier W H and Orteu J J 2008 The effect of out-of-plane motion on 2D and 3D digital image correlation measurements *Opt. Lasers Eng.* **46** 746–57
[12] Pan B, Yu L P and Wu D F 2013 High-accuracy 2D digital image correlation measurements with bilateral telecentric lens: error analysis and experimental verification *Exp. Mech.* **53** 1719–33
[13] Luo P F *et al* 1993 Accurate measurement of three-dimensional displacement in deformable bodies using computer vision *Exp. Mech.* **33** 123–32
[14] Garcia D, Orteu J J and Penazzi L 2002 A combined temporal tracking and stereo-correlation technique for accurate measurement of 3D displacements: application to sheet metal forming *J. Mater. Process. Technol.* **125** 736–42
[15] Bay B K *et al* 1999 Digital volume correlation: three-dimensional strain mapping using x-ray tomography *Exp. Mech.* **39** 217–26
[16] Bay B K 2008 Methods and applications of digital volume correlation *J. Strain Anal. Eng. Des.* **43** 745–60
[17] Helm J D, McNeil S R and Sutton M A 1996 Improved three-dimensional image correlation for surface displacement measurement *Opt. Eng.* **35** 1911–20
[18] Pan B, Xie H M, Gao J X and Asundi A 2008 Improved speckle projection profilometry for out-of-plane shape measurement *Appl. Opt.* **47** 5527–33
[19] Sun Z L, Lyons J S and McNeill S 1997 Measuring microscopic deformations with digital image correlation *Opt. Lasers Eng.* **27** 409–28
[20] Zhang D S, Luo M and Arola D D 2006 Displacement/strain measurements using an optical microscope and digital image correlation *Opt. Eng.* **45** 033605
[21] Berfield T A, Patel H K, Shimmin R G, Braun P V, Lambros J and Sottos N R 2006 Fluorescent image correlation for nanoscale deformation measurements *Small* **2** 631–5
[22] Sabate N *et al* 2006 Measurement of residual stresses in micromachined structures in a microregion *Appl. Phys. Lett.* **88** 071910
[23] Kang J *et al* 2005 Microscopic strain mapping using scanning electron microscopy topography image correlation at large strain *J. Strain Anal. Eng. Des.* **40** 559–70
[24] Wang H *et al* 2006 Error analysis of digital speckle correlation method under scanning electron microscope *Exp. Tech.* **30** 42–5
[25] Sutton M A *et al* 2006 Metrology in a scanning electron microscope: theoretical developments and experimental validation *Meas. Sci. Technol.* **17** 2613–22

[26] Chasiotis I and Knauss W G 2002 A new microtensile tester for the study of MEMS materials with the aid of atomic force microscopy *Exp. Mech.* **42** 51–7
[27] Knauss W G, Chasiotis I and Huang Y 2003 Mechanical measurements at the micron and nanometer scales *Mech. Mater.* **35** 217–31
[28] Chang S *et al* 2005 Nanoscale in-plane displacement evaluation by AFM scanning and digital image correlation processing *Nanotechnology* **16** 344–9
[29] Li X D *et al* 2006 Nanoscale deformation and cracking studies of advanced metal evaporated magnetic tapes using atomic force microscopy and digital image correlation techniques *Meas. Sci. Technol.* **22** 835–44
[30] Siebert T *et al* 2007 High-speed digital image correlation: error estimations and applications *Opt. Eng.* **46** 051004
[31] Tiwari V, Sutton M A and McNeill S 2007 Assessment of high speed imaging systems for 2D and 3D deformation measurements: methodology development and validation *Exp. Mech.* **47** 561–79
[32] Pankow M *et al* 2010 Three-dimensional digital image correlation technique using single high-speed camera for measuring large out-of-plane displacements at high framing rates *Appl. Opt.* **49** 3418–27
[33] Reu P 2014 All about speckles: speckle size measurement *Exp. Tech.* **38** 1–2
[34] Reu P 2015 All about speckles: speckle density *Exp. Tech.* **39** 1–2
[35] Reu P 2015 All about speckles: contrast *Exp. Tech.* **39** 1–2
[36] Reu P 2014 All about speckles: aliasing *Exp. Tech.* **38** 1–3
[37] Reu P 2015 All about speckles: edge sharpness *Exp. Tech.* **39** 1–2
[38] Pan B, Lu Z X and Xie H M 2010 Mean intensity gradient: an effective global parameter for quality assessment of the speckle patterns used in digital image correlation *Opt. Lasers Eng.* **48** 469–77
[39] Hua T, Xie H, Wang S, Hu Z, Chen P and Zhang Q 2011 Evaluation of the quality of a speckle pattern in the digital image correlation method by mean subset fluctuation *Opt. Laser Technol.* **43** 9–13
[40] Stoilov G, Kavardzhikov V and Pashkouleva D 2012 A comparative study of random patterns for digital image correlation *J. Theor. Appl. Mech.* **42** 55–66
[41] Crammond G, Boyd S W and Dulieu-Barton J M 2013 Speckle pattern quality assessment for digital image correlation *Opt. Lasers Eng.* **51** 1368–74
[42] Bomarito G F, Hochhalter J D, Ruggles T J and Cannon A H 2017 Increasing accuracy and precision of digital image correlation through pattern optimization *Opt. Lasers Eng.* **91** 73–5
[43] Bomarito G F, Hochhalter J D and Ruggles T J 2018 Development of optimal multiscale patterns for digital image correlation via local grayscale variation *Exp. Mech.* **58** 1–12
[44] Dong Y L and Pan B 2017 A review of speckle pattern fabrication and assessment for digital image correlation *Exp. Mech.* **57** 1–21
[45] Yoneyama S, Kikuta H, Kitagawa A and Kitamura K 2006 Lens distortion correction for digital image correlation by measuring rigid body displacement *Opt. Eng.* **45** 023602
[46] Pan B, Yu L P, Wu D F and Tang L Q 2013 Systematic errors in two-dimensional digital image correlation due to lens distortion *Opt. Lasers Eng.* **51** 140–7
[47] Ma S P, Pang J Z and Ma Q W 2011 The systematic error in digital image correlation induced by self-heating of a digital camera *Meas. Sci. Technol.* **23** 025403

[48] Dong Y, Kakisawa H and Kagawa Y 2013 Optical system for microscopic observation and strain measurement at high temperature *Meas. Sci. Technol.* **25** 025002
[49] Berke R B and Lambros J 2014 Ultraviolet digital image correlation (UV-DIC) for high temperature applications *Rev. Sci. Instrum.* **85** 045121
[50] LePage W S, Daly S H and Shaw J A 2016 Cross polarization for improved digital image correlation *Exp. Mech.* **56** 969–85
[51] Pan B, Yu L P and Wu D F 2014 High-accuracy 2D digital image correlation measurements with low-cost imaging lenses: implementation of a generalized compensation method *Meas. Sci. Technol.* **25** 025001
[52] Schreier H W and Sutton M A 2002 Systematic errors in digital image correlation due to undermatched subset shape functions *Exp. Mech.* **42** 303–10
[53] Lu H and Cary P D 2000 Deformation measurement by digital image correlation: implementation of a second-order displacement gradient *Exp. Mech.* **40** 393–400
[54] Xu X, Su Y and Zhang Q 2017 Theoretical estimation of systematic errors in local deformation measurements using digital image correlation *Opt. Lasers Eng.* **88** 265–79
[55] Wang B and Pan B 2015 Random errors in digital image correlation due to matched or overmatched shape functions *Exp. Mech.* **55** 1717–27
[56] Yu L and Pan B 2015 The errors in digital image correlation due to overmatched shape functions *Meas. Sci. Technol.* **26** 045202
[57] Poissant J and Barthelat F 2010 A novel 'subset splitting' procedure for digital image correlation on discontinuous displacement fields *Exp. Mech.* **50** 353–64
[58] Valle V, Hedan S, Cosenza P, Fauchille A L and Berdjane M 2015 Digital image correlation development for the study of materials including multiple crossing cracks *Exp. Mech.* **55** 379–91
[59] Turner D Z 2014 Peridynamics-based digital image correlation algorithm suitable for cracks and other discontinuities *J. Eng. Mech.* **141** 04014115
[60] Hassan G M, Dyskin A V, MacNish C K, Pasternak E and Shufrin I 2018 Discontinuous digital image correlation to reconstruct displacement and strain fields with discontinuities: dislocation approach *Eng. Fract. Mech.* **89** 273–92
[61] Han J and Pan B 2018 A novel method for measuring discontinuous deformation in digital image correlation based on partition and dividing strategy *Eng. Fract. Mech.* **204** 185–97
[62] Broggiato G B, Casarotto L, Del Prete Z and Maccarrone D 2009 Full-field strain rate measurement by white-light speckle image correlation *Strain* **45** 364–72
[63] Wang X, Liu X, Zhu H and Ma S 2017 Spatial-temporal subset based digital image correlation considering the temporal continuity of deformation *Opt. Lasers Eng.* **90** 247–53
[64] Chi Y and Pan B 2018 Spatial-temporal digital image correlation: a unified framework *arxiv:* 1812.04826
[65] Pan B, Xie H M and Wang Z Y 2010 Equivalence of digital image correlation criteria for pattern matching *Appl. Opt.* **49** 5501–9
[66] Tong W 2005 An evaluation of digital image correlation criteria for strain mapping applications *Strain* **41** 167–75
[67] Pan B, Asundi A, Xie H M and Gao J X 2009 Digital image correlation using iterative least squares and pointwise least squares for displacement field and strain field measurements *Opt. Lasers Eng.* **47** 865–74

[68] Huang J Y, Zhu T, Pan X Y, Qin L, Peng X L, Xiong C Y and Fang J 2010 A high-efficiency digital image correlation method based on a fast recursive scheme *Meas. Sci. Technol.* **21** 035101

[69] Press W H 2003 *C++ Numerical Algorithms* (Beijing: Publishing House of Electronics Industry)

[70] Schreier H W, Braasch J R and Sutton M A 2000 Systematic errors in digital image correlation caused by intensity interpolation *Opt. Eng.* **39** 2915–21

[71] Pan B and Li K 2011 A fast digital image correlation method for deformation measurement *Opt. Lasers Eng.* **49** 841–7

[72] Luu L, Wang Z, Vo M, Hoang T and Ma J 2011 Accuracy enhancement of digital image correlation with B-spline interpolation *Opt. Lett.* **36** 3070–2

[73] Ren M, Liang J, Tang Z, Hu H, Guo X and Li L 2014 Optimized interpolation filter for digital image correlation methods *J. Xi'an Jiaotong Univ.* **48** 65–70

[74] Ren M, Liang J and Wei B 2016 Accurate B-spline-based 3-D interpolation scheme for digital volume correlation *Rev. Sci. Instrum.* **87** 125114

[75] Su Y, Zhang Q, Xu X, Gao Z and Wu S 2018 Interpolation bias for the inverse compositional Gauss–Newton algorithm in digital image correlation *Opt. Lasers Eng.* **100** 267–78

[76] Baldi A and Bertolino F 2015 Experimental analysis of the errors due to polynomial interpolation in digital image correlation *Strain* **51** 248–63

[77] Pan B 2013 Bias error reduction of digital image correlation using Gaussian pre-filtering *Opt. Lasers Eng.* **51** 1161–7

[78] Mazzoleni P, Matta F, Zappa E, Sutton M A and Cigada A 2015 Gaussian pre-filtering for uncertainty minimization in digital image correlation using numerically-designed speckle patterns *Opt. Lasers Eng.* **66** 19–33

[79] Zhou Y, Sun C, Song Y and Chen J 2015 Image pre-filtering for measurement error reduction in digital image correlation *Opt. Lasers Eng.* **65** 46–56

[80] Wang D, Jiang Y, Wang W and Wang Y 2016 Bias reduction in sub-pixel image registration based on the anti-symmetric feature *Meas. Sci. Technol.* **27** 035206

[81] Bruck H A, McNeil S R, Sutton M A and Peters W H 1989 Digital image correlation using Newton–Raphson method of partial differential correction *Exp. Mech.* **29** 261–7

[82] Vendroux G and Knauss W G 1998 Submicron deformation field measurements: part 2. Improved digital image correlation *Exp. Mech.* **38** 86–92

[83] Pan B, Xie H M, Xu B Q and Dai F L 2006 Performance of sub-pixel registration algorithms in digital image correlation *Meas. Sci. Technol.* **17** 1615–21

[84] Sun Y *et al* 2005 Finite element formulation for a digital image correlation method *Appl. Opt.* **44** 7357–63

[85] Besnard G, Hild F and Roux S 2006 Finite-element displacement fields analysis from digital images: application to Portevin–Le Châtelier bands *Exp. Mech.* **46** 789–803

[86] Ma S, Zhao Z and Wang X 2012 Mesh-based digital image correlation method using higher order isoparametric elements *J. Strain Anal. Eng. Des.* **47** 163–75

[87] Pan B, Li K and Tong W 2013 Fast, robust and accurate digital image correlation calculation without redundant computations *Exp. Mech.* **53** 1277–89

[88] Gao Y, Cheng T, Su Y, Xu X, Zhang Y and Zhang Q 2015 High-efficiency and high-accuracy digital image correlation for three-dimensional measurement *Opt. Lasers Eng.* **65** 73–80

[89] Shao X, Dai X and He X 2015 Noise robustness and parallel computation of the inverse compositional Gauss–Newton algorithm in digital image correlation *Opt. Lasers Eng.* **71** 9–19
[90] Pan B and Wang B 2016 Digital image correlation with enhanced accuracy and efficiency: a comparison of two subpixel registration algorithms *Exp. Mech.* **56** 1395–409
[91] Pan B 2014 An evaluation of convergence criteria for digital image correlation using inverse compositional Gauss–Newton algorithm *Strain* **50** 48–56
[92] Jiang Z, Kemao Q, Miao H, Yang J and Tang L 2015 Path-independent digital image correlation with high accuracy, speed and robustness *Opt. Lasers Eng.* **65** 93–102
[93] Zhang L *et al* 2015 High accuracy digital image correlation powered by GPU-based parallel computing *Opt. Lasers Eng.* **69** 7–12
[94] Pan B 2009 Reliability-guided digital image correlation for image deformation measurement *Appl. Opt.* **48** 1535–42
[95] Pan B and Tian L 2015 Superfast robust digital image correlation analysis with parallel computing *Opt. Eng.* **54** 034106
[96] Pan B, Wang Z Y and Lu Z X 2010 Genuine full-field deformation measurement of an object with complex shape using reliability-guided digital image correlation *Opt. Express* **18** 1011–23
[97] Zhang Z F *et al* 2006 A novel coarse-fine search scheme for digital image correlation method *Measurement* **39** 710–8
[98] Chen D J, Chiang F P, Tan Y S and Don H S 1993 Digital speckle-displacement measurement using a complex spectrum method *Appl. Opt.* **32** 1839–49
[99] Zhao J Q, Zeng P, Lei L P and Ma Y 2012 Initial guess by improved population-based intelligent algorithms for large inter-frame deformation measurement using digital image correlation *Opt. Lasers Eng.* **50** 473–90
[100] Zhou Y, Pan B and Chen Y Q 2012 Large deformation measurement using digital image correlation: a fully automated approach *Appl. Opt.* **51** 7674–83
[101] Wang Z, Vo M, Kieu H and Pan T 2014 Automated fast initial guess in digital image correlation *Strain* **50** 28–36
[102] Pan B, Wang Y and Tian L 2017 Automated initial guess in digital image correlation aided by Fourier–Mellin transform *Opt. Eng.* **56** 014103
[103] Pan B, Wu D F and Xia Y 2012 Incremental calculation for large deformation measurement using reliability-guided digital image correlation *Opt. Lasers Eng.* **50** 586–92
[104] Zhou Y H and Chen Y Q 2012 Propagation function for accurate initialization and efficiency enhancement of digital image correlation *Opt. Lasers Eng.* **50** 1789–97
[105] Pan B, Xie H M, Guo Z Q and Hua T 2007 Full-field strain measurement using a two-dimensional Savitzky–Golay digital differentiator in digital image correlation *Opt. Eng.* **46** 033601
[106] Sutton M A, Turner J L, Bruck H A and Chao T A 1991 Full-field representation of discretely sampled surface deformation for displacement and strain analysis *Exp. Mech.* **31** 168–77
[107] Meng L B, Jin G C and Yao X F 2007 Application of iteration and finite element smoothing technique for displacement and strain measurement of digital speckle correlation *Opt. Lasers Eng.* **45** 57–63
[108] Yoneyama S 2011 Smoothing measured displacements and computing strains utilising finite element method *Strain* **47** 258–66

[109] Tong W 1997 Detection of plastic deformation patterns in a binary aluminum alloy *Exp. Mech.* **37** 452–9

[110] Wang C C B *et al* 2002 An automated approach for direct measurement of two-dimensional strain distributions within articular cartilage under unconfined compression *J. Biomech. Eng.* **124** 557–67

[111] Xiang G F *et al* 2007 Time-resolved deformation measurements of the Portevin–Le Chatelier bands *Scr. Mater.* **56** 721–4

[112] Han Y G, Kim D W and Kwon H J 2012 Application of digital image cross-correlation and smoothing function to the diagnosis of breast cancer *J. Mech. Behav. Biomed.* **14** 7–18

[113] Wattrisse B C, Muracciole A and Nemoz-Gaillard J M 2001 Analysis of strain localization during tensile tests by digital image correlation *Exp. Mech.* **41** 29–39

[114] Pan B and Tian L 2016 Advanced video extensometer for non-contact, real-time, high-accuracy strain measurement *Opt. Express* **24** 19082–93

[115] Lv Z, Xu X, Yan T, Cai Y, Su Y and Zhang Q 2018 High-accuracy optical extensometer based on coordinate transform in two-dimensional digital image correlation *Opt. Lasers Eng.* **100** 61–70

[116] Tian L, Yu L P and Pan B 2018 Accuracy-enhanced video extensometer with real-time strain error compensation *Opt. Lasers Eng.* **110** 272–8

[117] Hoult N A, Take W A, Lee C and Dutton M 2013 Experimental accuracy of two dimensional strain measurements using digital image correlation *Eng. Struct.* **46** 718–26

[118] Zhu F, Bai P, Shi H, Jiang Z, Lei D and He X 2016 Enhancement of strain measurement accuracy using optical extensometer by application of dual-reflector imaging *Meas. Sci. Technol.* **27** 065007

[119] Zhu F *et al* 2017 High-accuracy biaxial optical extensometer based on 2D digital image correlation *Meas. Sci. Technol.* **28** 085006

[120] Zhu F, Bai P, Gong Y, Lei D and He X 2018 Accurate measurement of elastic modulus of specimen with initial bending using two-dimensional DIC and dual-reflector imaging technique *Measurement* **119** 18–27

[121] Schreier H W, Garcia D and Sutton M A 2004 Advances in light microscope stereo vision *Exp. Mech.* **44** 278–88

[122] Hu Z X, Luo H Y, Du Y J and Lu H B 2013 Fluorescent stereo microscopy for 3D surface profilometry and deformation mapping *Opt. Express* **20** 11808–18

[123] Pankow M, Justusson B and Waas A M 2010 Three-dimensional digital image correlation technique using single high-speed camera for measuring large out-of-plane displacements at high framing rates *Appl. Opt.* **49** 3418–27

[124] Genovese K, Casaletto L, Rayas J A, Flores V and Martinez A 2013 Stereo-digital image correlation (DIC) measurements with a single camera using a biprism *Opt. Lasers Eng.* **51** 279–85

[125] Xia S, Gdoutou A and Ravichandran G 2013 Diffraction assisted image correlation: a novel method for measuring three-dimensional deformation using two-dimensional digital image correlation *Exp. Mech.* **53** 1–11

[126] Pan B and Wang Q 2013 Single-camera microscopic stereo digital image correlation using a diffraction grating *Opt. Express* **21** 25056–68

[127] Pan B, Wu D and Yu L 2012 Optimization of a three-dimensional digital image correlation system for deformation measurements in extreme environments *Appl. Opt.* **51** 4409–19

[128] Yan T, Su Y and Zhang Q 2018 Precise 3D shape measurement of three-dimensional digital image correlation for complex surfaces *Sci. China Technol. Sci.* **61** 68–73

[129] Nguyen H, Wang Z, Jones P and Zhao B 2017 3D shape, deformation, and vibration measurements using infrared Kinect sensors and digital image correlation *Appl. Opt.* **56** 9030

[130] Chen Z, Liao H and Zhang X 2014 Telecentric stereo micro-vision system: calibration method and experiments *Opt. Lasers Eng.* **57** 82–92

[131] Hu Z *et al* 2013 Fluorescent stereo microscopy for 3D surface profilometry and deformation mapping *Opt. Express* **21** 11808–18

[132] Hu Z, Xu T, Luo H, Gan R Z and Lu H 2016 Measurement of thickness and profile of a transparent material using fluorescent stereo microscopy *Opt. Express* **24** 29822

[133] Zhu T, Sutton M A, Li N, Orteu J J, Cornille N, Li X and Reynolds A P 2011 Quantitative stereovision in a scanning electron microscope *Exp. Mech.* **51** 97–109

[134] Gupta S, Parameswaran V, Sutton M A and Shukla A 2014 Study of dynamic underwater implosion mechanics using digital image correlation *Proc. R. Soc.* A **470** 20140576

[135] Leblanc J and Shukla A 2011 Dynamic response of curved composite panels to underwater explosive loading: experimental and computational comparisons *Compos. Struct.* **93** 3072–81

[136] Spranghers K *et al* 2012 Full-field deformation measurements of aluminum plates under free air blast loading *Exp. Mech.* **52** 1371–84

[137] Spranghers K, Vasilakos I, Lecompte D, Sol H and Vantomme J 2013 Numerical simulation and experimental validation of the dynamic response of aluminum plates under free air explosions *Int. J. Impact Eng.* **54** 83–95

[138] Spranghers K, Vasilakos I, Lecompte D, Sol H and Vantomme J 2014 Identification of the plastic behavior of aluminum plates under free air explosions using inverse methods and full-field measurements *Int. J. Solids Struct.* **51** 210–26

[139] Pankow M, Justusson B, Salvi A, Waas A M, Yen C and Ghiorse S 2011 Shock response of 3D woven composites: an experimental investigation *Compos. Struct.* **93** 1337–46

[140] Kumar P, Leblanc J, Stargel D S and Shukla A 2012 Effect of plate curvature on blast response of aluminum panels *Int. J. Impact Eng.* **46** 74–85

[141] Koohbor B, Kidane A, Lu W Y and Sutton M A 2016 Investigation of the dynamic stress–strain response of compressible polymeric foam using a non-parametric analysis *Int. J. Impact Eng.* **91** 170–82

[142] Xia S, Gdoutou A and Ravichandran G 2016 Diffraction assisted image correlation: a novel method for measuring three-dimensional deformation using two-dimensional digital image correlation *Exp. Mech.* **53** 755–65

[143] Pan B, Ma L J and Xia Y 2016 A new technique for measuring 3D deformation of single lap joint *Sci. China: Phys. Mech. Astron.* **88** 528–36

[144] Wu L, Zhu J and Xie H 2015 Single-lens 3D digital image correlation system based on a bilateral telecentric lens and a bi-prism: validation and application *Appl. Opt.* **54** 7842–50

[145] Wu L F *et al* 2018 Bi-prism-based single-lens three dimensional digital image correlation system with a long working distance: methodology and application in extreme high temperature deformation test *Sci. China Technol. Sci.* **61** 37–50

[146] Yu L P and Pan B 2016 Single-camera stereo digital image correlation system using a four-mirror adapter: optimized design and validation *Opt. Lasers Eng.* **87** 120–8

[147] Yu L and Bing P 2016 Structure parameter analysis and uncertainty evaluation for single-camera stereo-digital image correlation with a four-mirror adapter *Appl. Opt.* **55** 6936–46

[148] Yu L and Pan B 2017 Color stereo-digital image correlation method using a single 3CCD color camera *Exp. Mech.* **57** 649–57

[149] Yu L and Pan B 2017 Full-frame, high-speed 3D shape and deformation measurements using stereo-digital image correlation and a single color high-speed camera *Opt. Lasers Eng.* **95** 17–25

[150] Pan B, Yu L P and Zhang Q B 2018 Review of single-camera stereo-digital image correlation techniques for full-field 3D shape and deformation measurement *Sci. China Technol. Sci.* **61** 1–20

[151] Zhang Z Y 2000 A flexible new technique for camera calibration *IEEE Trans. Pattern Anal. Mach. Intell.* **22** 1330–4

[152] Xue T *et al* 2007 Complete calibration of a structure-uniform stereovision sensor with free-position planar pattern *Sensors Actuators* A **135** 185–91

[153] Hartley 1994 An algorithm for self calibration from several views *Proc. IEEE Conf. Comput. Vis. Pattern Recognit. CVPR-94* pp 908–12

[154] Maybank S J and Faugeras O D 1992 A theory of self-calibration of a moving camera *Int. J. Comput. Vis.* **8** 123–51

[155] Luong Q-T and Faugeras O D 1997 Self-calibration of a moving camera from point correspondences and fundamental matrices *Int. J. Comput. Vis.* **22** 261–89

[156] Abdel-Aziz Y I 1971 Direct linear transformation from comparator coordinates in close-range photogrammetry *Proc. American Society of Photogrammetry Symp. on Close-Range Photogrammetry (Falls Church, VA)* pp 1–19

[157] Tsai R Y and Versatile A 1987 Camera calibration technique for high-accuracy 3D machine vision metrology using off-the-shelf TV cameras and lenses *IEEE J. Robot. Autom.* **3** 323–44

[158] Zhang H, Zhang L Y, Chen J and Zhao Z P 2007 Field calibration of binocular stereo system based on planar template and free snapping *Acta Aeronaut. Astronaut. Sin.* **28** 695–701

[159] Reu P L 2013 A study of the influence of calibration uncertainty on the global uncertainty for digital image correlation using a Monte Carlo approach *Exp. Mech.* **53** 1661–80

[160] Vo M, Wang Z, Luu L and Ma J 2011 Advanced geometric camera calibration for machine vision *Opt. Eng.* **50** 110503

[161] Shao X, Dai X, Chen Z, Dai Y, Dong S and He X 2016 Calibration of stereo-digital image correlation for deformation measurement of large engineering components *Meas. Sci. Technol.* **27** 125010

[162] Tang Z, Liang J, Xiao Z and Guo C 2012 Large deformation measurement scheme for 3D digital image correlation method *Opt. Lasers Eng.* **50** 122–30

[163] Zhou Y, Sun C and Chen J 2014 Adaptive subset offset for systematic error reduction in incremental digital image correlation *Opt. Lasers Eng.* **55** 5–11

IOP Publishing

Full Field Optical Metrology and Applications

Fernando Mendoza-Santoyo, Manuel De la Torre-Ibarra, María del Socorro Hernández-Montes and Jorge Mauricio Flores Moreno

Chapter 13

Speckle-noise reduction in optical techniques by non-local means methods

Contributing authors: Manoj Kumar, Yassine Tounsi, Karmjit Kaur, Abdelkrim Nassim and Matoba Matoba

Within optical metrology, optical interferometric techniques are important non-invasive (or non-contact) research tools used in several fields—they are multi- and inter-disciplinary. These techniques encode the useful information of the object under study in the form of fringe patterns. However, due to speckle-noise the fringe patterns' quality is severely degraded and it may introduce errors in the measurements and, therefore, restrict their applications. Since the beginning of these techniques and to overcome the inherent speckle-noise, various methods have been proposed, including image-processing based algorithms, modified configurations of the set-ups, the use of low-coherence light sources or acquisitions of multiple object images (i.e. holograms), etc, and which one is chosen depends on the requirements and the understanding of the noise characteristics. In this chapter we will deal with image-processing based speckle-noise reduction methods based on non-local means (NLM). For this purpose, the performance of the NLM and its related adaptive kernel based filtering methods for speckle-noise reduction are presented in this chapter, emphasizing its application in fringe patterns obtained from digital speckle pattern interferometry (DSPI). In addition, methods based on NLM are validated, and their performance is evaluated on simulation data.

13.1 Introduction

Optical interferometric techniques keep on attracting increasing interest in scientific and technological research and development in many areas, mainly due to the continuous development in analog and digital electronics, electronic imaging devices (CCD and CMOS cameras, and the like), high-speed digital

doi:10.1088/978-0-7503-3027-5ch13

computers with growing capacities for data storage, and reliable image-processing software. These techniques have become automated, effective, and more user-friendly, and thus have found an enormous number of scientific, engineering, and industrial applications [1–29]. The interferometric techniques, including digital holography, speckle interferometry, shearography, DSPI, etc, encode diverse types of information from the object under study, and this data comes in the form of fringe patterns. However, the fringe patterns carry high-frequency speckle-noise due to the coherence of the light source used as the object illumination source. The presence of undesirable bright speckles in dark fringes (and vice versa) may lead to errors inherently introduced in the measurements. As seen in previous chapters, the speckle-noise is generated due to the self-interference of the scattered waves when coherent light waves pass through a randomly fluctuating medium or are reflected from a rough surface that consists of independent scattering areas [19]. Therefore, it is always desirable (required) to solve this speckle-noise in the recorded fringes before data processing, leading to the desired measurable quantity.

13.1.1 Basic principles of speckle denoising

Let us begin with one relationship that mathematically represents the problem of denoising:

$$I_n(x, y) = I(x, y)*n(x, y). \quad (13.1)$$

Here, $I_n(x, y)$ is the observed speckled image, $I(x, y)$ is the clean image, and $n(x, y)$ is the speckle-noise distribution. The objective of the denoising methods is to accurately estimate $I(x, y)$ given $I_n(x, y)$, with image details and edge preservation, as shown in figure 13.1.

Several attempts [30–39] that depend on the requirements and understanding of the noise characteristics have been made for a speckle-noise reduction in optical interferometric techniques, and the work is well in progress in this direction, i.e. it is an active research field. One very efficient method is to eliminate the speckle-noise as much as possible while preserving the fringe edges, which contain useful information.

The speckle-noise reduction methods are broadly classified into three groups [39]: (i) low-coherence and incoherent imaging systems; (ii) temporal integration by multiplexing holograms; and (iii) image-processing based speckle-noise reduction

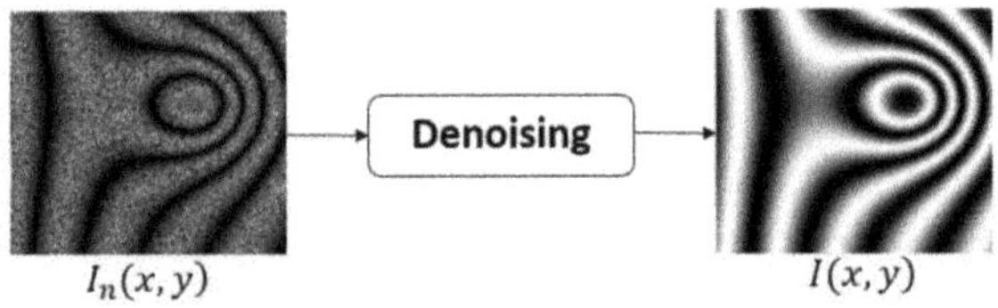

Figure 13.1. Schematic of denoising.

algorithms. Furthermore, the image-processing based denoising methods can be classified into the following four categories [39]:

1. Spatial-domain denoising methods.
2. Transform-domain denoising methods.
3. Learning based denoising methods.
4. Hybrid methods.

The first three image-processing based speckle-noise reduction algorithms will be presented in the following.

13.1.2 Spatial-domain denoising methods

The spatial-domain methods exploit the correlation among the pixels/patches in the noisy image to eliminate the noise by implementing an appropriate set of operations on the image matrix [39]. In a fringe pattern, the noise is accumulated randomly with a higher frequency, whereas the lower frequency is associated with the useful information which occurs repetitively. Therefore, the high-frequency components need to be removed, and this process can easily be achieved. However, the other important information, such as the object features and fringe edges, also carry higher frequency, and they are non-repetitive. Therefore, removing higher frequency components results in the blurring of useful feature information. The spatial-domain denoising methods help eliminate the higher frequency noise and not the higher frequency signal associated with useful information about the fringe patterns. Additionally, considering the large window size of filters and the use of the convolution process, these methods carry more computational complexity [39].

The spatial filters are classified into two categories: local and non-local filters. A filter is local if the spatial distance restricts the candidate selection process used for filtering. A filter is non-local if the candidate selection depends only on the similarity and is not restricted by the spatial distance. In local methods, the fringe pattern is denoised by taking the mean value of a group of pixels defined by the user, such as a window of 3×3 pixels surrounding the target pixel. In these methods, the spatial neighborhood of the target pixel and the filter coefficients are restricted by the spatial distance. Based on this concept, several local algorithms for noise reduction have been developed: mean filter, Gaussian filter, Wiener filter, least mean squares filter, bilateral filter, anisotropic filtering, etc. Figure 13.2 shows the schematic working principle of a local denoising method, the median filter, on an image. The local denoising methods remove the speckle-noise reasonably and are effective in terms of time computing complexity. However, these methods do not perform well when the noise level is high because the severe noise corrupts the correlations between neighboring pixels. Also, these methods fail to preserve the fringe edges.

The most popular and classical local filters exploited for image denoising are the Gaussian filter, least mean square filters, bilateral filter, Weiner filter, SUSAN filter, median filter, anisotropic diffusion filter, homomorphic filter, rank filter, steering kernel regression (SKR), metric steering kernel regression, and trained filters.

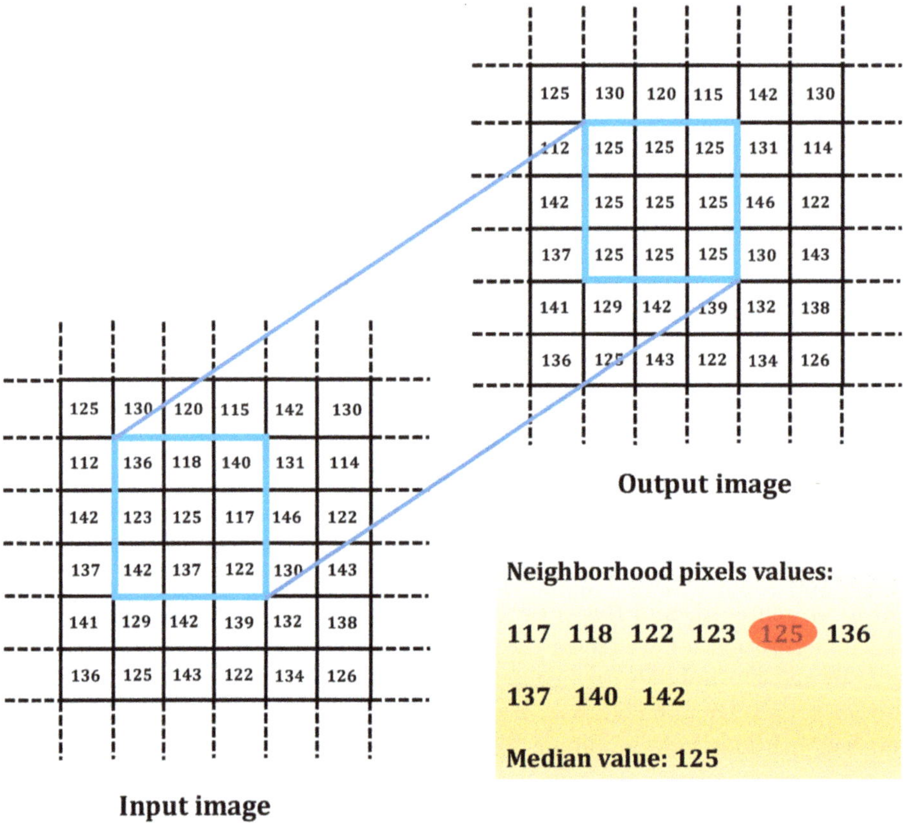

Figure 13.2. Working principle of the median filter on an image.

The local filter class is also divided into two families: linear and non-linear filters. Only some of these filters can reduce this multiplicative noise for speckle-noise reduction. Let us look at some.

13.1.2.1 The Weiner filter

The Weiner filter optimizes the overall mean square error (MSE) in filtering and noise smoothing. This filter is a linear estimation of the original noised image. In the Fourier domain, the Weiner filter is defined as

$$W(u_1, u_2) = \frac{H^*(u_1, u_2)S_o(u_1, u_2)}{|\,H(u_1, u_2)|^2\; S_o(u_1, u_2) + S_n(u_1, u_2)}, \tag{13.2}$$

where $H(u_1, u_2)$ is the spreading function that can be chosen to be unitary; while $S_o(u_1, u_2)$ and $S_n(u_1, u_2)$ are the estimated power spectra densities of the original image and the noise, respectively. Then, a median filter is applied to the noised image to obtain the original image. When the median filter is used, the noise estimation is calculated by the difference between the raw noise image and the filtered image. In particular, as an example, this filter was used to denoise ultrasound images [40] and to successfully reduce speckle-noise from speckled fringes [41].

13.1.2.2 The Gaussian filter

The Gaussian filter is a local linear filter and an anisotropic filter. This filter assumes that the neighborhood pixels are homogeneous and have smooth spatial variations. The well-known Gaussian kernel expression is defined as

$$G(x, y) = \frac{1}{2\pi\sigma^2} e^{-\frac{x^2+y^2}{2\sigma^2}}. \tag{13.3}$$

The principle of image denoising by Gaussian smoothing uses the two-dimensional distribution as a point spread function. The denoised output image is obtained by the results of the convolution product between the noised image and the above Gaussian kernel.

The majority of the local filters, including the Gaussian filter, do not preserve information at the edges of the denoised images (blurring edges). Now, let us look at some methods that overcome this blurring—they preserve edges, image details, and geometry.

13.1.2.3 Conventional anisotropic diffusion filter

Perona and Malik [42] introduced a non-linear diffusion method for image denoising. It is based on a diffusion process governed by a partial differential equation (PDE), where the diffusion amount depends on the image gradients. With this method, an image can be selectively smoothed, which gives good edge preservation. They apply an inhomogeneous process that reduces diffusivity at locations with a more significant likelihood of being an edge. This method is based on the following equation:

$$\frac{\partial f}{\partial t} = \mathrm{div}(g(|\nabla f(x, y, t)|^2)\nabla f(x, y, t)), \tag{13.4}$$

where f is the image obtained at an instant t, div is the divergence operator, ∇ is the gradient operator, and g is the diffusion coefficient, also called the edge-stopping function. It is important to mention that the diffusion coefficient $g(\nabla f)$ is a local image gradient. The magnitudes of the amount of diffusion and edges (gradient value) are inversely proportional. Within homogeneous regions, the gradient value is small; for this reason, the diffusion coefficient $g(\nabla f)$ approaches 1, hence smoothening or removing the uniform region, whereas in the case of near edges or boundaries $g(\nabla f)$ tends towards zero and then preserves edges [43].

The diffusion coefficient can take either of the two following forms:

$$g(\nabla f) = e^{-(\nabla f/r)^2} \tag{13.5}$$

or

$$g(\nabla f) = \frac{1}{1 + (\nabla f/r)^2}, \tag{13.6}$$

where r is the contrast parameter or gradient modulus threshold, which controls the amount of denoising.

13.1.2.4 Bilateral filter and weighted bilateral filter

Another filter was introduced for edge-preserving and image smoothing. It is called a bilateral filter (BF) [44]. The bilateral filter of a given image *f* is defined as

$$\mathrm{BF}[f]_{p_1} = \frac{1}{W_{p_1}} \sum_{p_2 \in S} G_{\sigma_s}(\| p_1 - p_2 \|) G_{\sigma_r}(f_{p_1} - f_{p_2}) f_{p_1}, \tag{13.7}$$

where W_{p1} is a normalization factor

$$\frac{1}{W_{p_1}} \sum_{p_2 \in S} G_{\sigma_s}(\| p_1 - p_2 \|) G_{\sigma_r}(f_{p_1} - f_{p_2}). \tag{13.8}$$

The parameters σ_s and σ_r are the standard deviations of the bilateral filter and represent, respectively, the spatial-domain standard deviation and the intensity domain standard deviation. The two parameters σ_s and σ_r determine the performance of the BF; σ_r controls the edge maintaining the ability and the smoothness. The lower the values for σ_r, the better the image edge preservation, whereas larger values for σ_s render the best degree of image smoothness. *G* represents the Gaussian kernel, and $G_{\sigma s}$ and $G_{\sigma r}$ represent a spatial-domain Gaussian that decreases the influence of distant pixels and a range Gaussian that decreases the influence of pixel f_{p1} with an intensity value different to f_{p2}, respectively.

The difference between the Gaussian filter and the BF is that the BF considers the variation of intensities to preserve edges. Later, the same authors presented a modified version of the BF called optimally weighted bilateral filters [45]. This version is introduced to make this filter robust at significant noise levels.

13.1.3 Transform-domain denoising methods

Transform-domain denoising methods exploit the property of sparsity, i.e. the signal can be represented by some number of non-zero coefficients, or the fringe pattern is represented as a linear expansion of high-valued coefficients. These methods are highly effective for speckle-noise reduction. The reason is that the speckle-noise is multiplicative and it is difficult to separate the noise components from the original fringe pattern. Therefore, a transform is applied to the fringe pattern and the speckle-noise appears in the form of additive noise. After that, a suitable denoising process is employed to remove the noise components and preserve the details of the fringe pattern. The transform-domain denoising methods are dependent on several parameters, which are selected based on the superior performance of the denoising method by trial. There is no explicit formula or scheme to determine these parameters. The selection and adjustment of the right parameters for the desired denoising results make these methods time-consuming and inconvenient for dynamic measurements or real-time applications, where typically large amounts of data are recorded.

Denoising in the transform-domain is based on two principal transforms, the Fourier and wavelet transform. The Fourier transform has been found to be effective for speckle-noise reduction [31, 46]. However, fringe pattern details may also be filtered. This limitation can be overcome by using the wavelet transform that was first introduced for signal processing. These functions provide a suitable representation of the local fringe pattern structure: they are obtained by translating, dilating, and orienting a mother wavelet transform. Several research papers demonstrate the performance of wavelet transform based methods for speckle denoising [32, 38, 47, 48]. The principle of the wavelet transform for speckle denoising is based on the decomposition of the fringe pattern into high and low-frequency components called wavelet coefficients, where high-frequency coefficients are called *details* and low-frequency coefficients are called *approximations.* Consequently, a thresholding method should be applied to remove the spectral contribution of the speckle-noise that is highly presented in detail. Finally, after the appropriate thresholding operation, an inverse wavelet transform is applied to reconstruct the coefficients and provide a denoised fringe pattern. This process is summarized in figure 13.3.

Different types of wavelet-basis functions have been used for fringe pattern denoising. We recommend reading a detailed study by Federico and Kaufmann that presents a comparative analysis of the performance of these different basis functions for speckle removal [32]. Also, Kemao proposed an efficient and elegant technique for fringe pattern denoising [49]. It is worth mentioning that the windowed Fourier transform (WFT) approach is similar to that of the wavelet transform thresholding method, i.e. an appropriate thresholding technique is applied to the obtained WFT coefficients of the speckled image in order to eliminate the spectral contribution of the speckle-noise.

The expression of the WFT basis function is

$$h_{u_1, u_2, \mu, \eta}(x, y) = h(x - u_1, y - u_2)\exp[\mathrm{i}(\mu x + \eta y)]. \tag{13.9}$$

The WFT of a given image intensity distribution is computed analytically as

$$\mathrm{WFT}_{u_1, u_2, \mu, \eta}(I) = \int_{-\infty}^{+\infty}\int_{-\infty}^{-\infty} I(x, y)h^{*}_{u_1, u_2, \mu, \eta}(x, y)\,\mathrm{d}x\,\mathrm{d}y. \tag{13.10}$$

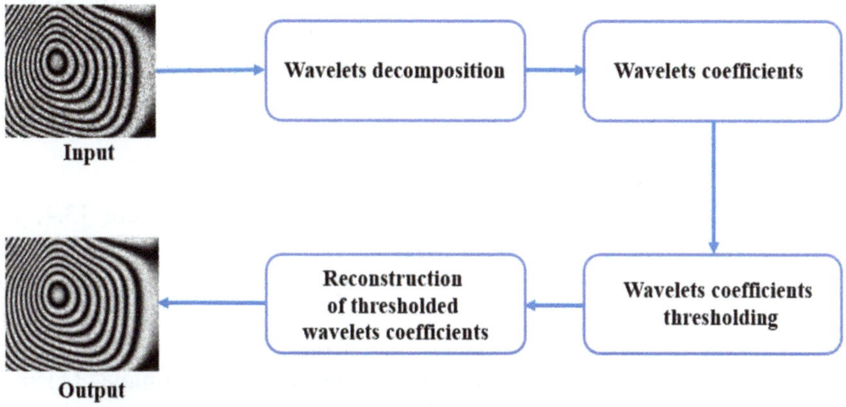

Figure 13.3. Principle of speckle denoising by the wavelet transform method.

As described previously, for the wavelet based methods, the obtained WFT coefficients undergo an appropriate thresholding value, where this value depends on the noise strength; the criteria to select this value are provided in [50]. Finally, an inverse WFT transform of the thresholded WFT coefficients is computed to obtain the denoised image. This algorithm has shown great results for speckle fringe patterns and phase map denoising [49–51].

13.1.4 Learning based denoising methods

Deep learning is based on the idea of artificial neural networks, which can handle large amounts of data by adding layers to the network. A deep-learning model can extract features from the raw data through multiple processing layers composed of multiple linear and non-linear transformations and learn about these features bit-by-bit through each layer with minimal human intervention. Deep-learning based approaches for noise removal have made significant breakthroughs, have been found to be more suitable, and continue progressing, so there are huge possibilities for further developments in the future and this is indeed an exciting area of research to pursue.

Apart from the methods presented, other image denoising methods, which are either based on the statistical domain or the combination of different domains, have been developed, for example, the probability based Bayesian model, Bayesian least squares estimator [52], total variational (TV) method [53], etc.

Next, the non-local means and related adaptive kernel based filtering methods for speckle-noise reduction will be presented.

13.2 Non-local means methods

The non-local denoising methods are based on the concept of self-similarity, i.e. exploiting the similar patches with similar structures in the entire image. These similar patches are estimated by creating the appropriate statistics from the similarly structured patches located all over the image (fringe pattern) [36, 37]. Then these patches are denoised by applying an operation. For example, in the NLM method, a weighted average is employed for all the similarity patches to reduce the noise in the entire image. These methods are not restricted by spatial distance. The NLM using adaptive kernel based methods has been proposed to reduce speckle-noise in digital speckle pattern interferometry and digital holography [36, 37].

13.2.1 Non-local means method

An example illustrates the concept of the NLM method in figure 13.4. Let us consider the square patch p, the window of interest, and the other square patches with the same dimensions, say q_1, q_2, and q_3, representing the similarity windows. Denoising of the central pixel of the patch p is realized by replacing it with the weighted average of surrounding pixel values from the pixels located in a more extensive search window, where the search window is the entire image. This process is carried out by dividing the search window into similarity windows such as q_1, q_2,

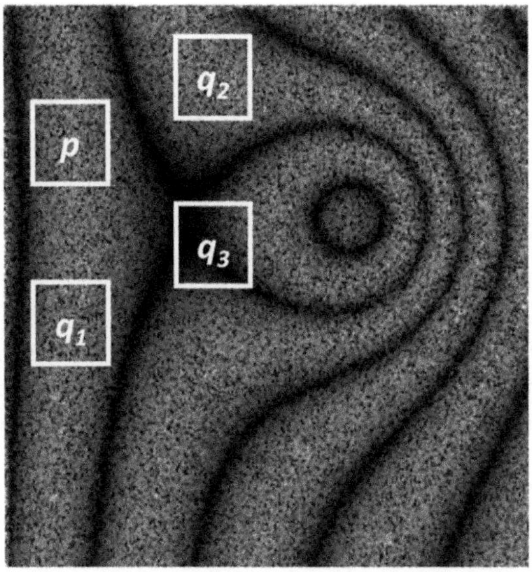

Figure 13.4. Illustration of the NLM denoising method [36].

and q_3. Then, the similarity window is compared to the window of interest to generate a weight, representing the similarity between these two windows. The weight is used to average the pixel values at the center of the similarity window with those in the center of the window of interest [36, 37]. A mathematical description of this method is explained fully in [36].

Each pixel of the noised image is denoised with the NLM method by computing the following formula [54]:

$$I_{\text{out}}(x_i, y_i) = \sum_{x_i, y_i \in \Omega} w(x_j, y_j).\ I(x_i, y_i), \tag{13.11}$$

where $I_{\text{out}}(x_i, y_i)$ is the denoised pixel value and $I(x_i, y_i)$ is the gray level of the pixel in the similarity window. $w(x_j, y_j)$ is the weight of the pixels in the similarity window centered around (x_i, y_i), which is calculated as [54]

$$w(x_j, y_j) = (1/Z_i) \cdot \exp\left[\frac{\left\| I(N_i) - I(N_j) \right\|_{2,\alpha}^2}{h^2}\right], \tag{13.12}$$

where $I(N_i)$ and $I(N_j)$ represent the gray level of the pixels in the window of interest and in the similarity windows, respectively. The expression $\left\| I(N_i) - I(N_j) \right\|_{2,\alpha}^2$ represents the Euclidean distance (gray level) between the similarity window and the window of interest, convolved with a Gaussian kernel with parameter α. The parameter Z_i is a normalizing factor used to ensure that the sum of the weights is equal to 1, and h is a parameter controlling the filter size.

Several parameters are responsible for evaluating the performance of the NLM denoising method. These parameters are:

1. *The size of the search window*—the NLM method shows better performance for the large size of the search window, meaning a large computational complexity for the algorithm. However, the performance of the denoising method becomes asymptotically constant beyond a specific size of the search window [36, 37]. The optimized size of the search window is between 40×40 and 70×70 pixels [39].
2. *The size of the similarity window*—if the size of the similarity window is too large, no similarity windows will be found. However, if it is too small, too many will be present. For the better performance of the denoising method, the size of the similarity window is typically chosen to be between 3×3 and 9×9 pixels [39].
3. *The filter size*—if the filter size is set too small, not enough noise will be removed from the image; on the other hand, if it is set too large, the denoising method will cause a blurry effect in the denoised image. Therefore, the proper filter size must be adapted to the noise level of the pixel of interest.
4. The Euclidean distance between these windows, i.e. the distance between the search window and the similarity window, is also essential in defining the method's performance.

The NLM method was developed initially for additive noise filtering. In the case of multiplicative noise, equation (13.12) needs to be modified accordingly [54]

$$w(x_j, y_j) = (1/Z_i) \cdot \exp\left[\frac{\left\|I(N_i) - I(N_j)\right\|_{2,\,a}^2}{I(N_j)^{2\gamma}}\right], \tag{13.13}$$

where γ is a noise factor depending on the imaging system parameters, with a nominal value of 0.5 [54].

13.2.2 Variant NLM methods

Now we will describe the other NLM-related methods, which differ only in the choice of weights $w(x_j,y_j)$.

The square patches used in the classical NLM method consume a considerable computation time (due to the complexity of the algorithm); also, the edges of the noised image will not be satisfactorily denoised. For this reason, other NLM based methods have been proposed, e.g. NLM-average, NLM with shapes adaptive patches (NLM-SAP), NLM based different reprojection, and NLM with local polynomial regression (NLM-LPR) [36, 37, 47]. Let us look at these briefly.

Non-local means average (NLM-av): One of the issues of the NLM is that it requires a significant computation time for measuring the Euclidean distances between patches. This problem can be significantly overcome by calculating the average of the pixels within each patch. The difference in the averages is used to estimate the distances between patches, and the weight is changed accordingly.

Non-local means local polynomial regression (NLM-LPR): In order to make full use of the structural redundancy in similar patches, an LPR can be used with the NLM framework, creating a new NLM method. The NLM-LPR method provides a higher order of smoothness in the filtered images and relatively better denoising performance near the boundaries.

Non-local means shape adaptive patch (NLM-SAP): Several arbitrary shapes can be exploited to compare pixel neighborhoods instead of square patches. Since the square patches do not provide satisfactory results around the edges because they introduce a noise halo around the edges of the image, the use of arbitrary patch shapes is needed. The NLM-SAP denoising method effectively reduces the noise halo produced around high-contrasted edges using the proper patch shape. In this method, the distance between two pixels is measured in patches, and the weighted Euclidean distance, hence the NLM-SAP weighted term, is modified. So, depending on the requirements, a collection of patch shapes can be exploited, such as square, Gaussian, disc, etc.

Non-local means reprojections (NLM-reprojection): In NLM methods, first a collection of patches is created in the patch domain, then each patch is denoised by employing the proper operation, and, finally, the denoised information is reprojected to the pixel space in order to recover the pixel estimators, as shown in figure 13.5. There are several reprojection procedures to recover the pixelated denoised information from the denoised

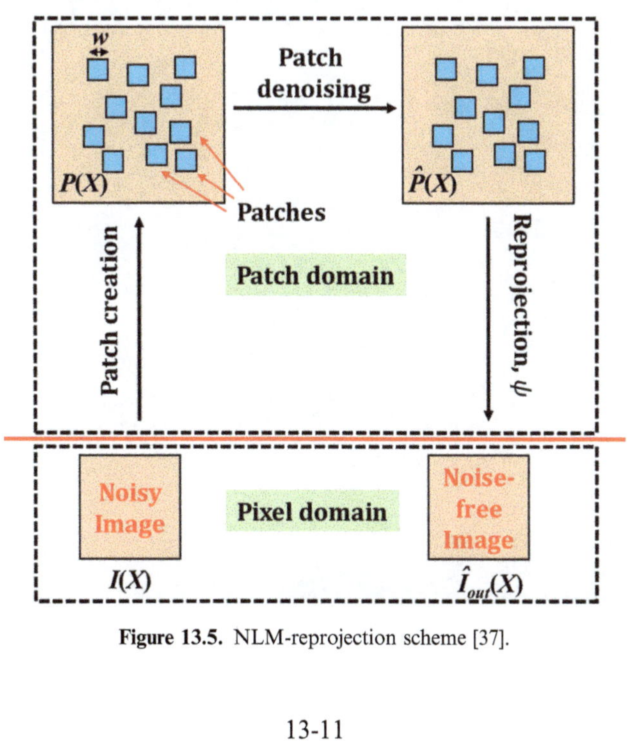

Figure 13.5. NLM-reprojection scheme [37].

patches, namely central reprojection, a uniform average of estimators reprojection, minimizing variance reprojection, minimizing variance with weighted average reprojection, etc, all of them discussed in detail in [37]. Further, the performance of all these techniques depends on the choice of the kernels, such as Gaussian, flat, box, or triangular kernels characterized by their bandwidth.

13.3 Examples

The speckle-noise reduction capability of the NLM methods and their performance is demonstrated on the computer-simulated speckle fringe pattern obtained from an out-of-plane digital speckle pattern interferometry layout. The computer used in the simulation process had a 32-bit Intel Pentium microprocessor with a CPU clock rate of 1.87 GHz. Figure 13.6 shows the fringe pattern on which various NLM methods will be employed to remove the speckle-noise.

Figures 13.7(a)–(c) show the NLM method's denoising results.

Figures 13.8(a)–(c) show the NLM-av method's denoising results.

Figures 13.9(a)–(c) show the denoising results obtained using the NLM-LPR method.

Figures 13.10(a)–(c) and 13.11(a)–(c) show the denoising results obtained by employing the NLM-SAP method for pie and rectangle shaped patches, respectively.

Figure 13.6. Input speckle fringe pattern.

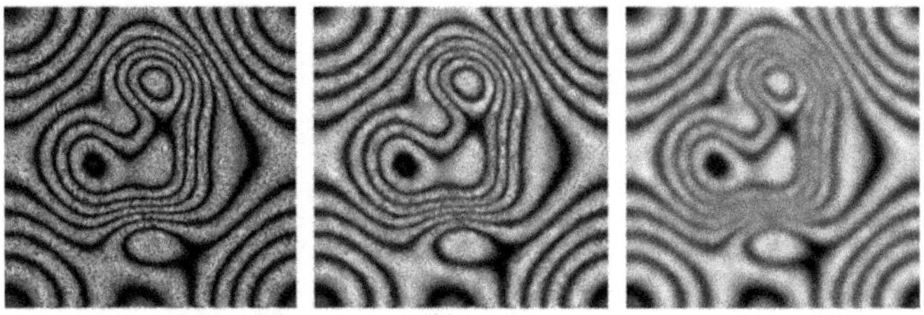

Figure 13.7. Speckle denoising results of NLM for (a) $\sigma = 25$, (b) $\sigma = 50$, and (c) $\sigma = 100$.

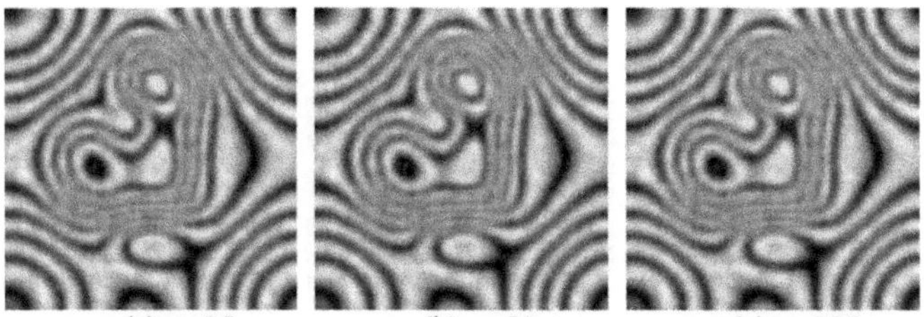

Figure 13.8. Speckle denoising results of the NLM-av method for (a) $\sigma = 25$, (b) $\sigma = 50$, and (c) $\sigma = 100$.

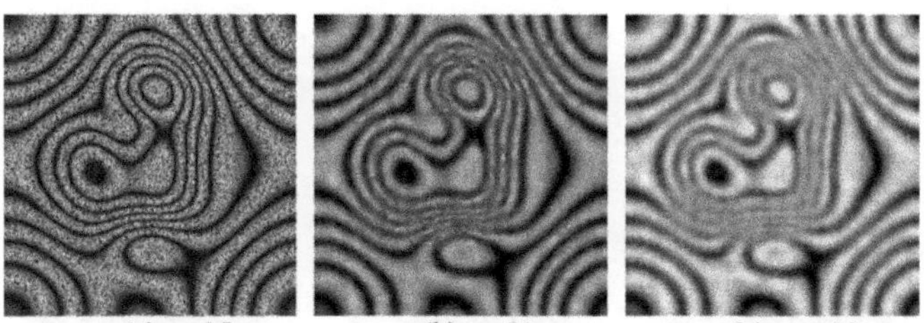

Figure 13.9. Speckle denoising results of the NLM-LPR method for (a) $\sigma = 25$, (b) $\sigma = 50$, and (c) $\sigma = 100$.

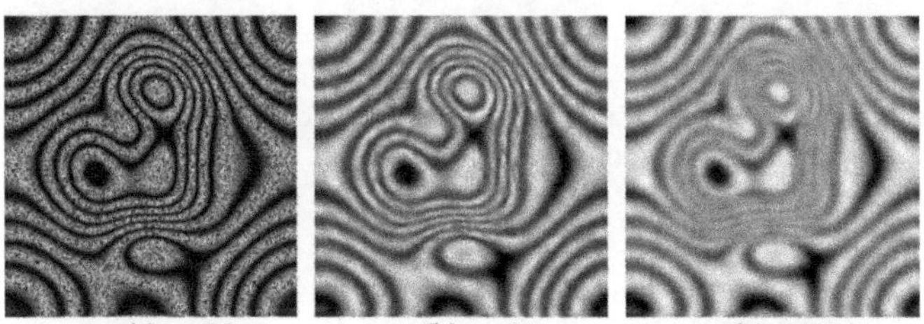

Figure 13.10. Speckle denoising results of the NLM-SAP method with pie shaped patches for (a) $\sigma = 25$, (b) $\sigma = 50$, and (c) $\sigma = 100$.

The denoising results of several schemes of the NLM-Rep method for different σ values are shown in figures 13.12–13.16. Figures 13.12(a)–(c) show the denoising results of central reprojection with a box kernel; figures 13.13(a)–(c) show the denoising results of central reprojection with a Gaussian kernel; figures 13.14(a)–(c)

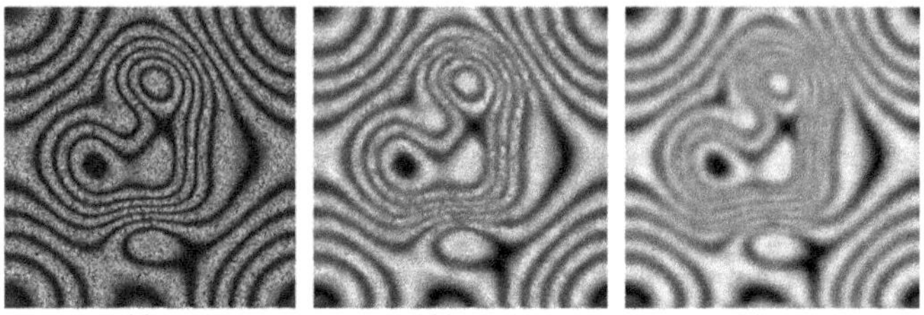

(a) σ=25 (b) σ=50 (c) σ=100

Figure 13.11. Speckle denoising results of the NLM-SAP method with rectangular shaped patches for (a) $\sigma = 25$, (b) $\sigma = 50$, and (c) $\sigma = 100$.

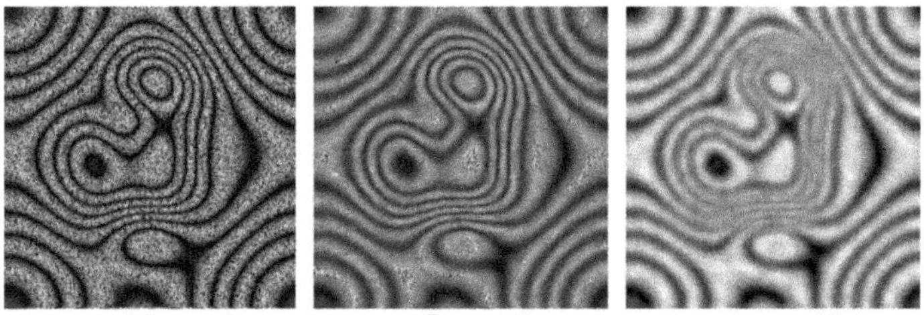

(a) σ=25 (b) σ=50 (c) σ=100

Figure 13.12. Speckle denoising results of the NLM-Rep method, central reprojection with box kernel, for (a) $\sigma = 25$, (b) $\sigma = 50$, and (c) $\sigma = 100$.

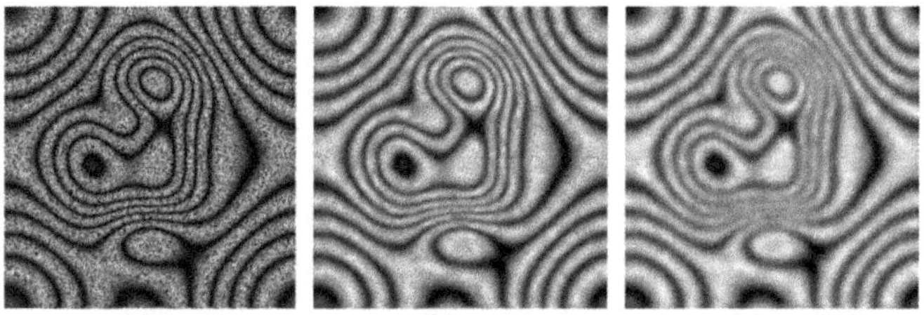

(a) σ=25 (b) σ=50 (c) σ=100

Figure 13.13. Speckle denoising results of the NLM-Rep method, central reprojection with a Gaussian kernel, for (a) $\sigma = 25$, (b) $\sigma = 50$, and (c) $\sigma = 100$.

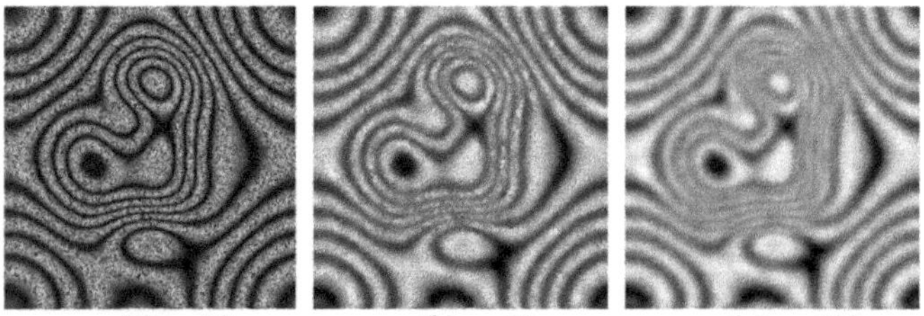

(a) σ=25 (b) σ=50 (c) σ=100

Figure 13.14. Speckle denoising results of the NLM-Rep method, uniform reprojection with box kernel, for (a) $\sigma = 25$, (b) $\sigma = 50$, and (c) $\sigma = 100$.

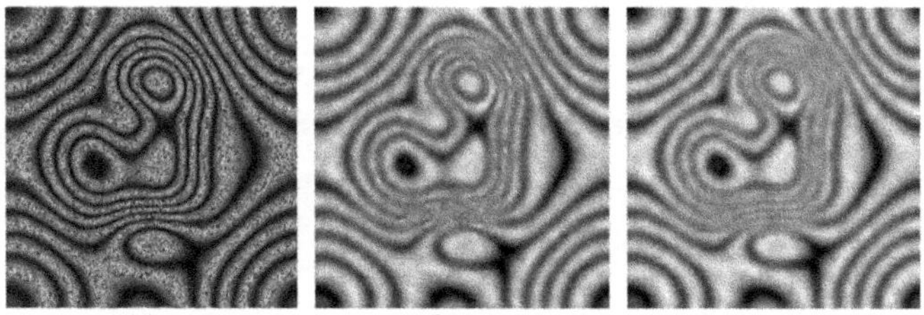

(a) σ=25 (b) σ=50 (c) σ=100

Figure 13.15. Speckle denoising results of the NLM-Rep method, minimizing variance reprojection with box kernel, for (a) $\sigma = 25$, (b) $\sigma = 50$, and (c) $\sigma = 100$.

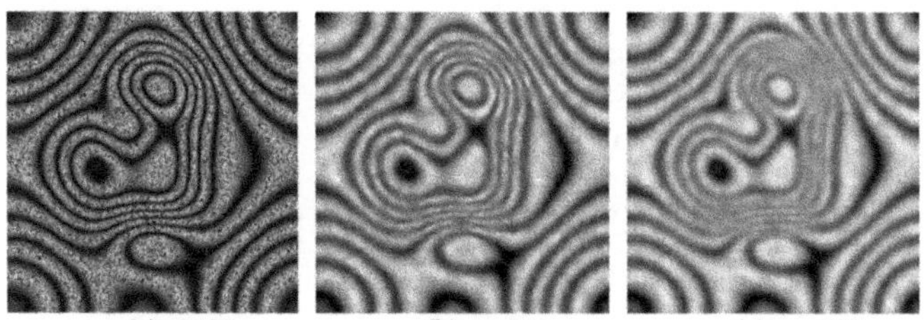

(a) σ=25 (b) σ=50 (c) σ=100

Figure 13.16. Speckle denoising results of the NLM-Rep method, weighted averaging reprojection based on the inverse variance with box kernel, for (a) $\sigma = 25$, (b) $\sigma = 50$, and (c) $\sigma = 100$.

show the denoising results of uniform reprojection with a box kernel; figures 13.15(a)–(c) show the denoising results of minimizing variance reprojection with a box kernel; and figures 13.16(a)–(c) show the denoising results of the weighted averaging reprojection based on the inverse variance with a box kernel.

13.3.1 Evaluation criteria

In order to evaluate the performance of any denoising method, some image quality metrics should be used [39], e.g. peak signal to noise ratio (PSNR), structure similarity index (SSIM), and edge preservation index (EPI). These metrics require two inputs, the reference and denoised image, i.e. the real or *perfect* result is known and the obtained denoised image is compared. The criteria evaluation is summarized in figure 13.17.

The expected results would be as follows: (a) the higher the PSNR value, the better the denoising quality; (b) as the maximum value for SSIM and EPI is 1, which means that the denoised fringe pattern and the perfect *clean* fringe pattern are similar, then, in general, good denoising is considered when the SSIM and EPI values are near to 1. With this in mind, these three metrics are computed for the results obtained with the NLM variant methods. The resulting data are presented as plots in figures 13.18–13.20, where PSNR, SSIM, and EPI were calculated using the different NLM methods for the images in figures 13.7–13.16. Further, these methods' computation time (seconds) was also evaluated and plotted for different σ parameters in figure 13.21.

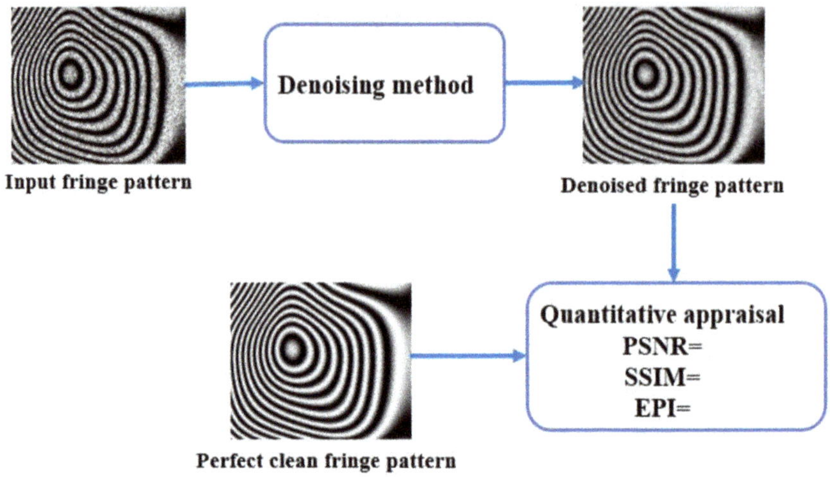

Figure 13.17. Process of quality evaluation for the denoising method.

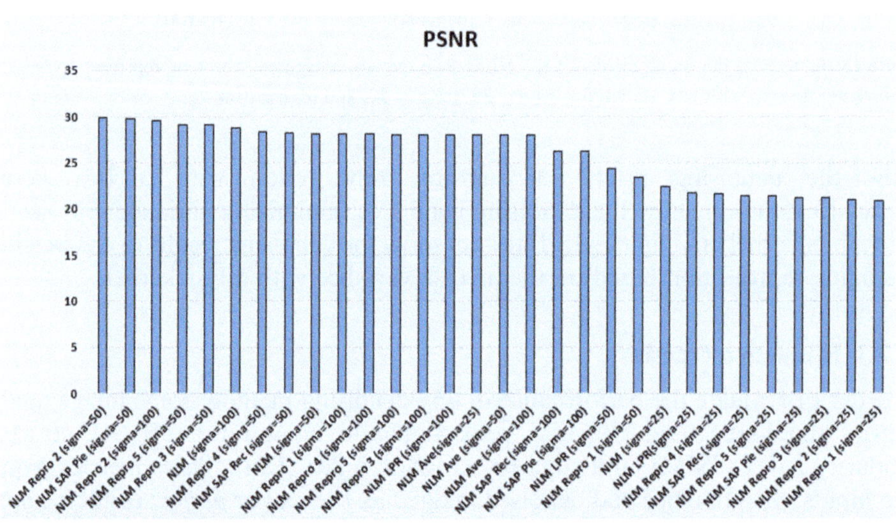

Figure 13.18. PSNR values calculated for the denoised speckled fringe pattern for various denoising methods.

A brief overview of the NLM and its variant kernel based denoising methods was presented for the speckle-noise reduction in fringe patterns. Based on the obtained filtered results as evaluated by some criteria and metrics, it can be concluded that these methods are effective in speckle-noise reduction and, therefore, can be established as powerful denoising tools in optical metrology and indeed in a wide range of other areas of science and engineering. As discussed in [36, 37, 39], these methods provide better results than the local denoising methods. It should be pointed out that we also observed that these methods introduce over-smoothing artifacts resulting in the loss of sharp edges in the final fringe pattern. Moreover, these methods are not effective when the self-similarity assumption fails. However, parameters such as patch size, suitable kernel, reprojection estimator scheme selection, speckle size, etc, should be optimized to obtain the best denoising results.

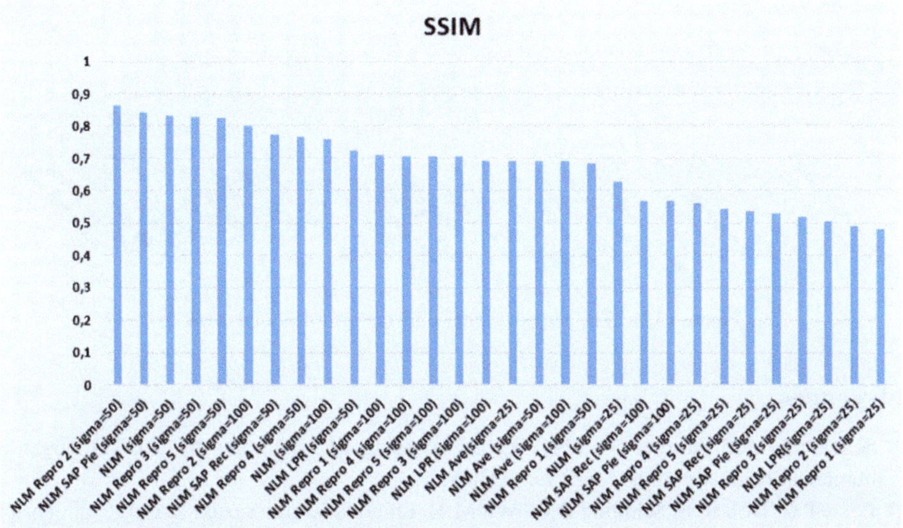

Figure 13.19. SSIM values calculated for the denoised speckled fringe pattern for various denoising methods.

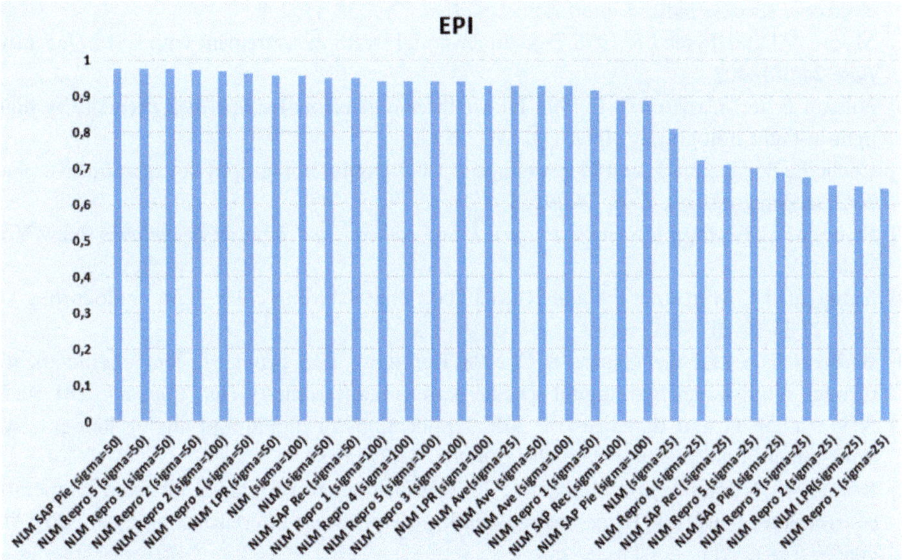

Figure 13.20. EPI values calculated for the denoised speckled fringe pattern for various denoising methods.

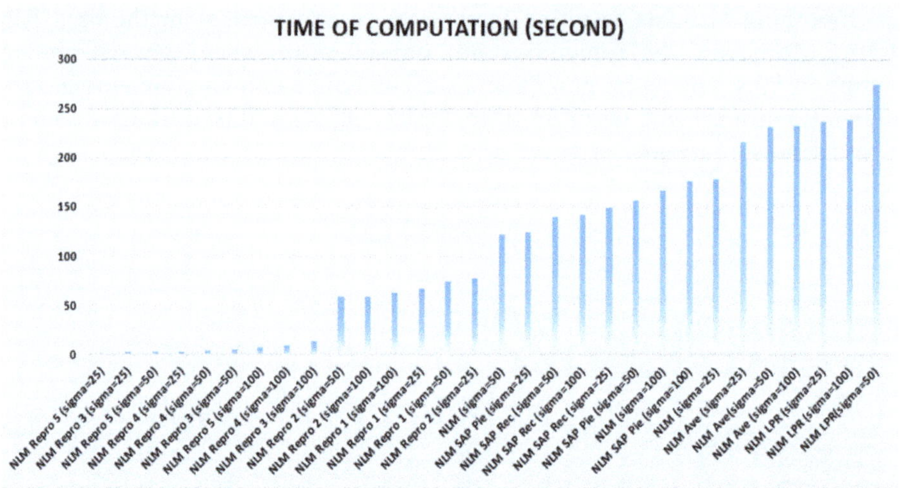

Figure 13.21. The computational time for each denoised method.

References

[1] Schnars U and Jüptner W 1994 Direct recording of holograms by a CCD target and numerical reconstruction *Appl. Opt.* **33** 179–81

[2] Poon T C, Doh K B, Schilling B W, Wu M H, Shinoda K and Suzuki Y 1995 3-dimensional microscopy by optical-scanning holography *Opt. Eng.* **34** 1338–44

[3] Lehman M M, Pomarico J A and Torroba R D 1995 Digital speckle pattern interferometry applied to a surface roughness study *Opt. Eng.* **34** 1148–53

[4] Wang W C, Hwang C H and Lin S Y 1996 Vibration measurement by the time-averaged electronic speckle pattern interferometry *Appl. Opt.* **35** 4502–9

[5] Moore A J and Tyrer J R 1996 Two-dimensional strain measurement with ESPI *Opt. Lasers Eng.* **24** 381–402

[6] Nilsson B and Carlsson T E 1998 Direct three-dimensional shape measurement by digital light-in-flight holography *Appl. Opt.* **37** 7954–9

[7] Cuche E, Bevilacqua F and Depeursinge C 1999 Digital holography for quantitative phase-contrast imaging *Opt. Lett.* **24** 291–3

[8] Rastogi P 2001 *Digital Speckle Pattern Interferometry and Related Techniques* (New York: Wiley)

[9] Yamaguchi I, Matsumura T and Kato J 2002 Phase-shifting color digital holography *Opt. Lett.* **27** 1108–10

[10] Barbosa E A and dos Santos S C 2008 Refractive and geometric lens characterization through multi-wavelength digital speckle pattern interferometry *Opt. Commun.* **281** 1022–9

[11] Nehmetallah G and Banerjee P P 2012 Applications of digital and analog holography in three-dimensional imaging *Adv. Opt. Photon.* **4** 472–553

[12] Kumar M, Kumar V and Shakher C 2013 Measurement of temperature and temperature distribution in diffusion flames using digital speckle pattern interferometry *Proc. SPIE* **9066** 9066Y

[13] Osten W, Faridian A, Gao P, Körner K, Naik D, Pedrini G, Singh A K, Takeda M and Wilke M 2014 Recent advances in digital holography *Appl. Opt.* **53** G44–63

[14] Kumar M and Shakher C 2015 Measurement of temperature and temperature distribution in gaseous flames by digital speckle pattern shearing interferometry using holographic optical element *Opt. Lasers Eng.* **73** 33–9

[15] Kumar M, Gaur K K and Shakher C 2015 Measurement of material constants (Young's modulus and Poisson's ratio) of polypropylene material using digital speckle pattern interferometry *J. Jpn Soc. Exp. Mech.* **15** s87–91

[16] Kumar M, Agarwal S, Kumar V, Khan G S and Shakher C 2015 Experimental investigation on butane diffusion flames under the influence of magnetic field by using digital speckle pattern interferometry *Appl. Opt.* **54** 2450–60

[17] Kumar M, Khan G S and Shakher C 2015 Measurement of elastic and thermal properties of composite materials using digital speckle pattern interferometry *Proc. SPIE* **9660** 966011

[18] Kumar M, Agarwal R, Bhutani R and Shakher C 2016 Measurement of strain distribution in cortical bone around miniscrew implants used for orthodontic anchorage using digital speckle pattern interferometry *Opt. Eng.* **55** 054101

[19] Kumar M 2016 Some novel applications of digital speckle pattern interferometry and digital holographic interferometry *Doctoral Dissertation* Indian Institute of Technology, New Delhi, India

[20] Kreis T 2016 Application of digital holography for nondestructive testing and metrology: a review *IEEE Trans. Ind. Inform.* **12** 240–7

[21] Kumar M and Shakher C 2016 Experimental characterization of the hygroscopic properties of wood during convective drying using digital holographic interferometry *Appl. Opt.* **55** 960–8

[22] Pedrini G, Martínez-García V, Weidmann P, Wenzelburger M, Killinger A, Weber U, Schmauder S, Gadow R and Osten W 2016 Residual stress analysis of ceramic coating by laser ablation and digital holography *Exp. Mech.* **56** 683–701

[23] Pandey P K, Kumar M, Kumar V and Shakher C 2017 Measurement of temperature and temperature profile of wick stabilized micro diffusion flame under the effect of magnetic field using digital speckle pattern interferometry *Opt. Eng.* **56** 014106

[24] Ruiz C G T, De La Torre-Ibarra M H, Flores-Moreno J M, Frausto-Reyes C and Mendoza Santoyo F 2018 Cortical bone quality affectations and their strength impact analysis using holographic interferometry *Biomed. Opt. Express* **9** 4818–33

[25] Quan X, Kumar M, Matoba O, Awatsuji Y, Hayasaki Y, Hasegawa S and Wake H 2018 Three-dimensional stimulation and imaging-based functional optical microscopy of biological cells *Opt. Lett.* **43** 5447–50

[26] Kumar M, Birhman A S, Kannan S and Shakher C 2018 Measurement of initial displacement of canine and molar in human maxilla under different canine retraction methods using digital holographic interferometry *Opt. Eng.* **57** 094106

[27] Tounsi Y, Kumar M, Siari A, Santoyo F M, Nassim A and Matoba O 2019 Digital four-step phase-shifting technique from a single fringe pattern using Riesz transform *Opt. Lett.* **44** 3434–7

[28] Kumar M, Quan X, Awatsuji Y, Cheng C, Hasebe M, Tamada Y and Matoba O 2020 Common-path multimodal three-dimensional fluorescence and phase imaging system *J. Biomed. Opt.* **25** 032010

[29] Kumar M, Quan X, Awatsuji Y, Tamada Y and Matoba O 2020 Digital holographic multimodal cross-sectional fluorescence and quantitative phase imaging system *Sci. Rep.* **10** 7580

[30] Lim J S and Nawab H 1981 Techniques for speckle-noise removal *Opt. Eng.* **20** 203472

[31] Kerr D, Mendoza Santoyo F and Tyrer J R 1989 Manipulation of the Fourier components of speckle image fringe patterns as a part of an interferometric analysis process *J. Mod. Opt.* **36** 195–203
[32] Federico A and Kaufmann G H 2001 Comparative study of wavelet thresholding methods for denoising electronic speckle pattern interferometry fringes *Opt. Eng.* **40** 2598–604
[33] Federico A and Kaufmann G H 2005 Local denoising of digital speckle pattern interferometry fringes using multiplicative correlation and weighted smoothing splines *Appl. Opt.* **44** 2728–35
[34] Bernini M B, Federico A and Kaufmann G H 2008 Noise reduction in digital speckle pattern interferometry using bidimensional empirical mode decomposition *Appl. Opt.* **47** 2592–8
[35] Trusiak M, Patorski K and Wielgus M 2012 Adaptive enhancement of optical fringe patterns by selective reconstruction using FABEMD algorithm and Hilbert spiral transform *Opt. Express* **20** 23463–79
[36] Tounsi Y, Kumar M, Nassim A and Mendoza Santoyo F 2018 Speckle-noise reduction in digital speckle pattern interferometric fringes by non-local means and its related adaptive kernel-based methods *Appl. Opt.* **57** 7681–90
[37] Tounsi Y, Kumar M, Nassim A, Mendoza-Santoyo F and Matoba O 2019 Speckle denoising by variant non-local means methods *Appl. Opt.* **58** 7110–20
[38] Zada S, Tounsi Y, Kumar M, Mendoza-Santoyo F and Nassim A 2019 Contribution study of monogenic wavelets transform to reduce speckle-noise in digital speckle pattern interferometry *Opt. Eng.* **58** 034109
[39] Kumar M, Tounsi Y, Kaur K, Nassim A, Mandoza-Santoyo F and Matoba O 2020 Speckle denoising techniques in imaging systems *J. Opt.* **22** 063001
[40] Baselice F, Ferraioli G, Pascazio V and Schirinzi G 2018 Enhanced Wiener filter for ultrasound image denoising *European Medical and Biological Engineering Conf., Nordic–Baltic Conf. on Biomedical Engineering and Medical Physics* vol 65H Eskola, O Väisänen, J Viik and J Hyttinen (Singapore: Springer), pp 65–8
[41] Muhire D, Tounsi Y, Kumar M, Siari A, Mendoza-Santoyo F, Matoba O and Nassim A 2020 Wavelets Teager–Kaiser Hilbert approach for AM–FM signal demodulation: application in the field of speckle metrology *Opt. Eng.* **59** 102407
[42] Perona P and Malik J 1990 Scale-space and edge detection using anisotropic diffusion *IEEE Trans. Pattern Anal. Mach. Intell.* **12** 629–39
[43] Tsiotsios C and Petrou M 2013 On the choice of the parameters for anisotropic diffusion in image processing *Pattern Recognit.* **46** 1369–81
[44] Tomasi C and Manduchi R 1998 Bilateral filtering for gray and color images *6th Int. Conf. on Computer Vision* (Piscataway, NJ: IEEE), pp 839–46
[45] Chaudhury K N and Rithwik K 2015 Image denoising using optimally weighted bilateral filters: a sure and fast approach *2015 IEEE Int. Conf. on Image Processing (ICIP)* (Piscataway, NJ: IEEE), pp 108–12
[46] Varman P and Wykes C 1982 Smoothing of speckle and moiré fringes by computer processing *Opt. Lasers Eng.* **3** 87–100
[47] Tounsi Y, Ghlaifan A, Kumar M, Mendoza-Santoyo F, Matoba O and Nassim A 2019 Fringe pattern analysis in wavelet domain *Holographic Materials and Applications* (London: IntechOpen)
[48] Yassine T, Ahmed S and Abdelkrim N 2017 Speckle-noise reduction in digital speckle pattern interferometry using Riesz wavelets transform *2017 Int. Conf. on Advanced Technologies for Signal and Image Processing (ATSIP)* (Piscataway, NJ: IEEE), pp 1–4

[49] Kemao Q 2004 Windowed Fourier transform for fringe pattern analysis *Appl. Opt.* **43** 2695–702

[50] Kemao Q 2007 Two-dimensional windowed Fourier transform for fringe pattern analysis: principles, applications and implementations *Opt. Lasers Eng.* **45** 304–17

[51] Kemao Q, Wang H and Gao W 2008 Windowed Fourier transform for fringe pattern analysis: theoretical analyses *Appl. Opt.* **47** 5408–19

[52] Wong A, Mishra A, Bizheva K and Clausi D A 2010 General Bayesian estimation for speckle-noise reduction in optical coherence tomography retinal imagery *Opt. Express* **18** 8338–52

[53] Rudin L I, Osher S and Fatemi E 1992 Non-linear total variation based noise removal algorithms *Physica* D **60** 259–68

[54] Uzan A, Rivenson Y and Stern A 2013 Speckle denoising in digital holography by non-local means filtering *Appl. Opt.* **52** A195–200

IOP Publishing

Full Field Optical Metrology and Applications

Fernando Mendoza-Santoyo, Manuel De la Torre-Ibarra, María del Socorro Hernández-Montes and Jorge Mauricio Flores Moreno

Chapter 14

Overview of recent applications using DHI

The word hologram was introduced by Denis Gabor during the 1940s [1]. At that time the practical demonstration of the concept of holography remained largely unexplored due to the lack of either photon or electron coherent sources. This lack of sources is why holography lay dormant for almost twenty years, waiting for a new scientific and technological event that would trigger an awakening to catapult it unexpectedly and powerfully into use. A major invention and a development in imaging systems occurred in the 1960s, namely, the invention of the laser and the development of the first video cameras. At the time, 3D images were reconstructed from holograms, as reported by Leith and Upatnieks [2, 3], on high-resolution photographic plates that needed a lengthy and sometimes cumbersome wet chemical process. These pioneers introduced the so-called off-axis configuration that removed Gabor's twin images which were present in his in-line configuration, an issue that caused overlap and obstruction of the original object image upon observation of images reconstructed from holograms.

14.1 Introduction

The principal theoretical aspects of holographic interferometry (HI) were first published in the 1960s [4, 5]. The idea consists (see in particular chapters 6 and 8) of comparing the optical path difference produced by two object states, a reference or base state and another state when the object changes (e.g. a displacement or deformation, static or dynamic). Thus two events were captured on the same holographic plate at two different times. First, the object base state overlapped with the reference beam, and the hologram plate was developed and placed back at the same place; then, a second image of the object was captured on the same plate, but for a displaced object state overlapped with the reference beam. This process produced a full field of view interference fringes that reflected the object displacement uniquely. The use of HI allows the study of diffusively reflecting objects as compared to the more common interferometers that, at the time, were used to measure optically

doi:10.1088/978-0-7503-3027-5ch14

polished surfaces (or also called specular surfaces because of their optical quality). The ability to measure a wide range of physical and mechanical variables from objects with diffuse surfaces is perhaps the most important characteristic of HI.

The wet chemical process for the hologram plates was often cumbersome and time-consuming, and sometimes more critical was the fact that, due to, for instance, environmental disturbances, the hologram of the object was not correctly registered, researchers then needed to go back and repeat the experiments. Thus the introduction of video cameras to replace the hologram plates represented a major step forward for using HI under different conditions, even outside the controlled environments of laboratories. An interferometer emerged late in the 1960s called the electronic speckle pattern interferometer (ESPI). The first ESPI systems did not directly register the holograms electronically. Instead, they scanned and externally digitized, via a digital memory frame grabber, an object image (hologram) captured by a so-called vidicon camera, whose resolution capacities were far from those reached by the photographic emulsion (typically of several thousands of lines per millimeter). The latter meant that, in its early stages, ESPI had a low spatial resolution and when the speckle noise was added poor visibility/contrast fringe patterns were obtained, which is not an advantage for ESPI when compared to the high resolution and quality HI fringe patterns acquired with holographic films.

As with HI and other interferometrical methods, the primary variable to obtain from ESPI was the optical phase, which is directly related to the optical path difference between the object states. Since the beginning, researchers working with ESPI (and also now with digital ESPI (DSPI) and with digital HI (DHI)) have tried to obtain a better spatial resolution for the fringe data and have also placed particular attention on reducing the speckle noise (see chapter 13, on speckle denoising). The aim is to extract with more accuracy the optical phase, a subject dealt with in chapter 3.

Eventually, during the 1970s, the vidicon camera was replaced by digital cameras, and DSPI and DHI replaced the old image capturing technologies. Let us concentrate from here onwards on DHI, a method very close to DSPI—the former is classified as a spatial carrier method since it uses a frequency carrier to retrieve the optical phase, one of the significant differences with DSPI where this experimentally introduced carrier frequency is not present; instead in DSPI the optical phase is obtained with phase stepping methods, typically using Pzt elements in the optical layout.

Today it is possible to search for and find many different optical configurations in the literature (papers and books). They are designed and constructed for specific applications. Many show advances in resolution, speed, and complexity of the set-ups and the are continued upgrades of electronic devices and faster computers for big data processing. As a result, DHI set-ups are configured as fully digital set-ups combining different methodologies [6]. Also, in the literature one can find full-digital, real-time, and high-resolution DHI systems for a wide range of applications, with a continuous input of new research results and contributions by an ever-growing number of groups worldwide. Also, in the last decades, we have witnessed the introduction of DHI into biomedical and biological applications, many under demanding experimental conditions. DHI has demonstrated that it is a reliable and mature measuring method. So, more specific and high-impact applications are demanding the use of DHI, which means that newer and more robust experimental configurations are needed—DHI is far from being an exhausted research topic. Indeed, it has plenty of years to continue improving and influencing several new research areas and topics. An excellent example of these is part of the research examples presented in chapter 9, related to digital holographic microscopy (DHM).

14.2 Basic applications of DHI

DHI requires recording the interference between the scattered light from an object and a smooth reference beam. The background theory for DHI may be found in chapters 6, 8, and 10. This chapter will describe some interesting and valuable applications of DHI and, as such, contains information that will evolve and change with time. It is a chapter that will be updated every time we produce a new version of this book. These applications include displacement, strain–stress, vibrations, flow, refracting index, and surface profilometry measurements.

14.2.1 Displacement measurements

One of the first commercially available digital systems based on the hologram concept was the Holocamera (Newport HC130), designed to measure the displacement maps from mechanical deformations on cantilevers and honeycomb structures [7]. Chang *et al* used a Pzt in the reference beam arm and a phase stepping method to retrieve the optical phase map. DHI was also applied to compare the relative phase difference between two reflecting surfaces [8]: a steel ball worked as a master surface profile and was used for comparison with a glass-eye surface. Figure 14.1 shows the resulting optical phase difference in the form of an astigmatic pattern.

By looking at the scale in this figure, it is seen that this system had low resolution, mainly because the camera sensor had a limited size of 756 × 581 pixels (with a pixel

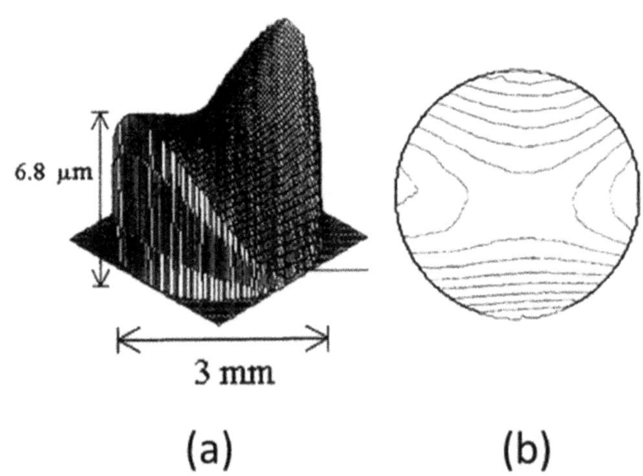

Figure 14.1. Optical phase difference resulting from comparing the shapes of a steel ball, the master, and a glass-eye surface: (a) pseudo-3D representation and (b) contour map. Reproduced with permission from [8]. Copyright 1995 Elsevier.

size of 11 μm [9]), which was used to observe and analyze an area of up to 3 × 3 mm. Beyond this area, the fringe density was too high, making it difficult to obtain useful phase information.

As imaging and semiconductor technology advanced, digital sensors with higher spatial resolution started to be used in DHI, making it possible to reduce the speckle size to a few pixels [10, 11]. Then, the displacement maps, derived from the optical phase difference, could analyze sub-nanometer variations using the so-called average windowed holograms [12]. Let us look into this. Consider that each section of a hologram contains phase information regarding the complete hologram. The authors propose a process by which the hologram is divided into n windowed holograms, see figure 14.2(a), each generating an intensity image representing the

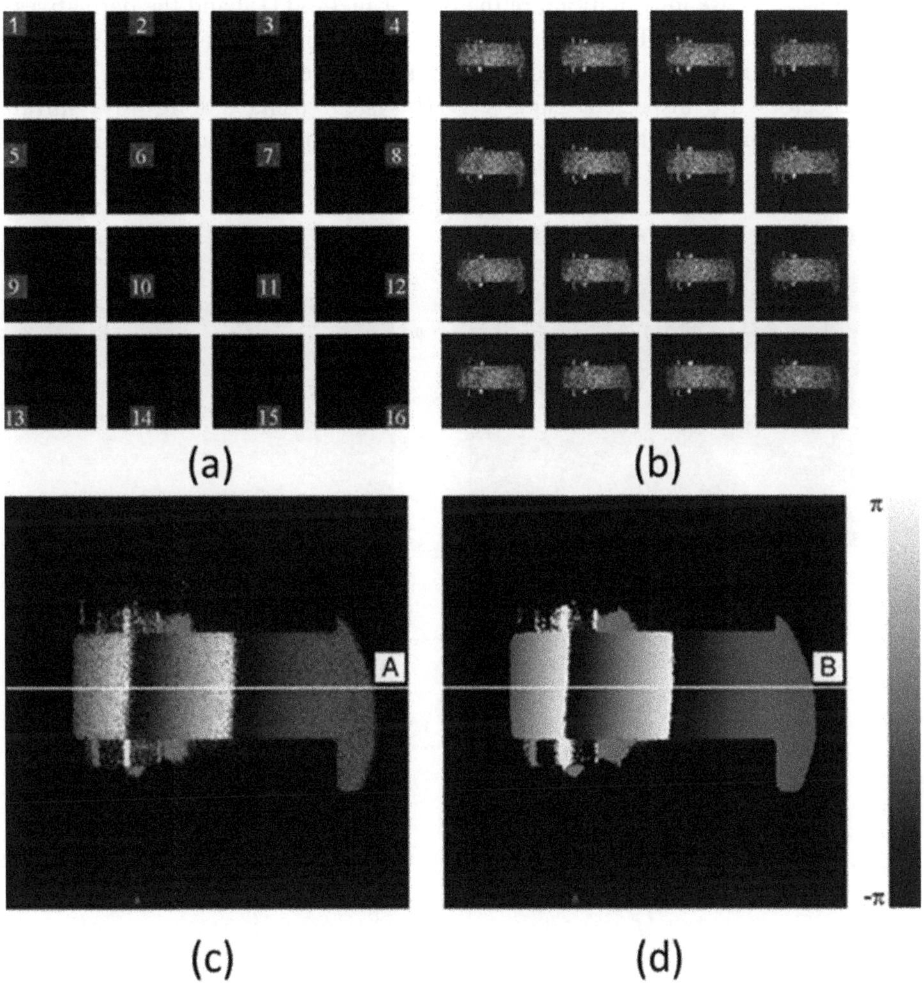

Figure 14.2. Windowed holograms: (a) distribution; (b) resulting images; (c) phase differences from the entire hologram $n = 1$; and (d) a windowed divided holograms with $n = 16$. Reproduced with permission from [12]. Copyright 2007 SPIE.

entire illuminated object, see figure 14.2(b). This technique improves the accuracy of the measurements since it provides a speckle noise reduction. A comparison between the optical phases is presented in figures 14.2(c) and (d), where the complete hologram and 16 windowed holograms are used, respectively. The speckle noise reduction in the smooth phase allows the quantification of tiny measurements.

Continuing with the examples, the same optical phase was used to measure the radius of curvature in solid samples [13] and the propagation of shock waves in polycrystalline structures [14]. In the latter, the shock waves were generated on the structure with pulses from a Q-switched Nd:YAG laser, while a laser at 532 nm was used to register the holograms of the provoked deformations, see figure 14.3. There, it is possible to identify regions for the evaporated material (1), compressed air (2), the shock wavefront (3), and the surrounding undisturbed air (4). More applications using the basic phase information are the phasor method [15] and the derivatives of the phase [16, 17]. A different approach for displacement measurements is when large areas are inspected in a remote and non-invasive form [18–20]. In [19] large space reflectors are inspected using a DHI system using a long-wave infrared illumination source. The latter allows for detecting more significant displacements,

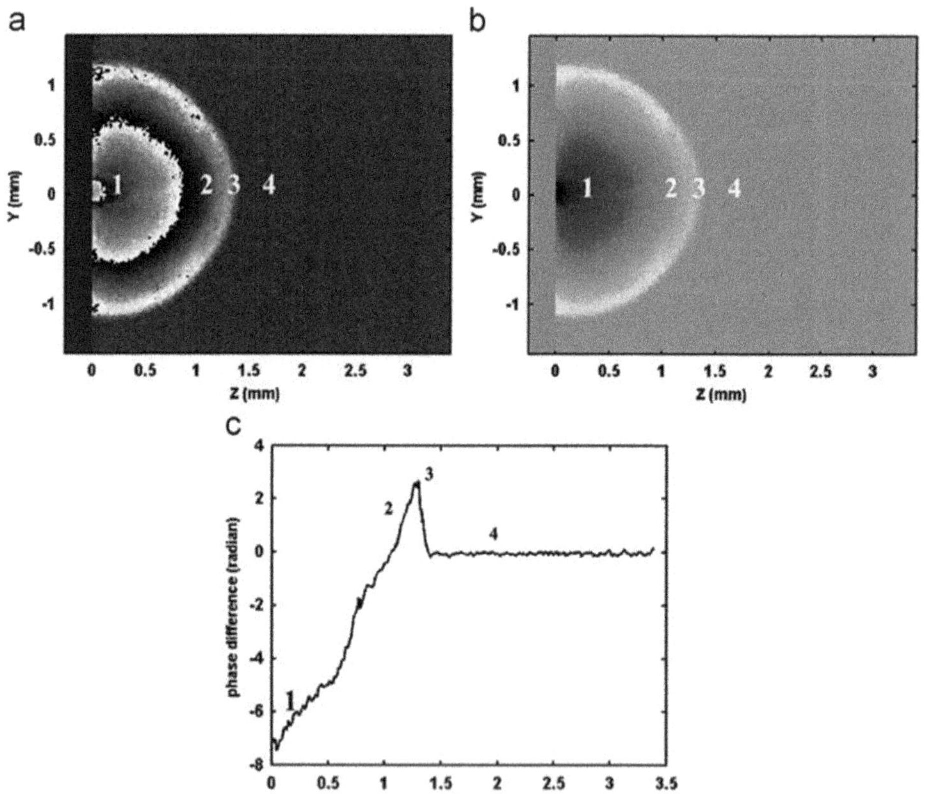

Figure 14.3. Shock wave propagation: (a) wrapped phase map; (b) unwrapped phase map; and (c) phase profile. Reproduced with permission from [14]. Copyright 2009 Elsevier.

which simulate thermal changes when the objects are in an Earth orbit. In figure 14.4 some results of this long-wave DHI system are presented.

A large area involves a high variation of the interferometer sensitivity vector, but the reflector's smooth optical phase change is achieved even in this demanding inspection. On the other hand, thanks to the method's versatility, another application involves the study of graded-index optical fibers, where their profile is analyzed using the refractive index variations [21]. The difference between a graded-index fiber and a regular fiber is the smooth rounded profile due to the refractive index variation around the center, as figure 14.5 shows. This condition complicates the measurement of the fiber's profile, as it is not a single on–off value remaining unaltered in a region. The profile indicates the fiber's parabolic or non-parabolic refractive index fields (a phase-shifting method is used to retrieve the optical phase). The results are seen in figure 14.6.

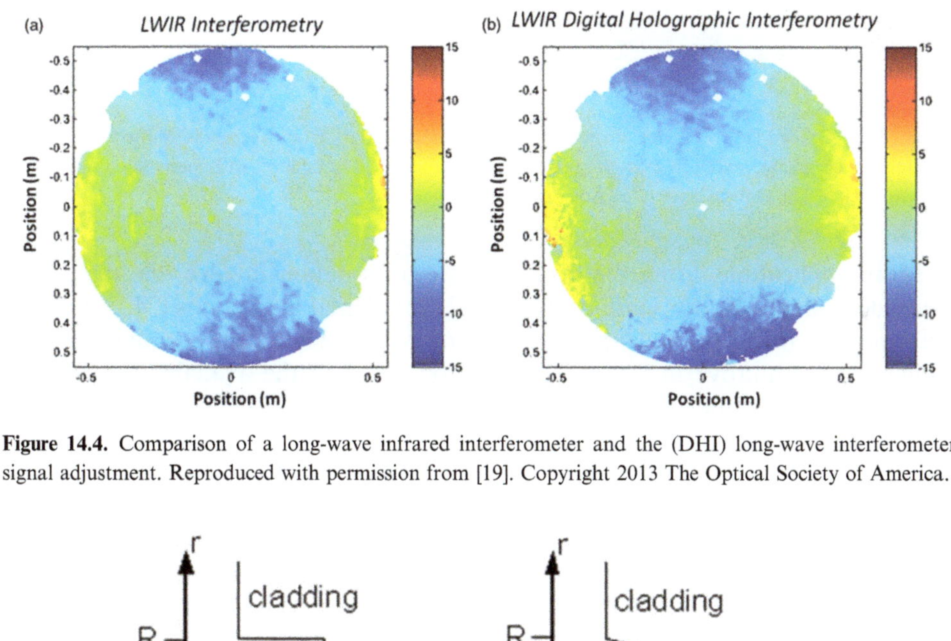

Figure 14.4. Comparison of a long-wave infrared interferometer and the (DHI) long-wave interferometer signal adjustment. Reproduced with permission from [19]. Copyright 2013 The Optical Society of America.

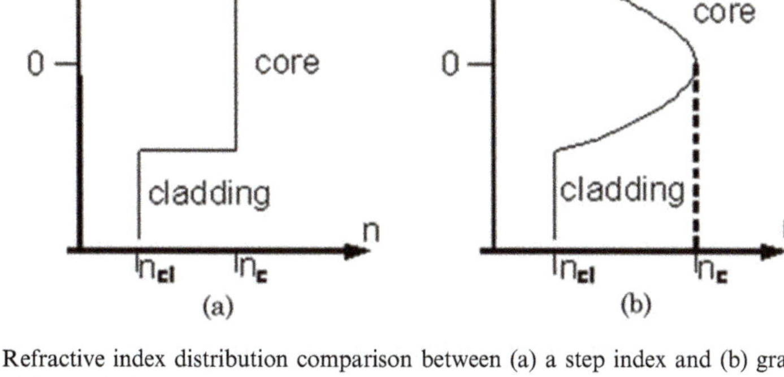

Figure 14.5. Refractive index distribution comparison between (a) a step index and (b) graded-index fibers. Reproduced with permission from [21]. Copyright 2009 The Optical Society of America.

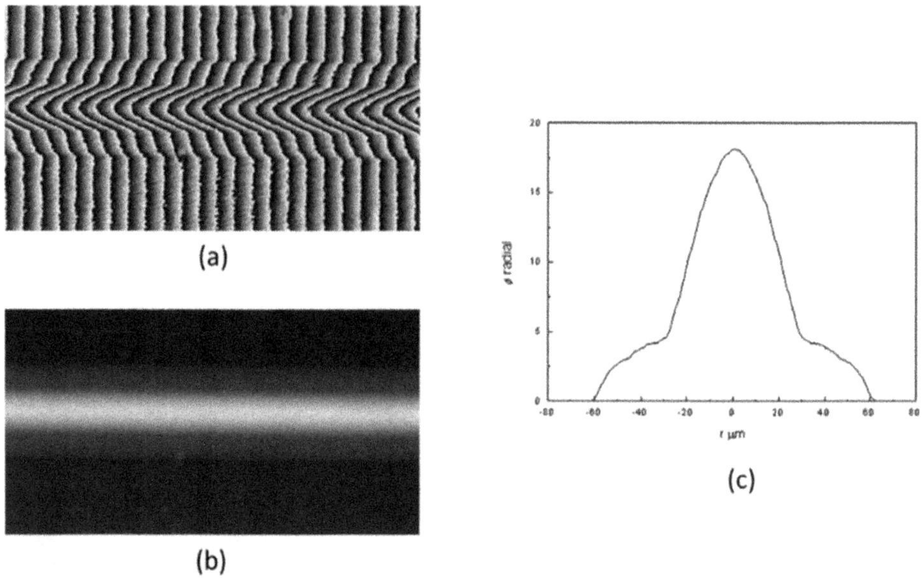

Figure 14.6. Graded-index optical fiber: (a) wrapped phase map from the phase-shifted holograms, (b) unwrapped phase, and (c) cross-section of the fiber core. Reproduced with permission from [21]. Copyright 2009 The Optical Society of America.

More applications using the optical phase information involve the analysis of beam deflection [21], transparent objects [22, 23], dental inspections [24, 25], micro deformations in electronic circuit boards [26], rough material inspections [27], and techniques to improve the current calculation speed of the available information coming from the optical phase [28–30].

14.2.2 Strain–stress measurements

Even when the displacement information is useful, there are applications where it is not sufficient to fully describe the mechanical behavior of an object. An object's mechanical response to specific conditions is typically studied with more conventional methods. Several destructive and well-standardized (conventional) tests can be replicated with DHI, so the same (or even more) information can be obtained in a non-destructive (or non-invasive) fashion. An example of this is the strain–stress value that gives mechanical information about the sample (see figure 14.7). Both variables are linked to the elasticity modulus (Young's modulus), which describes the material's mechanical response under specific load conditions. The strain–stress curve is unique for each material and determines, for instance, its elastic and plastic range. If the deformation load is removed, the material restores its initial conditions in the elastic range. However, the deformation is permanent in the plastic range and does not disappear even if the load is removed. In addition, this curve gives information about the yield point where the transition between the two states occurs. Finally, there is the fracture point where the material fails and breaks.

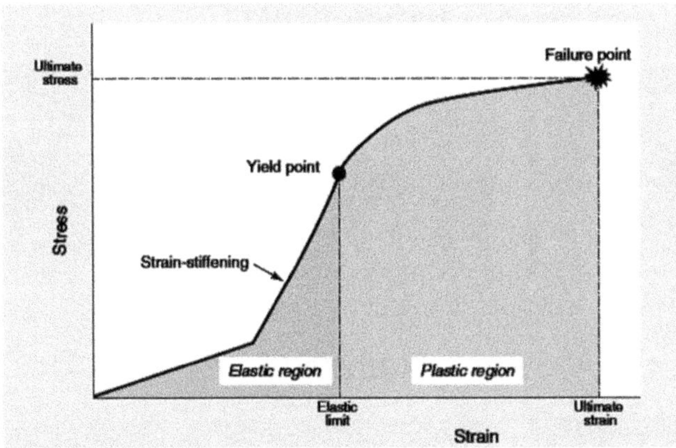

Figure 14.7. A general stress–strain curve related to a non-linear material behavior. At small strains, the stress increases linearly. Then, the elastic modulus can be calculated from the slope of the line. The strain will go back to zero in the elastic region if the deformation's load is removed. At larger strains, a non-linear material deforms easily. If the load is removed, the strain does not fully recover in the plastic region. At some point, the strain is larger and the non-linear material breaks (as viscous flow). The curve's magnitudes and shapes will change if the strain rate force changes.

One of the advantages of optical non-destructive tests relies on their high sensitivity, i.e. the tests do not require destroying the object, and even a very small deformation can give enough mechanical information. The strain–stress information is retrieved from the displacement maps calculated from the optical phase maps [31]. A mathematical model converts the measured full-field displacement maps, for each axis, into the principal strain components [32]. This process involves an approximation that makes it possible to have a finite range where the partial derivatives result in a useful strain or stress map. A clear example of this is reported in [33], where a transparent sample (a plaque with a central hole) made with acrylic (poly-methyl-methacrylate (PMMA)) was tested under a vertical tension load. The optical system uses a Mach–Zehnder interferometer configuration, and figure 14.8 shows the experimental and simulated stress values (σ_{00}) for different tension values.

In this case, the sample may be considered as an optically flat surface since it does not add speckle noise to the fringe patterns, a favorable fact that results in smooth stress maps. The authors of [34] report an elastic membrane (made with latex) deformed due to a controlled vibration, where DHI retrieves its elasticity modulus. As the experimental conditions were known, it was possible to acquire a series of holograms at various points (amplitudes) of the sinusoidal vibration and then calculate the strain–stress average values. The elasticity modulus is obtained for this material and compared with information available in the literature. Figure 14.9(a) presents the wrapped phase maps of the elastic membrane at different points during the applied sinusoidal signal. Figure 14.9(b) shows a table indicating the elasticity modulus of the membrane according to the different sound pressures applied to the membrane. From these results, it is possible to see that the high sensitivity of DHI can detect displacements on the order of tenths of nanometers, which modifies the Young's modulus value in hundredths.

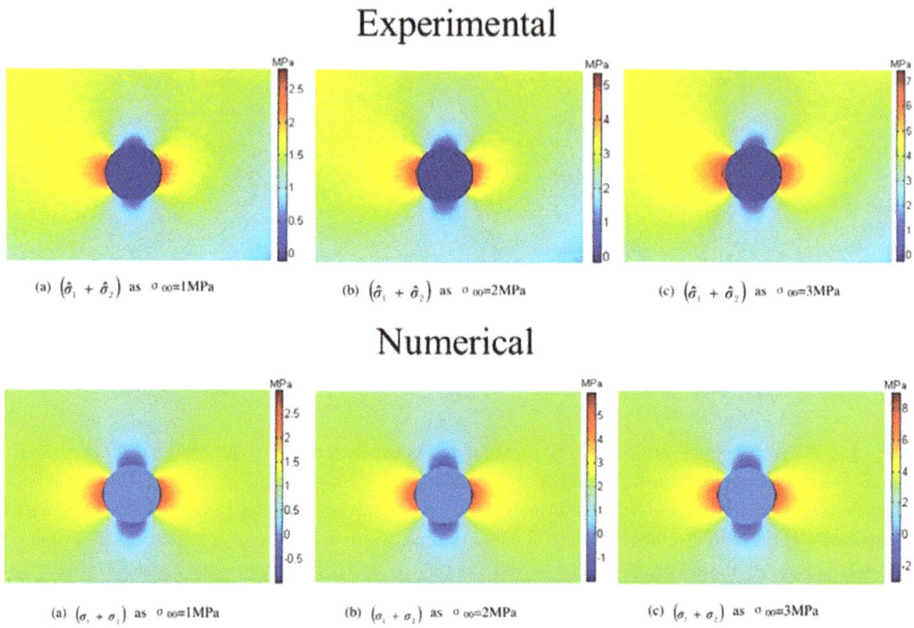

Figure 14.8. Experimental and numerical full-field stress comparison in a transparent acrylic sample (that has a hole in its middle). Reproduced with permission from [33]. Copyright 2016 Elsevier.

The study of the mechanical behavior of objects using DHI is not limited to engineering samples; it also has been applied to biological samples such as bones. This natural composite material offers a complex anisotropic response when its two phases (organic and inorganic) interact during a deformation. In [35] two different regions of a bovine femoral bone were analyzed. The sample was compressed under controlled conditions to simulate physiological and overload conditions. A series of holograms was recorded during the compression to retrieve the strain field map at different stages. The tests were performed in the cortical zone (outer region) and a transversal cut that exposed its yellow marrow. The experimental conditions for these two sections are presented in figures 14.10 and 14.11, respectively.

After the compression test, the holograms were processed in the manner indicated in chapter 3, i.e. using the spatial frequency carrier method (Fourier method). Then, the displacement data were processed and converted into the x and y principal strain components and the xy shear values. The resulting strain maps are shown in figures 14.12 and 14.13 for the cortical and transversal bone sections, respectively. In these figures, it is possible to see the remarkable mechanical differences between these two sections of the bone. The highly inorganic cortical bone has a rigid response, while the transversal one, due to the yellow marrow, is highly deformed. These differences can be explained with respect to the optical phase in each case. The cortical region is inorganic, while the transversal one has a strong organic component. The possibility of having the full-field strain map information for material like the bone helps to understand better the mechanical dependency

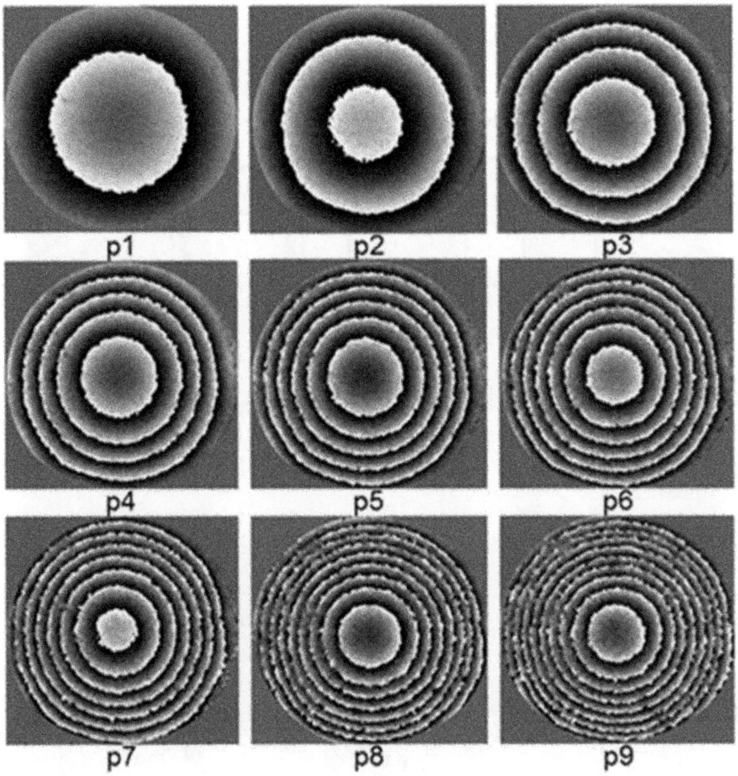

(a)

Sound Intensity (dB)	Pressure Max. (Pa)	Deformation z (μm)	Young's Modulus (MPa)
82.6	0.41	0.49	1.74
84.9	0.53	0.65	1.71
86.8	0.66	0.92	1.5
88.9	0.84	1.13	1.56
90.5	1.00	1.39	1.52
91.6	1.20	1.62	1.48
92.6	1.30	1.89	1.43
93.3	1.40	2.13	1.37
94.2	1.60	2.21	1.47

(b)

Figure 14.9. (a) Wrapped phase maps at several instants of the vibration cycle and (b) the elasticity modulus at different sound intensities. Reproduced with permission from [34]. Copyright 2011 The Optical Society of America.

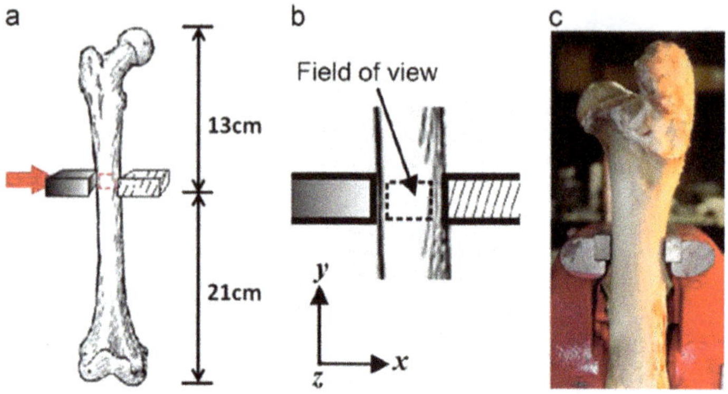

Figure 14.10. Schematic representation of the bone's loading: (a) the red arrow on the left shows where the compression was applied while the right side of the object is fixed; (b) the observed area shown by dotted lines; and (c) a picture of the bone fixed in place during the first test. Reproduced with permission from [35]. Copyright 2014 Elsevier.

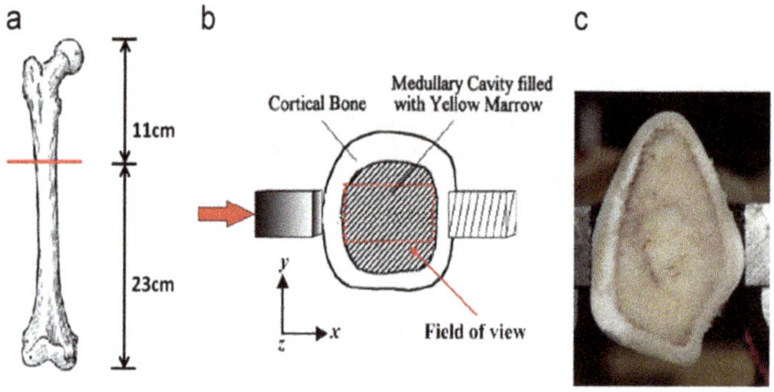

Figure 14.11. (a) Schematic representation of the bone where the red line indicates the cut line; (b) transversal bone region with the FOV indicated within the red dotted line; and (c) photograph of the bone fixed in place for the second test. Reproduced with permission from [35]. Copyright 2014 Elsevier.

between its organic and inorganic phases. The full field, high sensitivity, and resolution of DHI for mechanical inspections are continuously used and applied to different materials and conditions [36–40].

14.2.3 Vibrational analysis with DHI and ESPI

Another essential feature of materials in many engineering applications is their behavior when subjected to vibrations. Due to its many capabilities (i.e. full field of view, measuring accuracy, high sensitivity, and resolution), DHI has been extensively proved to be suitable for measuring vibrations on materials. With only one image from a hologram, it is possible to measure mechanical vibrations or analyze the frequency response of several materials. This vibrational information is useful since it gives maps or data to prevent fatigue failure in various materials and objects. There are reports on several methods to analyze vibrations with HI. For example,

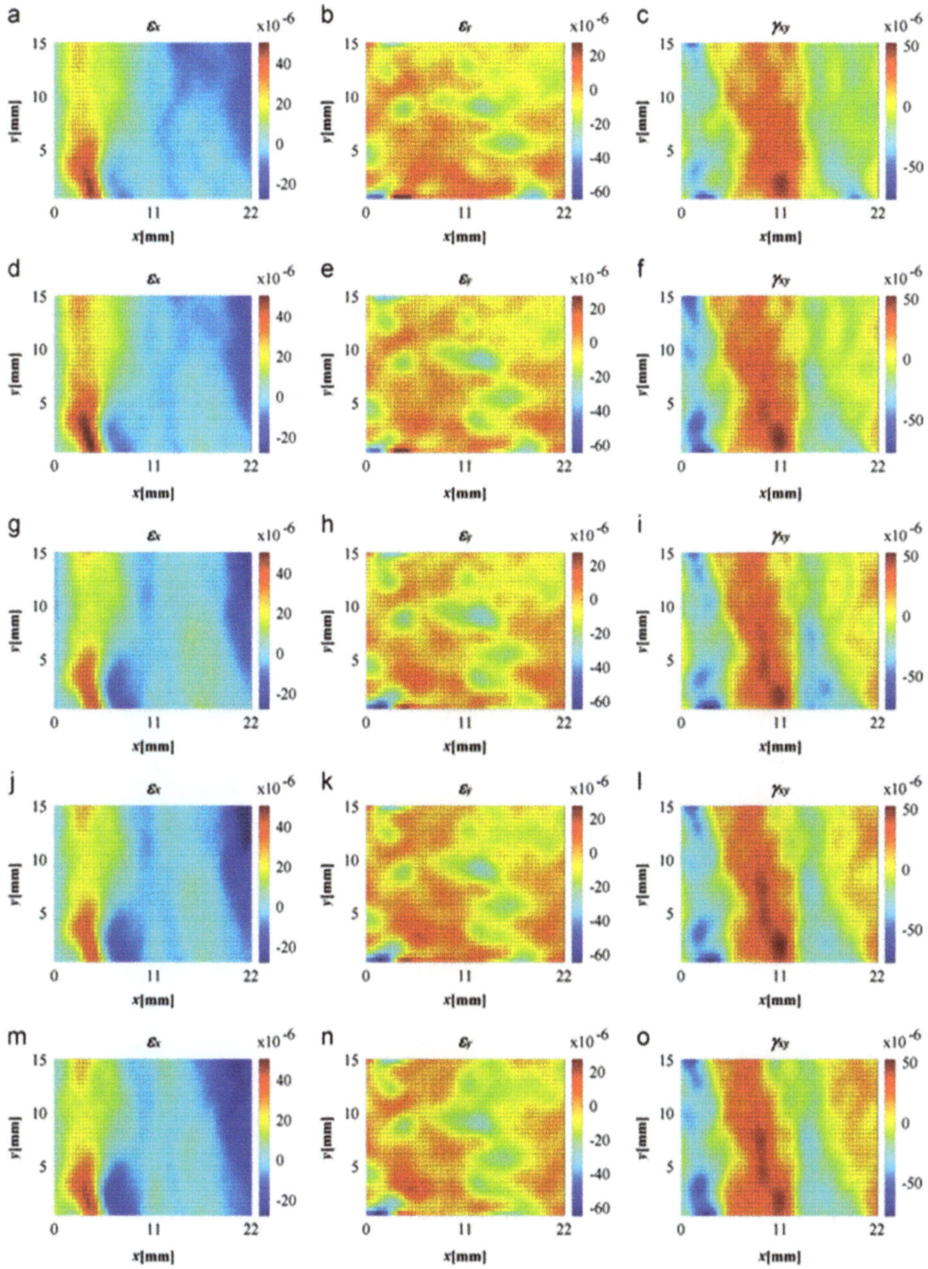

Figure 14.12. Experimental strain maps (ε) from the cortical (longitudinal) test: (a), (d), (g), (j), (m) for *x*; (b), (e), (h), (k), (n) for *y*; and (c), (f), (i), (l), (o) for the *xy* shear strain values. Reproduced with permission from [35]. Copyright 2014 Elsevier.

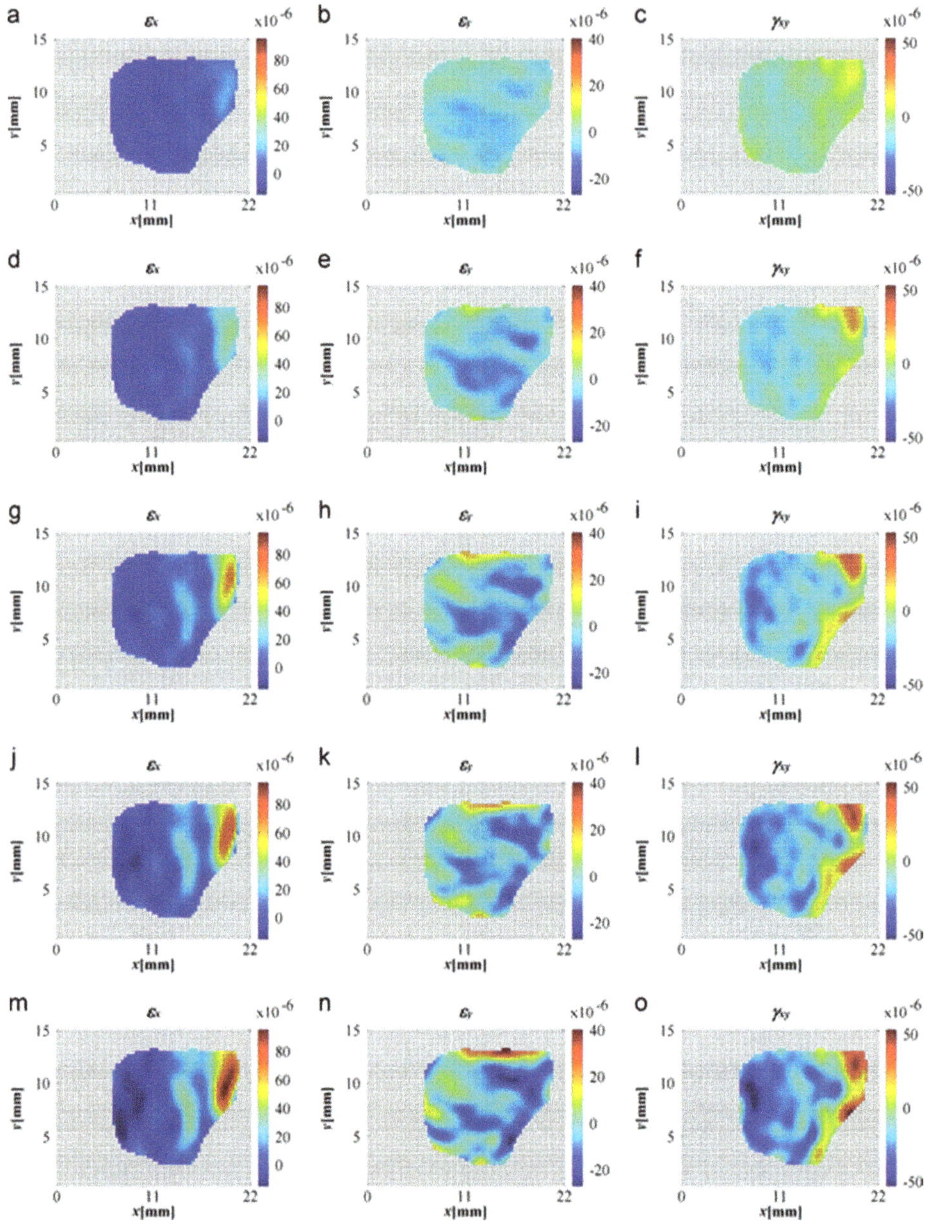

Figure 14.13. Experimental strain maps (ε) from the transversal test: (a), (d), (g), (j), (m) for x; (b), (e), (h), (k), (n) for y; and (c), (f), (i), (l), (o) for the xy shear strain values. Reproduced with permission from [35]. Copyright 2014 Elsevier.

the work in [41] gives a general classification of the methods based on HI to measure vibrations. Currently, the techniques mostly used are the time-average and pulsed or stroboscopic DHI (see chapters 6 and 8). Each has certain advantages over the others, depending on the specific vibration measurement applications. The basic principle of pulsed or stroboscopic DHI techniques is the same. The difference remains in using a sequence of pulses as illuminating beams. This property helps to overcome the problem of recording a hologram of an object vibrating simultaneously at several frequencies. By using a well-defined sequence of pulses, the different mode shapes can be separated. This technique has been used when recording events that happen during short times [42]. A suitable illumination source for these applications is a pulsed laser, where a high-intensity beam with femto or even attosecond duration can be obtained. These pulses must be synchronized with short exposure camera times. If a pulsed laser is not an option (sometimes the budget is not large enough), electro-optic modulators are also used for stroboscopic illumination [43]. Exact time synchronization between the strobed laser, the camera, and the vibrating signal applied to the sample is needed. Pulsed or strobed illumination has been applied to analyze rotating objects [44], vibrational characteristics of noisy brackets [45], biological tissues under vibration [42], and vibrating systems in general. Figure 14.14 shows the instantaneous phase change provoked by the spark discharge on atmospheric pressure air conditions [46], using a femtosecond light pulse and a very high-speed frame-rate camera.

According to reference [47], pulsed digital holography usually records two states of the vibrating object: the first one on its maximum displacement and the second on its maximum negative displacement. The resulting fringes are like those in double exposure holography (the first demonstration of HI [4, 5]). The interference phase at each point is given by the scalar product of the sensitivity and displacement vectors (see for instance chapter 8).

The most common images formed in vibration analysis are related to sinusoidal or Bessel-shaped fringe patterns. In the first case, the pattern is related to a load showing an equal brightness across the image, without regard to the fringe order.

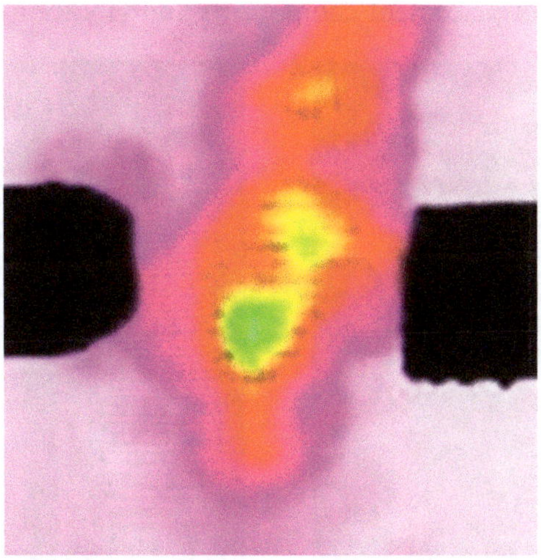

Figure 14.14. Phase-shifting digital holography using a single femtosecond light pulse. The author used as an object the set spark discharge in atmospheric pressure air induced by applying a high voltage between two electrodes. Reproduced with permission from [46]. Copyright 2012 The Optical Society of America.

The second case, the Bessel pattern, is usually associated with object vibrational loads and is characterized by the maximum brightness for the zeroth-order fringe and its decrease for the higher fringe orders. Figure 14.15 depicts both situations.

In time-average DHI, the object under investigation vibrates at an angular frequency ω, while the camera sensor opens for a period of time T much greater than one period of vibration cycle of the object. The hologram effectively records an ensemble of object images corresponding to the time-averaged complex amplitude of the light scattered. The object image comprises a set of interference fringes of unequal brightness (figure 14.15, left panel) that vary according to the square of the zeroth-order Bessel function of the first kind. The experimental se-up for recording a time-average hologram is the same for double exposure holography.

An exciting application has been reported in [49], where the authors study the vibrations on cutting tools using time-averaged amplitude fluctuation (AF) based on ESPI, see figure 14.16. This experiment uses a Michelson-type interferometer with ESPI for static and out-of-plane vibration displacement measurements.

As pointed out before, ESPI is a non-invasive full field of view optical technique that enables interferometric measurements of surface displacements and deformations. In reality, its underlying working mechanisms are very similar to those in DHI (remember that we stated at the beginning of this chapter that a major difference is that the optical phase in ESPI is retrieved via phase stepping methods, while DHI uses the fringe carrier approach, the Fourier method). Both use the speckle pattern produced when an object with a rough surface is illuminated with a laser. Then, a speckle pattern correlation is applied (usually a subtraction) before and after an object's deformation. ESPI measures the by evaluating the phase difference of two recorded speckle interferograms under different loading conditions. The basis of ESPI for measuring displacement components is to assess the relative phase difference between the object and reference beams [50]. Using ESPI, it is possible to measure different displacement components and vibration modes.

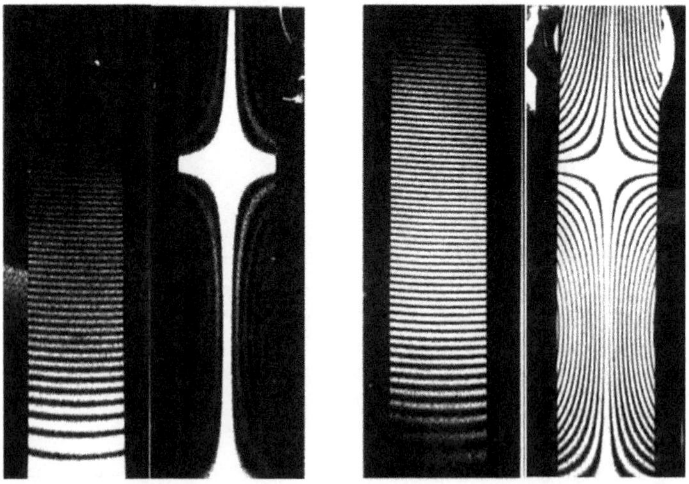

Figure 14.15. Fringes of zeroth and higher orders for a vibrational load (left) and static load (right). Reproduced with permission from [48]. Copyright 2001 Springer.

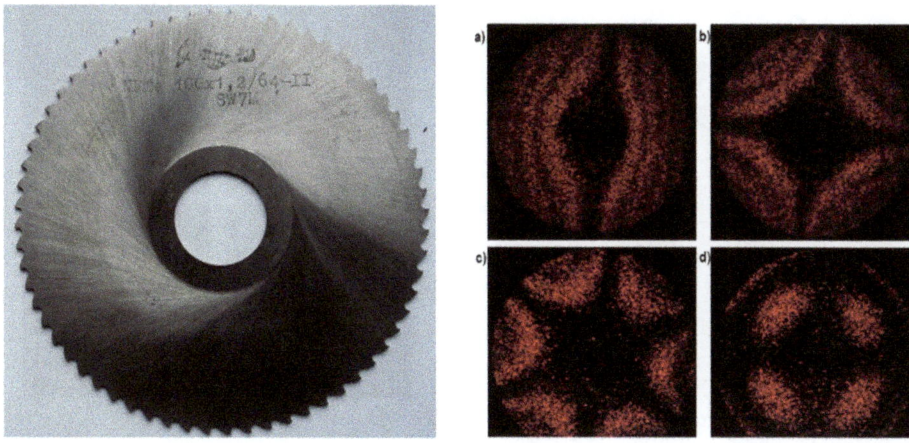

Figure 14.16. Circular slitting saw (left) and its mode shapes (right) at different resonant frequencies: (a) f = 607 Hz, (b) f = 840 Hz, (c) f = 1495 Hz, and (d) f = 5190 Hz. Reproduced with permission from [49]. Copyright 2018 De Gruyter Poland.

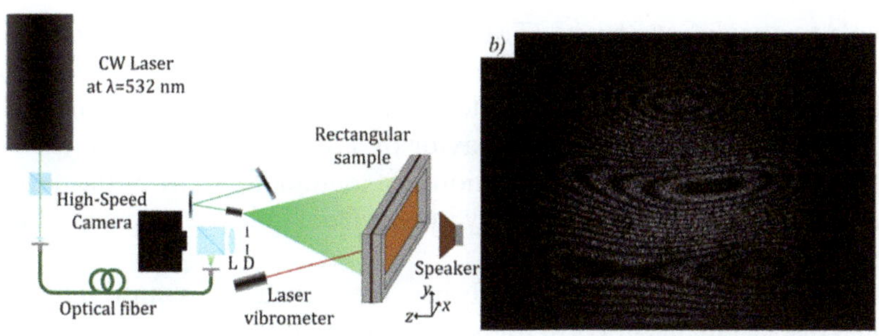

Figure 14.17. Experimental set-up to measure the vibration in a rectangular leather sample. The electronic speckle pattern interferometer uses a high-speed camera to observe the full-field deformation, while the laser vibrometer quantifies the deformation of the central point of the membrane. L: lens with f = 60 mm. D: diaphragm to control the size of the speckle. ESPI patterns for the resonant modes: (a) $m = n = 1$ and (b) $m = 1$, $n = 3$ for a representative sample. Reproduced with permission from [51]. Copyright 2018 Elsevier.

As an additional example, ESPI was used to measure the resonant frequencies of tanned leather. See figure 14.17, where the optical layout used is shown. It combines the single-point measurement of a laser vibrometer with the full-field observation of ESPI [51].

It is important to remark that both time-averaged DHI and ESPI help in studying vibrating objects even though there is a need to know *a priori* the direction of the movement of the sample (due to the lack of the optical phase appearing in the argument of the Bessel function). Otherwise, pulsed or stroboscopic DHI or ESPI present a solution to overcome this situation [52].

However, another interesting area of opportunity for vibration analysis is the vibrating characterization of biological samples. For example, figure 14.18 shows the dynamic excitation by acoustically induced motions on a postmortem

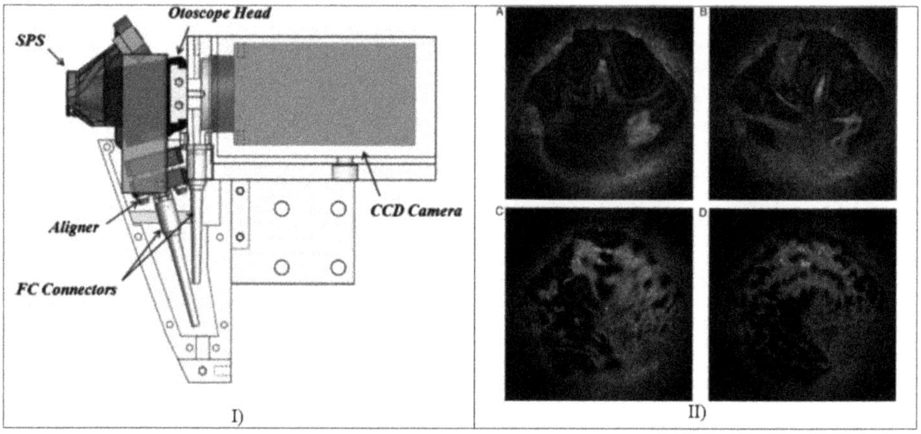

Figure 14.18. (I) A HI system packaged in a custom design (sound presentation system (SPS)) and (II) time-average information of deformations of a postmortem chinchilla TM subjected to a controlled acoustic signal: (A) 0.776 kHz, (B) 1.085 kHz, (C) 3.489 kHz, and (D) 7.675 kHz. Reproduced with permission from [53]. Copyright 2011 Wiley.

chinchilla's tympanic membrane (TM) using an ad hoc DHI otoscope [53]. Here, qualitative and quantitative data from the deformations of a TM subjected to controlled acoustic signals were reported, giving the hearing specialists important information about the mechanical behavior of TMs at different frequencies. Other applications of DHI for 3D reconstruction of a mammalian TM or the development of robust algorithms to recover intensity images from holograms acquired with DHI have been reported in [54, 55].

It us worth mentioning that DHI has been used to quantify the dynamic vibrational displacements in vocal folds in mammals and recently for the vibrational study of skin rigidity variation under UV radiation [56, 57].

14.2.4 Flow measurements

DHI's versatility is continuously exploited in new and more demanding fields, such as the study of fluid flows. The applications include various flow conditions such as speed, density, transparency, temperature, etc. Most of the DHI set-ups use a Mach–Zehnder configuration with separate paths for the object and the reference beams, where the object under study is located along the object beam's path (in-line mode). One of the first reports using DHI in a flow is in [58], where the heat from a solar flat plate collector is analyzed. The work also deals with the signal filtering as this in-line configuration conveys all the object beams to the camera sensor. In addition to the phase information, these beams carry noise such as parasitic rings, glares, and excessive speckle noise.

An example can be found in [59], where the authors describe the axisymmetric flow temperatures produced from an alcohol flame. A mathematical approximation to retrieve the radial temperature distribution is proposed. Figure 14.19 shows the flame image and the resulting fringe pattern.

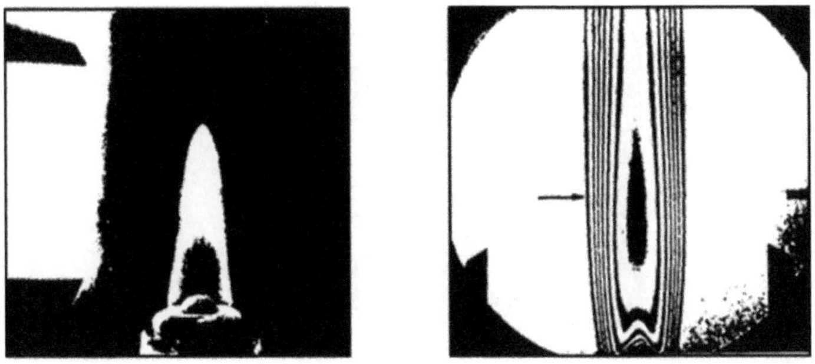

(a) (b)

Figure 14.19. (a) A flame is produced by burning alcohol and (b) the resulting DHI fringe pattern. Reproduced with permission from [59]. Copyright 1988 SPIE.

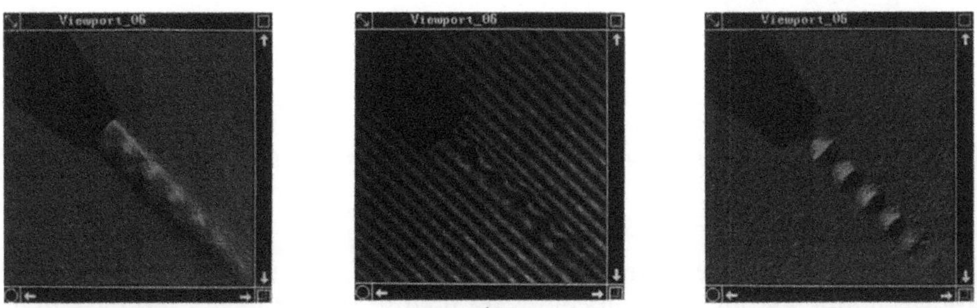

(a) (b) (c)

Figure 14.20. (a) Interferogram of an axisymmetric jet of air; (b) interferogram showing reference fringes parallel to the jet axis; (c) computed image (knife-edge normal to the jet axis). Reproduced with permission from [60]. Copyright 1990 Elsevier.

Faster speed flows have been analyzed with DHI [60, 61], see figure 14.20(a), where examples of images from [60] are presented. A jet flow was observed interferometrically, and the resulting data were processed digitally. At this time, digital data post-processing started to play a relevant role in this kind of test as more information could be obtained, see figures 14.20(b) and (c).

An example of the importance of image post-processing is presented in figure 14.21, obtained when the flow was observed during a gas quenching process [62]. A pulsed laser is used as the illumination source as it is impossible to isolate the hot and moving object. After the data were processed, the flow is shown in figure 14.21(a) as a wrapped phase map, and from it a pseudo-3D displacement map, a mesh plot, can be seen in figure 14.21(b).

In heat flows such as those coming from flames, the combustion process modifies the surrounding refractive index of the air. This feature has been used to estimate the temperature distribution in several research reports [63, 64]. In the latter, a comparison between the refractive index values and the optical phase was studied, and part of the results are presented in figure 14.22, at different heights of the flame.

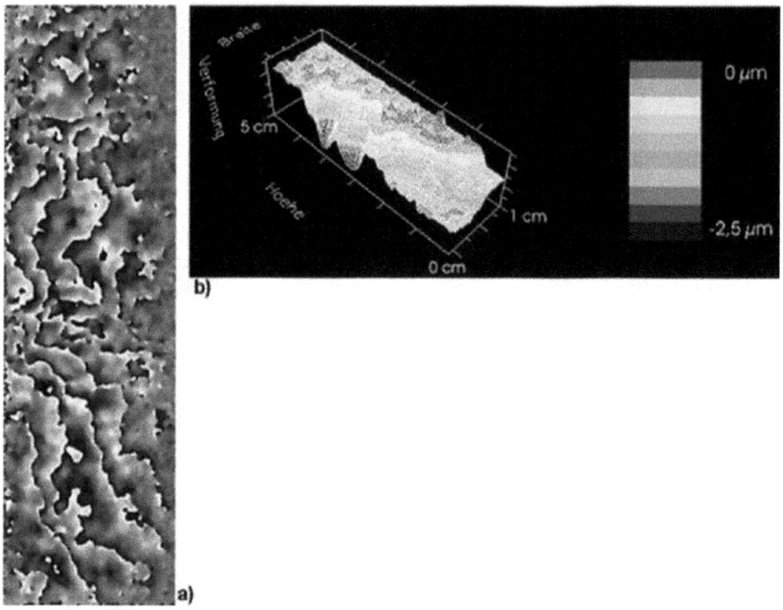

Figure 14.21. (a) Phase difference and (b) displacement field. Reproduced with permission from [62]. Copyright 2006 Wiley.

Some flow studies are carried out with alterations in the optical set-up, as is the case when different illumination colors are used to inspect it. For instance, in [65] the authors simultaneously use three different lasers (blue, green, and red) to detect the refractive index variations from a candle flame. Also, a diffusion flame has been studied to see how a magnetic field affects it [66]. The flame's temperature variations are related to the optical phase behaving as the magnetic field increases, decreases, and remains at a steady state, as seen in figure 14.23.

It was observed that a flow modifies its surroundings; however, this perturbation could be generated not only from a flame but also by an electric arc which can create a measurable perturbation in the medium [67]. This kind of study requires high-speed hardware to register the fast event. An example of this is reported in [68], where the ignition process from a butane–air mixture flame is recorded. Thanks to a high-speed camera, it is possible to track the ignition process within the first milliseconds after the electric arc appears. The optical system recorded holograms at 1000 fps, and it was possible to observe different stages of the combustion process. Figures 14.24 and 14.25 show an image sequence at 100 fps and some selected unwrapped phase maps of the ignition process recorded at 3000 fps, respectively.

Studying fast flows with DHI allows registering supersonic flows and their interaction with the surroundings [69] or even ultra-fast flows constrained in narrow channels [70]. In this application, the boundary conditions of the channels accelerate the flame as the channel reduces its transversal area. In figure 14.26 the schematic experimental set-up is shown, while in figure 14.27, some test results are presented.

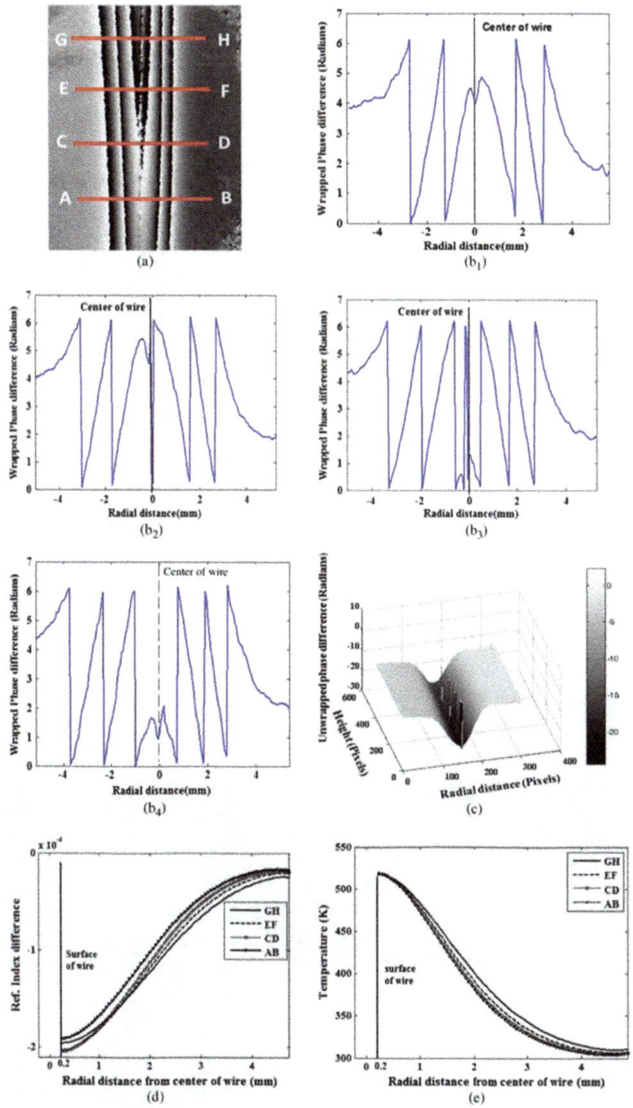

Figure 14.22. (a) Phase map showing the difference between air without and with heat; the object was a wire of diameter 0.4 mm, and the power applied to heat the wire was 70 W. (b1)–(b4) Line profile of the phase difference along the lines (as marked in (a)) AB, CD, EF, and GH, respectively. (c) 3D phase difference mesh plot calculated from (a). (d) Refractive index difference distribution along with the AB, CD, EF, and GH lines. (e) Temperature profiles of the heated air along with the AB, CD, EF, and GH lines. Reproduced with permission from [64]. Copyright 2014 The Optical Society of America.

These figures make it possible to see the flow expansion within the channel and its acceleration in only a few microseconds.

At this point, the flow has been analyzed from a 2D perspective, but a particular case occurs when the flow is analyzed in 3D as a tomography. Here, the

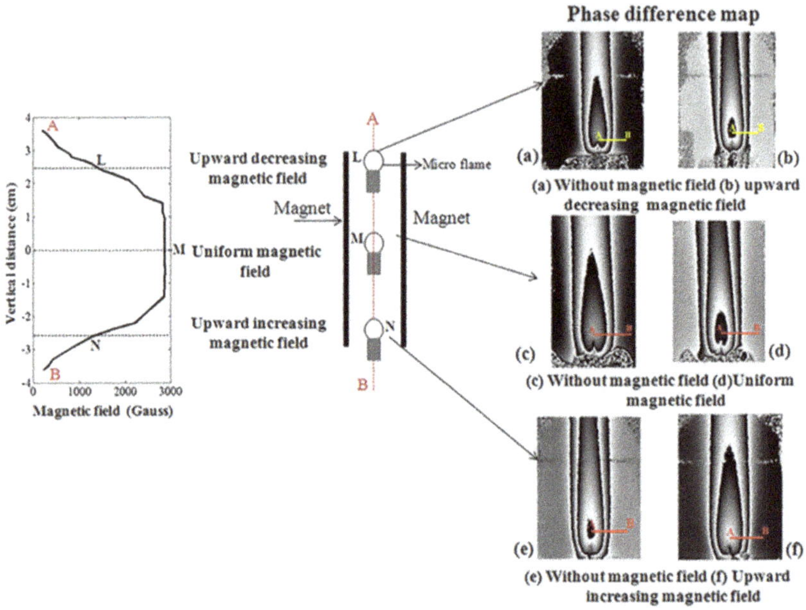

Figure 14.23. Schematic of the interaction between three different settings for the magnetic field and the difference between the air in the absence and presence of a micro flame (shown as wrapped phase maps). Reproduced with permission from [66]. Copyright 2018 Elsevier.

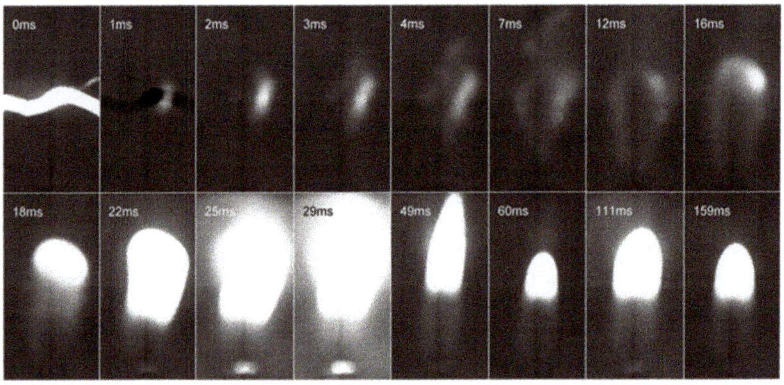

Figure 14.24. A sequence of the ignition process images was captured with the CCD camera. Some representative frames are shown and labeled with the time elapsed since the electric arc appears. Reproduced with permission from [68]. Copyright 2017 The Optical Society of America.

interferometer's sensitivity needs to be increased on more than one axis. The authors of [71] present the temperature profile for a flow coming out of three orifices of a synthetic jet actuator. The DHI flow measurements were compared with constant current anemometry (CCA) information. Figure 14.28 shows the 3D flow representation and the temperature profiles for each flow column at different planes. The

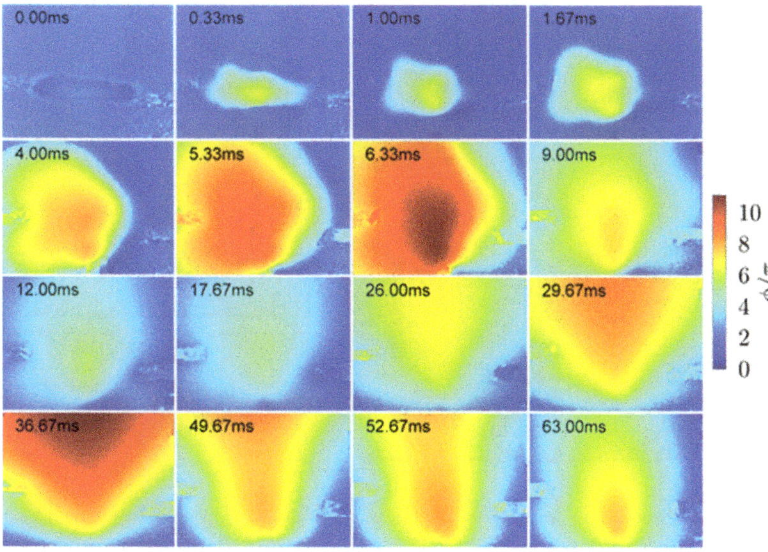

Figure 14.25. Ignition process sequence from DHI data, showing unwrapped optical phase maps. Some representative frames are shown and labeled with the time elapsed since the electric arc appears. Reproduced with permission from [68]. Copyright 2017 The Optical Society of America.

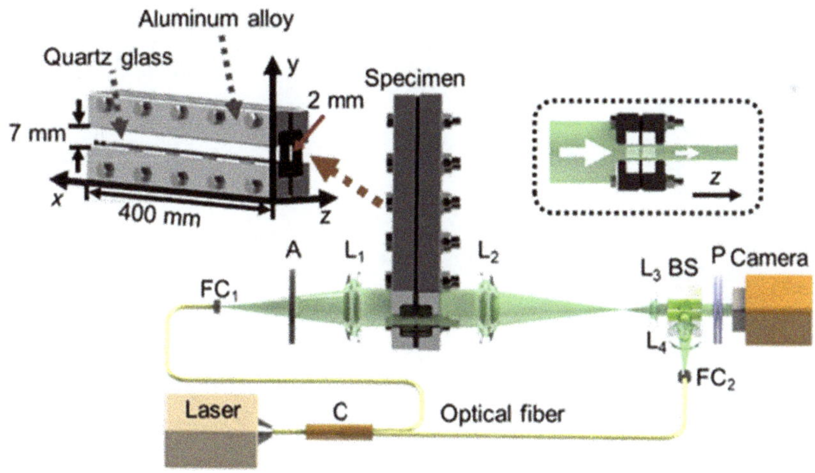

Figure 14.26. Experimental DHI set-up for measuring the combustion process in a narrow channel. C: fiber coupler; FC_1 and FC_2: fiber connectors; A: aperture; L_1, L_2, L_3, and L_4: lenses; BS: beam splitter; P: polarizer. Reproduced with permission from [70]. Copyright 2018 The Optical Society of America.

3D data were matched with the temperature values, and a full-field inspection was reached with this optical configuration.

A particular case of study in flows is in liquids and the interactions therein for which several reports use DHI [72–77]. An interesting example is presented in [78], where the phase information obtained can detect concentration variations in the

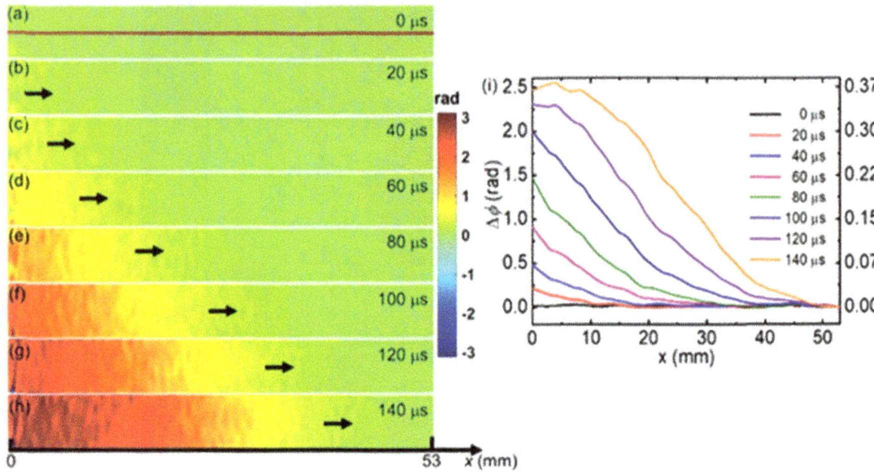

Figure 14.27. (a)–(h) Phase images of the compression wave. (i) Density variation from 0 to 140 μs along the red line in (a). Reproduced with permission from [70]. Copyright 2018 The Optical Society of America.

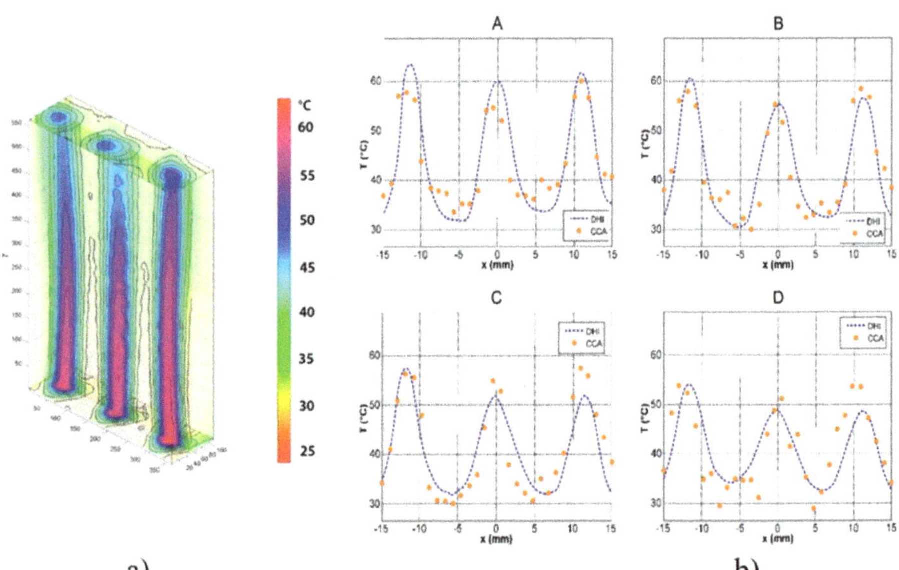

Figure 14.28. (a) Tomographic reconstruction of the temperature field for a steady flow and (b) temperature profiles measured over a distance of 5 mm (A), 10 mm (B), 20 mm (C), and 40 mm (D) from the orifices. Reproduced with permission from [71]. Copyright 2019 MDPI.

crystallization process of liquids. For this purpose, a cuvette with the solution is situated in the object's arm, and a temporal recording process is performed. The results show the phase variations according to the crystallization stage of the liquid. See figure 14.29, which shows a fringe number increase after 4750 min compared to 200 min. Another example of liquid concentration distribution is in [79], where an

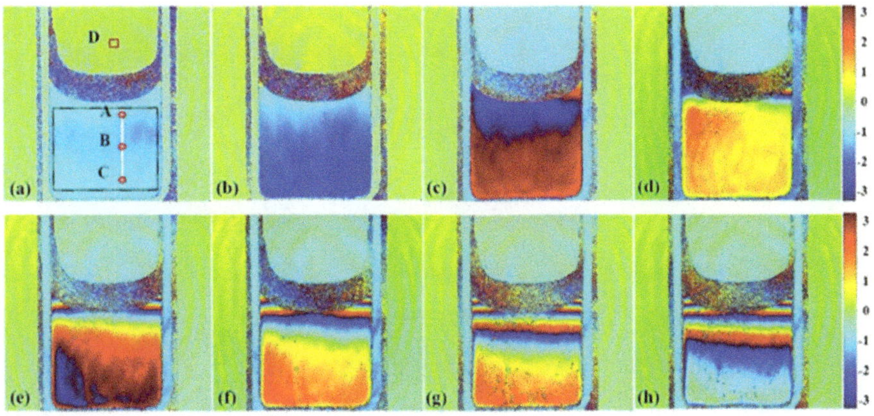

Figure 14.29. Reconstructed two-dimensional wrapped phase distributions of the solution region at different times (t): (a) $t = 200$ min; (b) $t = 1000$ min; (c) $t = 1550$ min; (d) $t = 2350$ min; (e) $t = 3050$ min; (f) $t = 3700$ min; (g) $t = 4200$ min; and (h) $t = 4750$ min. Reproduced with permission from [78]. Copyright 2012 The Optical Society of America.

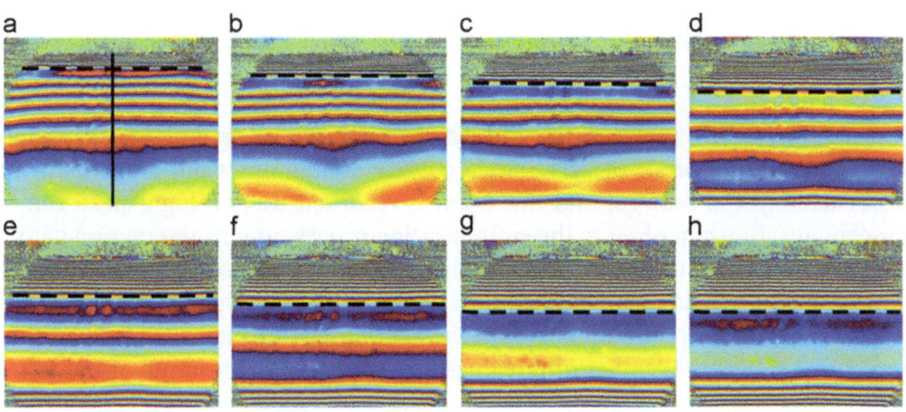

Figure 14.30. Wrapped phase maps representing the concentration distributions at different times during a reaction–diffusion process: (a) 100 s; (b) 200 s; (c) 300 s; (d) 400 s; (e) 500 s; (f) 600 s; (g) 700 s; and (h) 727 s. Reproduced with permission from [79]. Copyright 2014 Elsevier.

unstable reaction–diffusion process is studied. Figure 14.30 shows a time-lapse when this reaction occurs, and equal optical phase layers within the liquid are observed. The use of DHI makes it possible to detect the motion of drops by thermo-capillary [80]. This work uses a layer of a liquid and drops from another type of liquid, i.e. both have different viscosity, and their interaction during the tests registers the migration of the drops (see figure 14.31).

The available information using DHI in these tests contains essential data and represents a growing area of research and continuous development. A recent work substitutes the in-line Mach–Zehnder configuration for one where the object beam is not directly illuminating the camera's sensor [22]. This configuration eliminates the parasitic patterns, speckle noise, and illumination saturation. In addition, a neutral

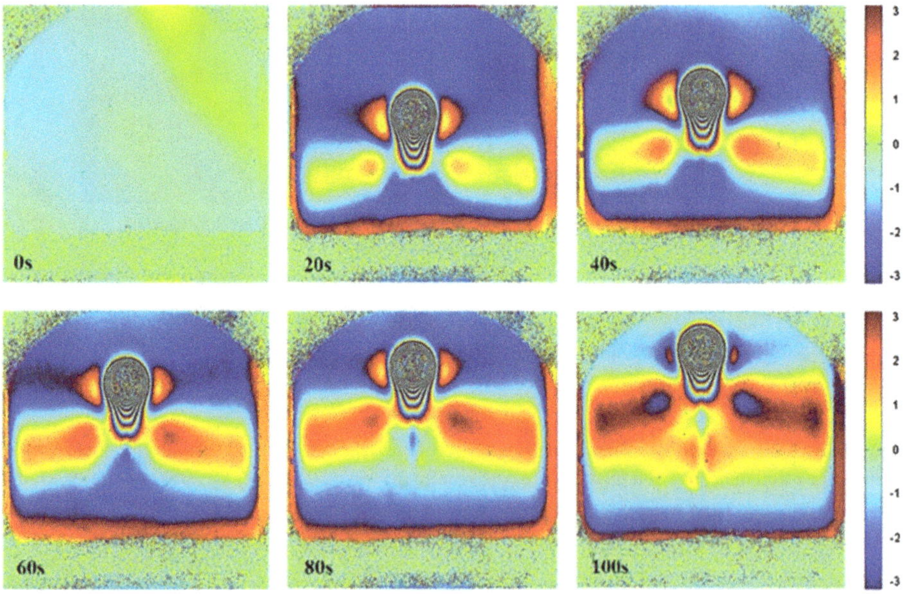

Figure 14.31. Reconstructed two-dimensional wrapped phase distributions at different times of a drop migration. Reproduced with permission from [80]. Copyright 2016 Springer.

phase screen (NPS) replaces the diffuse screen used in several Mach–Zehnder interferometers. As proof of principle, a cuvette containing water is heated with a resistor fixed at the bottom, and this event is recorded using a high-speed camera. This configuration can observe the entire heating process and its interaction with the liquid as the temperature rises. The simplicity of the optical set-up is tested further by mixing liquids with temperature differences <0.1 °C. Figure 14.32 shows the phase maps of a temporal lapse during the heating test of the cuvette's water, and figure 14.33 shows the temperature values for some selected instants of the test.

14.2.5 Measuring refracting index changes

Phase objects are transparent or translucent, meaning that the amplitude of a wavefront is little affected while passing through, but phase changes are not negligible. In these applications, the optical path difference varies due to the changes in the refractive index in the path that light travels through the sample. A typical HI configuration to measure the refractive index change of a phase object is based on an off-axis set-up. Using the double exposure method we can replace the dot product of the displacement and sensitivity vector by

$$\Delta\varphi(x, y) = \left(\frac{2\pi}{\lambda}\right) \int \Delta n(x, y, z) \mathrm{d}z, \tag{14.1}$$

where $\Delta n(x, y, z)$ represents the refractive index change through the object. Then, quantitative measurements of refractive index distributions on phase objects can be calculated.

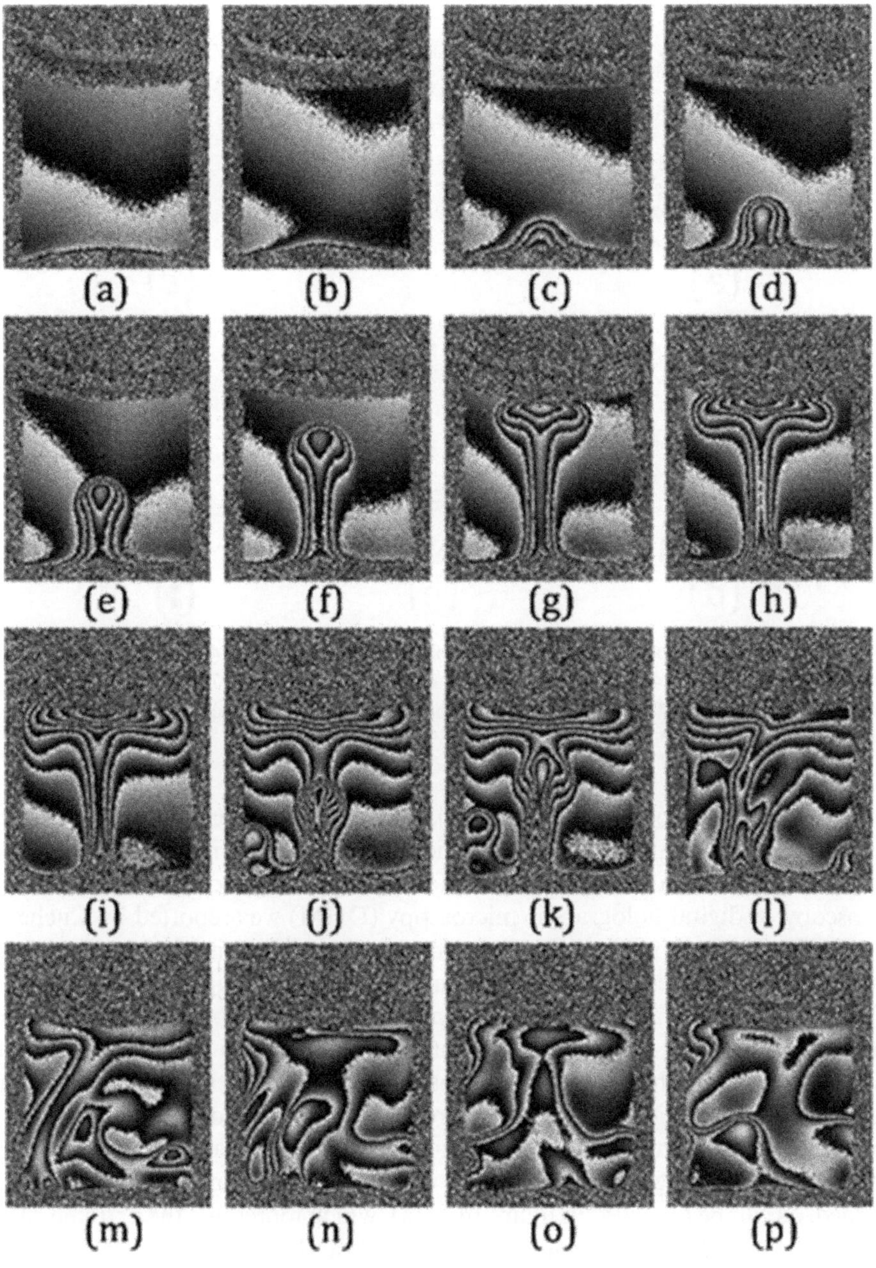

Figure 14.32. Wrapped phase maps at different times while a resistor heats the bottom of a water-filled cuvette. The field of view was 22.60 × 19.06 mm with a camera resolution of 2560 × 2160 pixels. Reproduced with permission from [22]. Copyright 2018 The Optical Society of America.

Measurements of changes in refraction index in microscopic transparent or semi-transparent samples with DHI have been addressed using interference microscopy techniques such as the Mirau or Linnik interference microscopes. These are based on

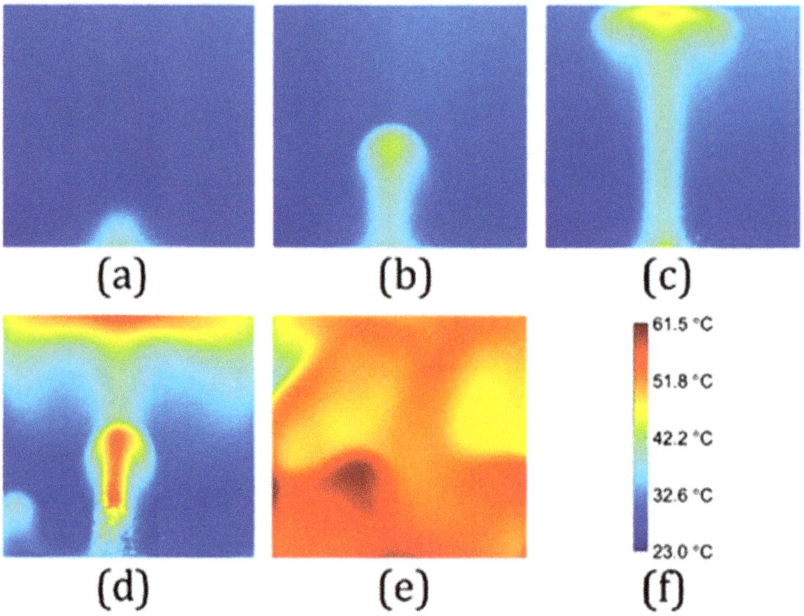

Figure 14.33. Temperature values from the wrapped phase maps of (a) figure 14.32(c); (b) figure 14.32(e); (c) figure 14.32(g); (d) figure 14.32(j); and (e) figure 14.32(p), which was at 47.46 s after the resistor was disconnected; (f) color bar for temperature scale (for reference). Reproduced with permission from [22]. Copyright 2018 The Optical Society of America.

spatial carrier fringe-pattern analysis [81], and their developments are a version of the classical Zernike and NorMarski optical phase-contrast microscopy [82].

A step forward and crucial advance for biological applications based on DHI microscopy or digital holographic microscopy (DHM) was reported by Cuche *et al* in 1999 [82]. The authors used a reconstruction method that simultaneously provides amplitude and phase-contrast information using a single hologram. A second report [83] implements these algorithms based on the Fresnel approximation to recover an epithelial cell's amplitude and phase-contrast image, shown in figure 14.34. Their algorithm involved the following reconstruction parameters: the reconstruction distance, the calculation of the digital reference wave, and the calculation of a digital phase mask; these must be adjusted to compensate for the wavefront curvature of the holograms. If these are not well adjusted, no information can be recorded, as can be observed in figure 14.34(a), which represents the extreme case on an image obtained without a digital phase mask and a digital reference wave; the remaining cases of figure 14.34 shown some variations of the parameters involved in the algorithm.

Several biological samples have been studied using DHI microscopy that follow the above ideas and results. Next, we will cite some recent applications to study bio-samples with DHM. Shown in figure 14.35, Bianco *et al* [84] report a hybrid system consisting of an optical tweezer coupled to DHM and used to quantify the bio-volume of red blood cells (RBCs). Shown in figure 14.36, Rastogi *et al* [85] show the

Figure 14.34. Five phase-contrast images of an epithelial cell change the fundamental parameters for numerical reconstruction: (a) without either a right digital phase mask or a digital reference wave; (b) with a digital phase mask and without a digital reference wave; (c) and (d) with the correct phase mask and the wrong reference wave, at different values; and (e) the correct digital phase mask and reference wave. Reproduced with permission from [83]. Copyright 1999 The Optical Society of America.

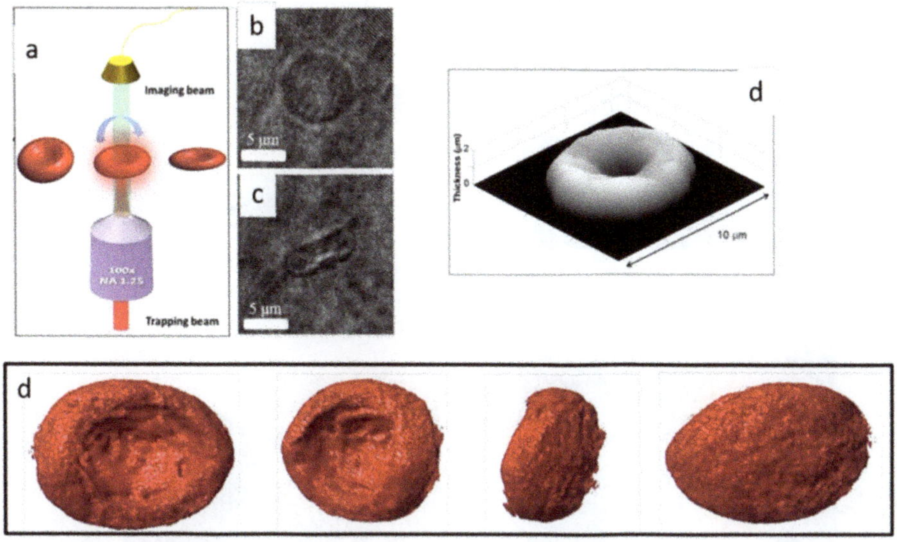

Figure 14.35. Quantitative estimation of the bio-volume of RBCs. Reproduced with permission from [84]. Copyright 2019 SPIE.

reconstructed 2D phase profile of an *E. coli* bacterium where its thickness is calculated. As can be seen in figure 14.37, Barroso *et al* [86] present a multispectral quantitative phase imaging system with DHM to characterize a dissected murine retina by a label-free refractive index measurement. The wavelength is tuned from 800 to 850 nm at steps of 10 nm.

The authors of [87] present interesting results acquired using a commercial DHM from Lyncée Tec., showing the advanced stage of DHI techniques. Figure 14.38 depicts reconstructed holographic images of hippocampal cells acquired with a 50× objective at 666 nm. The authors of [88] present a clever combination between a smartphone, a DVD optical head, and DHM algorithms to perform imaging of cells. The results are shown in figure 14.39, where we can see the reconstructed image from a smear of blood cells. By using partially spatial coherent light with DHM, Dubey *et al* [89] showed how sperm cell motility is affected when exposed to different levels of oxidative stress, see figure 14.40. Brodoline *et al* [90] show a complete 4D (time taken as the fourth dimension) characterization of RBCs inside

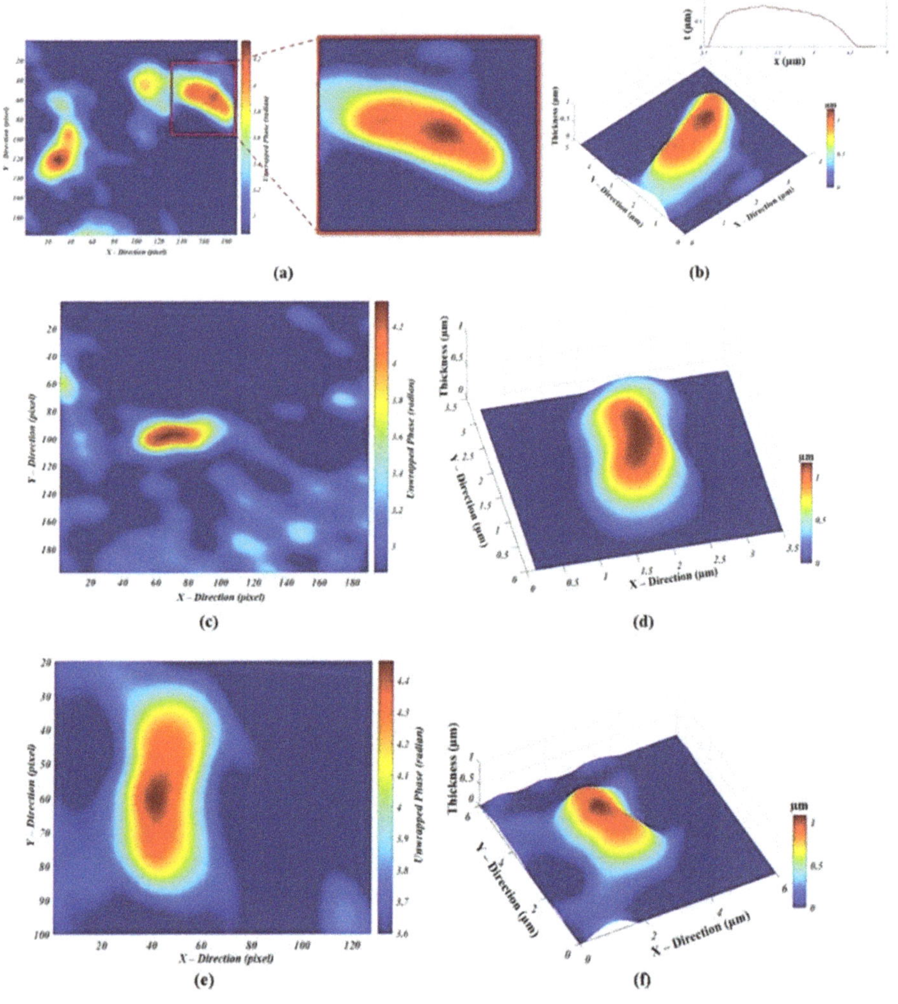

Figure 14.36. *E. coli* bacterium DHM imaging: (a) 2D phase profile; (b) thickness distribution of a single *E. coli* bacterium; (c) and (e) another 2D reconstruction of the same culture; and (d) and (f) their corresponding thickness distribution. Reproduced with permission from [85]. Copyright 2019 SPIE.

the vascular system of a zebrafish specimen, see figure 14.41. Figure 14.42 shows another interesting contribution by Mangal *et al* [91] that uses the capabilities of DHM with standard bright-field microscopy. Here some morphological parameters of the cell nuclei in cervical cells are reported, identifying suspicious cells for further examination by the pathologist using a methodology that complements information provided by the bright-field and quantitative phase images. Another example is shown in figure 14.43, where Frausto *et al* [92] use transmission DHI (t-DHI) to measure the dimensions of pollen grains from a Lantana flower and compare its phase magnitude measurement with the acquired image obtained with standard confocal microscopy. The t-DHI is a novel optical configuration that can retrieve

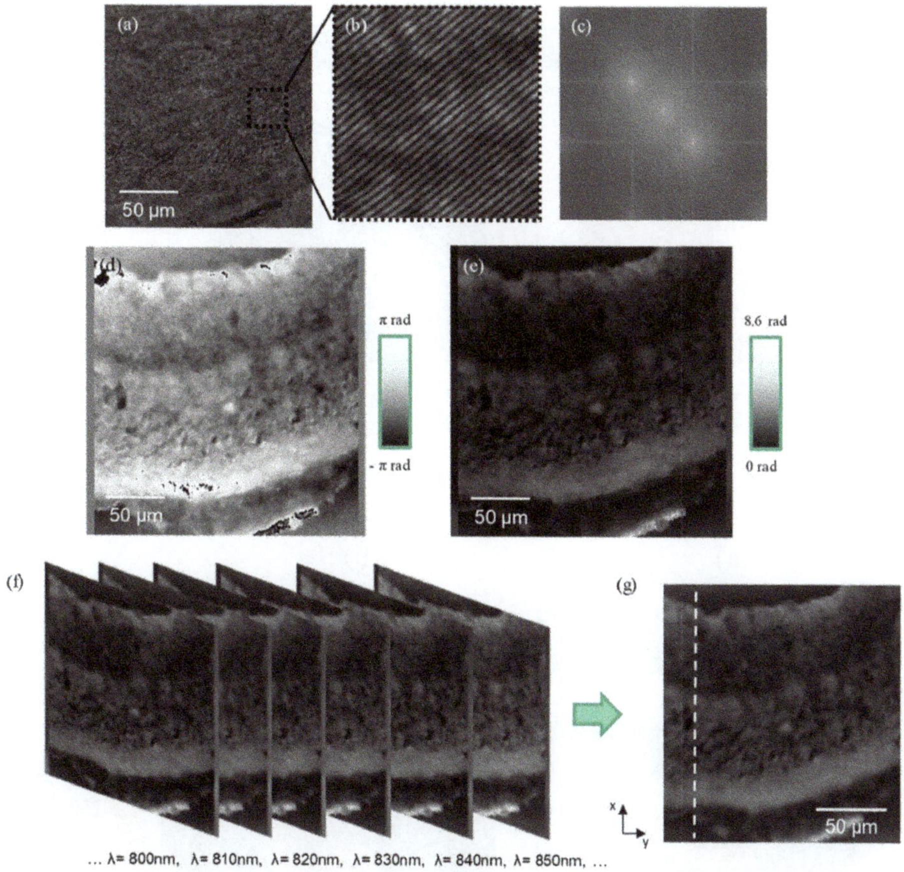

Figure 14.37. Multi-wavelength DHM images of a dissected mouse retina: (a) digital off-axis hologram of a mouse retina section; (b) enlarged spatial carrier fringe pattern of (a); (c) its Fourier spectrum; (d) quantitative phase-contrast image; (e) unwrapped phase; (f) stack of unwrapped phase images at different wavelengths for the same region of the retina; (g) averaged phase imaging of the stack of the unwrapped images, where the dashed line is a cross-section chosen for refractive index determination. Reproduced with permission from [86]. Copyright 2019 SPIE.

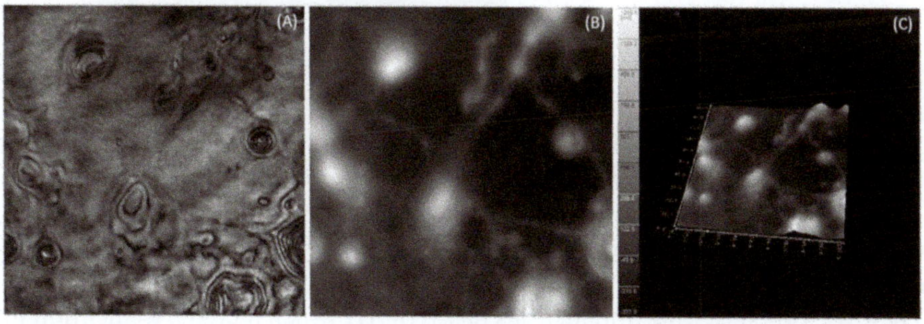

Figure 14.38. DHM images of hippocampal cells: (A) intensity image; (B) spatial phase image; and (C) optical path length scale in nanometers. Reproduced with permission from [87]. Copyright 2019 SPIE.

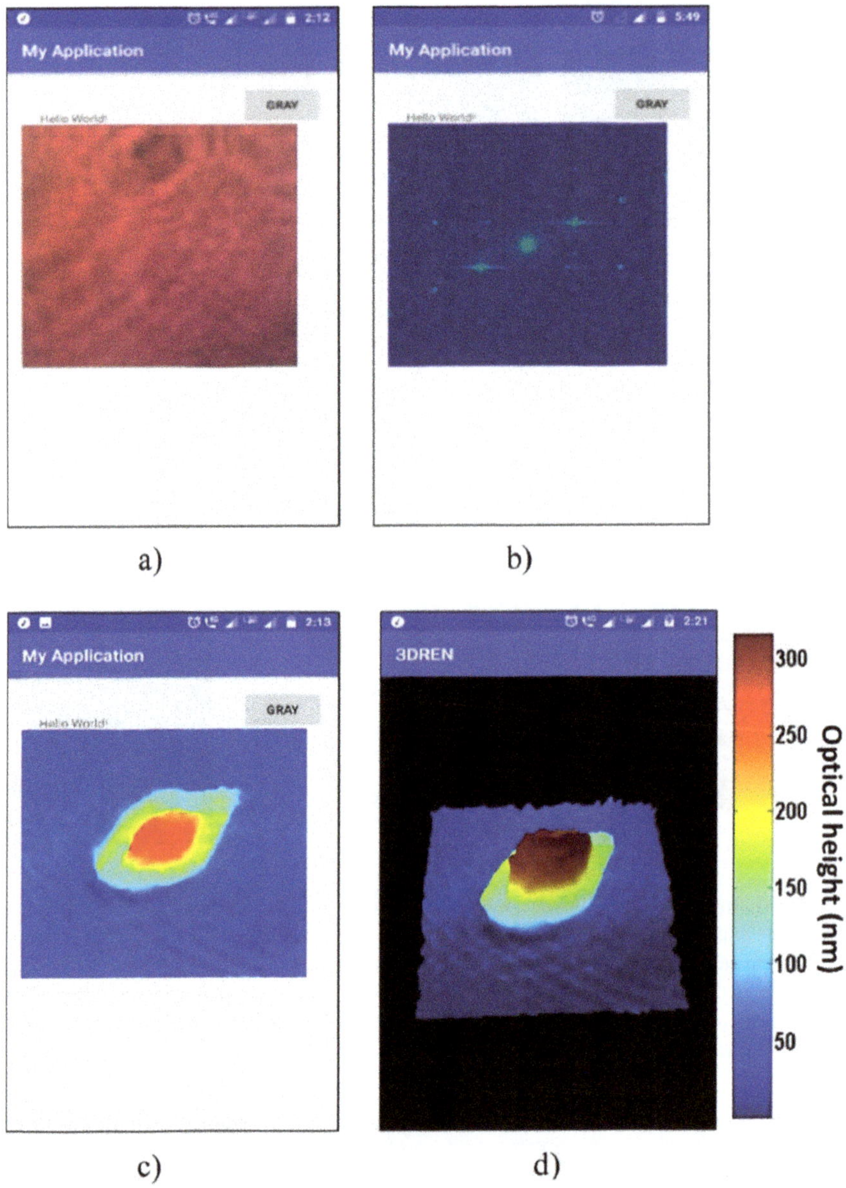

Figure 14.39. Screenshots of (a) an interferogram acquired with a smartphone, (b) a 2D Fourier spectrum of (a), (c) a color-coded phase image of a smear blood cell, and (d) a rendered phase image of the interferogram. Reproduced with permission from [88]. Copyright 2019 Elsevier.

useful information from semi-transparent samples as pollen grains by using phase information and comparing the information with confocal images.

There are other exciting applications based on HI which include, to mention just a couple, optical scanning holography (OSH) [93–98] and heterodyne holography (HH), which is considered an extension of phase-shifting digital holography [99].

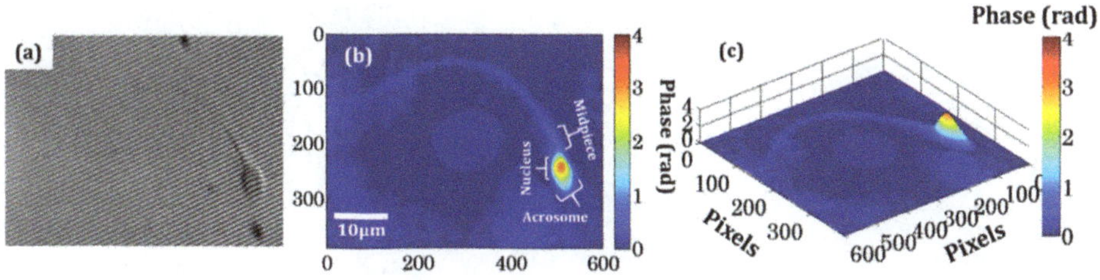

Figure 14.40. DHM imaging reconstruction process: (a) hologram; (b) reconstructed phase map of a sperm cell; and (c) pseudo-3D map. Reproduced with permission from [89]. Copyright 2019 Science.

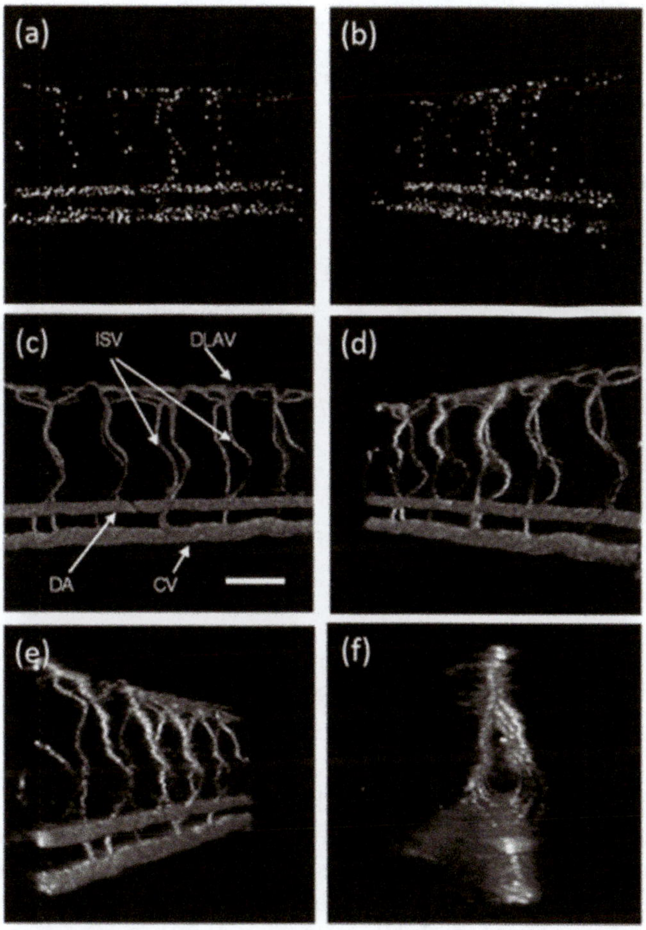

Figure 14.41. (a) and (b) Representation of diffusers (RBCs) considering a single 3D reconstruction. Views at 0° and +50°+50°. (c)–(f) Representation of the vascular system considering a summation of a sequence of 100 holograms. Views at 0°, +44°+44°, −48°−48°, and −90°−90°. In (c), DA: dorsal aorta; CV: caudal vein; ISV; intersegmental vessels; DLAV: dorsal longitudinal anastomotic vessels. The scale bar is 100 μm. Reproduced with permission from [90]. Copyright 2019 The Optical Society of America.

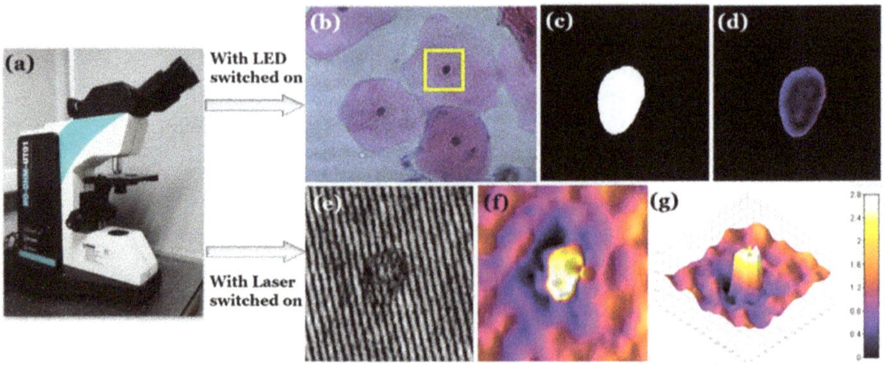

Figure 14.42. Processing steps to image the cervical cell parameters: (a) a commercial DHM; (b) bright-field image of the cell nuclei; (c) binary mask; (d) resulting image after multiplying the mask with the bright-field image; (e) hologram; (f) pseudo-color phase image; and (g) 3D phase image rendering the nucleus. Reproduced with permission from [91]. Copyright 2019 Wiley.

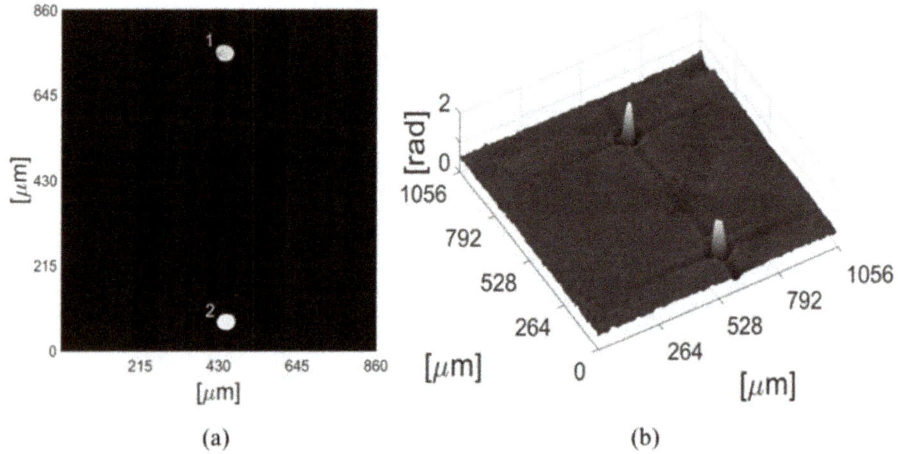

Figure 14.43. Comparison between the pollen images recorded with (a) confocal microscopy and (b) a phase magnitude t-DHI system. Reproduced with permission from [92]. Copyright 2019 The Optical Society of America.

Also, it is important to stress the fact that the DHI systems have reached a high level of confidence, making them innovative methodologies for the characterization of bio-samples, i.e. DHI is becoming a standard technique in disciplines such as biology, chemistry, tissue engineering, pharmaceutics, and others. The latter makes it possible to have DHM and HI system manufacturers such as LynceeTec, Holoxica Limited, Zebra Imaging, Eon Reality, and others [100–102]; surely, this list will grow. Currently, DHM systems are present as straightforward, outlined products that incorporate applications, blueprints, industry chain structures, and definitions. At this time, there are new start-up ventures and affiliations with purposeful macroeconomic signs. There are various aspects to be considered:

label-free interferometric techniques, quantitative phase imaging, and applications in pharmaceutical and biotechnology companies, hospitals, medical centers, research organizations, and many others.

There are other breakthrough developments based on interferometric techniques, including HI. For example, we can mention the LIGO project, the giant laser interferometer built with 4 km long arms. This interferometer is particularly important in the search for gravitational waves since the longer the arms, the farther the laser travels, and this increases the instrument's sensitivity [103].

Due to its actual importance and potential in the future, an important application of DHM as a tool for surface profilometry was presented in chapter 9. Applications for profilometry are broad and diverse in surface metrology using low coherence illumination light sources such as LEDs and white light; low coherence holography has shown significant novel advances in recent years, mainly because it presents a significant reduction in coherent noise. This aspect boosts applications in the characterization of cells and tissues.

References

[1] Gabor D 1972 Holography, 1948–71. Nobel lecture *Proc. IEEE* **60** 655–68

[2] Leith E N and Upatnieks J 1963 Wavefront reconstruction with continuous tone objects *J. Opt. Soc. Am.* **53** 1377–81

[3] Leith E N and Upatnieks J 1964 Wavefront reconstruction with diffused illumination and three-dimensional objects *J. Opt. Soc. Am.* **54** 1295

[4] Brooks R E, Heflinger L O and Wuerker R F 1965 Interferometry with a holographically reconstructed comparison beam *Appl. Phys. Lett.* **7** 248

[5] Brown G, Grant R and Stroke G 1966 Theory of holographic interferometry *J. Acous. Soc. Am.* **45** 1166–79

[6] Frausto-Rea G, de la Torre-Ibarra M H, Muñoz-Huerta R F and Casillas F J 2020 Mechanical test study in composites using digital holographic interferometry and optical coherence tomography simultaneously *Appl. Opt.* **59** 857–65

[7] Chang M, Hu C P, Lam P and Wyant J 1985 High precision deformation measurement by digital phase shifting holographic interferometry *Appl. Opt.* **24** 3780

[8] Pedrini G, Zou Y L and Tiziani H J 1995 Comparison of two reflecting surfaces by using digital holographic interferometry *Opt. Commun.* **118** 186–92

[9] Yaroslavsky L P and Merzyalov N S 1980 *Methods of Digital Holography* (New York: Consultant Bureau)

[10] Pedrini G, Schedin S and Tiziani H J 1999 Lensless digital-holographic interferometry for the measurement of large objects *Opt. Commun.* **171** 29–36

[11] Demoli N, Vukicevic D and Torzynski M 2003 Dynamic digital holographic interferometry with three wavelengths *Opt. Express* **11** 767

[12] Morimoto Y 2007 Subnanometer displacement measurement by averaging of phase difference in windowed digital holographic interferometry *Opt. Eng.* **46** 025603

[13] Anand A and Chhaniwal V K 2007 Measurement of parameters of simple lenses using digital holographic interferometry and a synthetic reference wave *Appl. Opt.* **46** 2022–6

[14] Amer E, Gren P and Sjödahl M 2009 Laser-ablation-induced refractive index fields studied using pulsed digital holographic interferometry *Opt. Lasers Eng.* **47** 793–9

[15] Tay C J, Quan C and Chen W 2009 Dynamic measurement by digital holographic interferometry based on complex phasor method *Opt. Laser Technol.* **41** 172–80
[16] Quan C, Tay C J and Chen W 2009 Determination of displacement derivative in digital holographic interferometry *Opt. Commun.* **282** 809–15
[17] Rajshekhar G, Gorthi S S and Rastogi P 2010 Estimation of displacement derivatives in digital holographic interferometry using a two-dimensional space-frequency distribution *Opt. Express* **18** 18041
[18] De la Torre M H, Flores-Moreno J M, Aguayo D D, Hernandez-Montes M S, Perez-Lopez C and Mendoza Santoyo F 2014 Displacement measurements over a square meter area using digital holographic interferometry *Opt. Eng.* **53** 092009
[19] Georges M P *et al* 2013 Digital holographic interferometry with CO_2 lasers and diffuse illumination applied to large space reflector metrology *Appl. Opt.* **52** A102–16
[20] Vandenrijt J F, Thizy C, Queeckers P, Dubois F, Doyle D and Georges M P 2014 Long-wave infrared digital holographic interferometry with diffuser or point source illuminations for measuring deformations of aspheric mirrors *Opt. Eng.* **53** 112309
[21] Wahba H H and Kreis T 2009 Characterization of graded index optical fibers by digital holographic interferometry *Appl. Opt.* **48** 1573–82
[22] De la Torre M H, Mendoza Santoyo F and Hernandez-Montes M S 2018 Transmission out-of-plane interferometer to study thermal distributions in liquids *Opt. Lett.* **43** 871
[23] Xia H T, Guo R X, Fan Z B, Cheng H M and Yang B C 2012 Non-invasive mechanical measurement for transparent objects by digital holographic interferometry based on iterative least-squares phase unwrapping *Exp. Mech.* **52** 439–45
[24] Brozović J, Demoli N, Farkaš N, Sušić M, Alar Ž and Gabrić Pandurić D 2014 Properties of axially loaded implant-abutment assemblies using digital holographic interferometry analysis *Dent. Mater.* **30** 17–27
[25] Kumar M, Birhman A S and Kannan S 2018 Measurement of initial displacement of canine and molar in human maxilla under different canine retraction methods using digital holographic interferometry *Opt. Eng.* **57** 1
[26] Monroy-Ramirez F and Garcia-Sucerquia J 2014 Monitoring micro-mechanical changes in electronic circuit boards with digital holographic interferometry *Optik* **125** 2113–6
[27] Vincitorio F, Bahuer L, Fiorucci M P, López A J and Ramil A 2018 Improvement of crack detection on rough materials by digital holographic interferometry in combination with non-uniform thermal loads *Optik* **163** 43–8
[28] Vishnoi A and Rajshekhar G 2019 Rapid deformation analysis in digital holographic interferometry using graphics processing unit accelerated Wigner–Ville distribution *Appl. Opt.* **58** 4420
[29] Kulkarni R and Rastogi P 2015 Three-dimensional displacement measurement from phase signals embedded in a frame in digital holographic interferometry *Appl. Opt.* **54** 3393
[30] Vishnoi A, Ramaiah J and Rajshekhar G 2019 Phase recovery method in digital holographic interferometry using high-resolution signal parameter estimation *Appl. Opt.* **58** 1485
[31] De la Torre-Ibarra M, Mendoza Santoyo F, Pérez-López C and Saucedo-A T 2005 Detection of surface strain by three-dimensional digital holography *Appl. Opt.* **44** 27–31
[32] Cloud G 1995 *Optical Methods of Engineering Analysis* 1st edn (New York: Cambridge University Press)

[33] Xia H, Guo R, Yan F, Cheng H, Lin Z, Ma Q and Wang F 2016 Simultaneous measurement of stress-optic constant and stress field of transparent plate by digital holographic interferometry *Optik* **127** 11974–81

[34] Solís S, Hernández-Montes M S and Mendoza Santoyo F 2011 Measurement of Young's modulus in an elastic material using 3D digital holographic interferometry *Appl. Opt.* **50** 3383–8

[35] Alvarez A S, De La Torre M H, Mendoza Santoyo F and Anaya T S 2014 Strain determination in bone sections with simultaneous 3D digital holographic interferometry *Opt. Lasers Eng.* **57** 101–8

[36] Rajshekhar G, Gorthi S S and Rastogi P 2009 Strain, curvature, and twist measurements in digital holographic interferometry using pseudo-Wigner–Ville distribution based method *Rev. Sci. Instrum.* **80** 093107

[37] Morimoto Y and Fujigaki M 2011 Three-dimensional displacement and strain measurements by windowed phase-shifting digital holographic interferometry *Advanced Holography —Metrology and Imaging* (London: InTech Open) pp 29–52

[38] De la Torre-Ibarra M *et al* 2006 Surface strain distribution on thin metallic plates using 3-D digital holographic interferometry *Opt. Eng.* **45** 105603

[39] Pantelic D V, Grujic D and Vasiljevic D M 2014 Single-beam, dual-view digital holographic interferometry for biomechanical strain measurements of biological objects *J. Biomed. Opt.* **19** 127005

[40] Gorthi S S and Rastogi P 2009 Simultaneous measurement of displacement, strain and curvature in digital holographic interferometry using high-order instantaneous moments *Opt. Express* **17** 17784

[41] Pryputniewicz R J 1985 Time average holography in vibration analysis *Opt. Eng.* **24** 245843

[42] Schedin S, Pedrini G and Tiziani H J 2000 Pulsed digital holography for deformation measurements on biological tissues *Appl. Opt.* **39** 2853–7

[43] Hariharan P, Oreb B F and Freund C H 1987 Stroboscopic holographic interferometry: measurements of vector components of a vibration *Appl. Opt.* **26** 3899–903

[44] Pérez-López C, Mendoza-Santoyo F and Guerrero J A 2003 Decoupling the x, y and z displacement components in a rotating disc using three-dimensional pulsed digital holography *Meas. Sci. Technol.* **14** 97–100

[45] Fieldhouse J D and Newcomb T P 1996 Double pulsed holography used to investigate noisy brakes *Opt. Lasers Eng.* **25** 455–94

[46] Kakue T *et al* 2012 Single-shot femtosecond-pulsed phase-shifting digital holography *Opt. Express* **20** 20286

[47] Kreis T 2005 *Handbook of Holographic Interferometry: Optical and Digital Methods* (Weinheim: Wiley)

[48] Rastogi P K 1994 *Holographic Interferometry* vol 68 (Berlin: Springer)

[49] Mrozek P, Mrozek E and Werner A 2018 Electronic speckle pattern interferometry for vibrational analysis of cutting tools *Acta Mech. Autom.* **12** 135–40

[50] Yang L *et al* 2014 Review of electronic speckle pattern interferometry (ESPI) for three dimensional displacement measurement *Chin. J. Mech. Eng.* **27** 1

[51] Sanchez Preciado J, Perez Lopez C, del Socorro Hernández-Montes M, De la Torre-Ibarra M, Flores Moreno J M, Tavera Ruiz C, Mendoza Santoyo F and Galan M 2018 Laser vibrometry in the quality control of the break of tanned leather *Opt. Lasers Eng.* **104** 78–83

[52] Pedrini G, Zou Y L and Tiziani H J 1996 Digital double-pulsed holographic interferometry for vibration analysis *Shock Vib.* **42** 367–74
[53] Flores-Moreno J M *et al* 2011 Holographic otoscope for nanodisplacement measurements of surfaces under dynamic excitation *Scanning* **33** 342–52
[54] Hernandez-Montes M S *et al* 2015 Surface strain-field determination of tympanic membrane using 3D-digital holographic interferometry *Opt. Lasers Eng.* **71** 42–50
[55] Flores-Moreno J M, Mendoza Santoyo F and Estrada J C 2013 Holographic otoscope using dual-shot-acquisition for the study of eardrum biomechanical displacements *Appl. Opt.* **52** 1731–42
[56] Hernández-Montes M S *et al* 2016 Quantification of the vocal folds' dynamic displacements *J. Phys. D: Appl. Phys.* **49** 175401
[57] Silva-Acosta L *et al* 2020 Study of skin rigidity variations due to UV radiation using digital holographic interferometry *Opt. Lasers Eng.* **126** 105909
[58] Choudry A 1981 Digital holographic interferometry of convective heat transport *Appl. Opt.* **20** 1240
[59] Lei H, Shi-Ping H and Xiao-Ping W 1988 Holographic interferometry using digital image processing technique for the measurement of 3-D axisymmetric temperature field *Hologr. Appl.* **0673** 192
[60] Lanen T A W M 1990 Digital holographic interferometry in flow research *Opt. Commun.* **79** 386–96
[61] Lanen T A W M, Bakker P G and Bryanston-Cross P J 1992 Digital holographic interferometry in high-speed flow research *Exp. Fluids* **13** 56–62
[62] Kreis T, Müller J, Von-Kopylow C and Jüptner W 2006 Noncontacting measurement of distortion by digital holographic interferometry *Materialwiss. Werkstofftech.* **37** 76–80
[63] Sharma S, Sheoran G and Shakher C 2012 Investigation of temperature and temperature profile in axi-symmetric flame of butane torch burner using digital holographic interferometry *Opt. Lasers Eng.* **50** 1436–44
[64] Kumar V, Kumar M and Shakher C 2014 Measurement of natural convective heat transfer coefficient along the surface of a heated wire using digital holographic interferometry *Appl. Opt.* **53** G74
[65] Desse J-M, Picart P and Tankam P 2008 Digital three-color holographic interferometry for flow analysis *Opt. Express* **16** 5471
[66] Agarwal S, Kumar V and Shakher C 2018 Temperature measurement of wick stabilized micro diffusion flame under the influence of magnetic field using digital holographic interferometry *Opt. Lasers Eng.* **102** 161–9
[67] Demoli N, Torzynski M and Vukicevic D 2007 Enhanced sensitivity digital holographic interferometry *Opt. Express* **15** 10672
[68] Pérez-Huerta J S, Saucedo-Anaya T, Moreno I, Ariza-Flores D and Saucedo-Orozco B 2017 Digital holographic interferometry applied to the investigation of ignition process *Opt. Express* **25** 13190
[69] Desse J M and Olchewsky F 2018 Dual-reference digital holographic interferometry for analyzing high-density gradients in fluid mechanics *Opt. Lett.* **43** 1635
[70] Xi T, Di J, Li Y, Dai S, Ma C and Zhao J 2018 Measurement of ultrafast combustion process of premixed ethylene/oxygen flames in narrow channel with digital holographic interferometry *Opt. Express* **26** 28497

[71] Dancova P, Psota P and Vit T 2019 Measurement of a temperature field generated by a synthetic jet actuator using digital holographic interferometry *Actuators* **8** 27
[72] Guerrero-Mendez C *et al* 2016 Real-time measurement of the average temperature profiles in liquid cooling using digital holographic interferometry *Opt. Eng.* **55** 121730
[73] Guerrero-Méndez C *et al* 2017 Measurements of concentration differences between liquid mixtures using digital holographic interferometry *Metrol. Meas. Syst.* **24** 19–26
[74] Kuzmina T B, Andreeva N V, Ismagilov A O and Andreeva O V 2018 A study of fractionation of milk as a typical biological fluid by digital holographic interferometry *Opt. Spectrosc.* **125** 977–81
[75] Fernández-Sempere J, Ruiz-Beviá F, García-Algado P and Salcedo-Díaz R 2010 Experimental study of concentration polarization in a crossflow reverse osmosis system using digital holographic interferometry *Desalination* **257** 36–45
[76] He M G, Zhang S, Zhang Y and Peng S G 2015 Development of measuring diffusion coefficients by digital holographic interferometry in transparent liquid mixtures *Opt. Express* **23** 10884
[77] Anand A, Chhaniwal V K and Narayanamurthy C S 2006 Diffusivity studies of transparent liquid solutions by use of digital holographic interferometry *Appl. Opt.* **45** 904–9
[78] Zhang Y, Zhao J, Di J, Jiang H, Wang Q, Wang J and Yin D 2012 Real-time monitoring of the solution concentration variation during the crystallization process of protein-lysozyme by using digital holographic interferometry *Opt. Express* **20** 18415
[79] Wang J, Zhao J, Di J, Rauf A and Hao J 2014 Dynamically measuring unstable reaction-diffusion process by using digital holographic interferometry *Opt. Lasers Eng.* **57** 1–5
[80] Zhang S, Duan L and Kang Q 2016 Experimental research on thermocapillary migration of drops by using digital holographic interferometry *Exp. Fluids* **57** 1–13
[81] Brand U and Hillmann W 2016 Calibration of step height standards for nanometrology using interference microscopy and stylus profilometry *Precis. Eng.* **17** 22–33
[82] Cuche E, Bevilacqua F and Depeursinge C 1999 Digital holography for quantitative phase-contrast imaging *Opt. Lett.* **24** 291–3
[83] Cuche E, Marquet P and Depeursinge C 1999 Simultaneous amplitude-contrast and quantitative phase-contrast microscopy by numerical reconstruction of Fresnel off-axis holograms *Appl. Opt.* **38** 6994–7001
[84] Bianco V *et al* 2019 3D imaging in microfluidics: new holographic methods and devices *Proc. SPIE* **10875** 108750U
[85] Rastogi V, Gadkari R, Agarwal S, Kumar S and Shakher C 2019 Digital holographic interferometric *in-vitro* imaging of *Escherichia coli* (*E. coli*) bacteria *Proc. SPIE* **11030** 1103011
[86] Barroso P A, Ketelhut S, Heiduschka P, Nettels-Hackert G, Schnekenburger J and Kemper B 2019 Refractive index properties of the retina accessed by multi-wavelength digital holographic microscopy *Proc. SPIE* **10883** 108830X
[87] Romero-Sansano C, González D, Madrigal R, Fernández E and Fimia A 2019 Digital holographic microscopy for the study of light-cell interaction in real time *Proc. SPIE* **11030** 1103014
[88] Karthik G B, Shinde D D, Udupa D V, Murali K C, Divakar R K and Sahoo N K 2019 Low cost digital holographic microscope for 3-D cell imaging by integrating smartphone and DVD optical head *Opt. Lasers Eng.* **114** 1–6

[89] Dubey V *et al* 2019 Partially spatially coherent digital holographic microscopy and machine learning for quantitative analysis of human spermatozoa under oxidative stress condition *Sci. Rep.* **9** 3564

[90] Brodoline A, Rawat N, Alexandre D, Cubedo N and Gross M 2019 4D compressive sensing holographic microscopy imaging of small moving objects *Opt. Lett.* **44** 2827–30

[91] Mangal J *et al* 2019 Unsupervised organization of cervical cells using bright-field and single-shot digital holographic microscopy *J. Biophotonics* **12** e201800409

[92] Frausto-Rea G, De la Torre M H, Flores J M, Silva L, Briones-R M and Mendoza S F 2019 Micrometric size measurement of biological samples using a simple and non-invasive transmission interferometric set up *Opt. Express* **27** 26251–63

[93] Poon T C 1985 Scanning holography and two-dimensional image-processing by acousto-optic 2-pupil synthesis *J. Opt. Soc. Am.* A **2** 521–7

[94] Poon T C, Doh K B, Schilling B W, Wu M H, Shinoda K and Suzuki Y 1995 Three-dimensional microscopy by optical scanning holography *Opt. Eng.* **34** 1338–44

[95] Poon T C, Kim T, Indebetouw G, Schilling B W, Wu M H, Shinoda K and Suzuki Y 2000 Twin-image elimination experiments for three-dimensional images in optical scanning holography *Opt. Lett.* **25** 215–7

[96] Swoger J, Martinez-Corral M, Huisken J and Stelzer E H K 2002 Optical scanning holography as a technique for high resolution three-dimensional biological microscopy *J. Opt. Soc. Am.* A **19** 1910–8

[97] Poon T C, Doh K B, Schilling B, Shinoda K, Suzuki Y and Wu M H 1997 Holographic three-dimensional display using an electron-beam-addressed spatial light modulator *Opt. Rev.* **4** 567–71

[98] Schilling B W, Poon T C, Indebetouw G, Storrie B, Shinoda K, Suzuki Y and Wu M H 1997 Three-dimensional holographic fluorescence microscopy *Opt. Lett.* **22** 1506–8

[99] Le-Clerc F, Collot L and Gross M 2000 Numerical heterodyne holography with two dimensional photodetector arrays *Opt. Lett.* **25** 716–8

[100] Lyncée Tec www.lynceetec.com (Accessed: 21 October 2021) (Accessed: 21 October 2021)

[102] EON Reality www.eonreality.com (Accessed: 21 October 2021)

[103] LIGO www.ligo.caltech.edu/page/ligos-ifo (Accessed: 21 October 2021)

Fernando Mendoza-Santoyo, Manuel De la Torre-Ibarra, María del Socorro Hernández-Montes and Jorge Mauricio Flores Moreno

Chapter 15

A couple of trending applications related to holography

Over the years, holography has become a healthy, mature, and active field in optics with a wide variety of applications in many areas such as materials, biology, medicine, etc. Its diverse approaches include but are not limited to holographic interferometry, digital holography, digital holographic microscopy, white-light holography, fluorescence holography, etc. Recently, thanks to the advanced technology in photodetectors and object-interrogation sources, two configurations have aroused particular interest, electron holography and quantum holography, two innovative and promising 'hot topics'. In this final chapter, we will provide a sneak peek into these two techniques, as there are still a lot of developments and research to come. Interestingly, most of the technology and methods developed for holography with light sources can be used or adapted easily to these two particular approaches.

15.1 Electron holographic interferometry

15.1.1 Introduction

Electron holographic interferometry (EHI) is a measurement method developed for transmission electron microscopes (TEMs), which have been used to characterize the physical and mechanical properties of a large number of nanomaterials and their structures, giving an extraordinary opportunity to study a wide variety of physical parameters in, on, and around an object.

Initially, when Dennis Gabor published his work on electron holography (EH) the proposed design consisted of recording a combination of in-line or on-axis object and reference wavefronts (a hologram). The object and reference beams originated from the same 'optical' axis; upon hologram reconstruction, the object image was seen on this axis, but it was obstructed by the reference beam meaning that the object

doi:10.1088/978-0-7503-3027-5ch15

appeared to be *shadowed* by the reference, making it almost impossible to observe and thus retrieve the object information [1, 2]. The reconstruction of the object wavefront involved an all-optical set-up that included illuminating the hologram, recorded on the film, with a partially coherent quasi-monochromatic beam working in the visible part of the electromagnetic spectrum. This quasi-monochromatic beam was later replaced by a laser, which meant that a brighter and fully coherent beam of light could be used to reconstruct the wavefront of the object embedded in the random structure of the interferogram recorded in the hologram. The lack of a reliable and coherent source of electrons and the fact that holograms do not always result in a well-defined object image limited its applications. Sometimes no object image was recorded at all due to various circumstances, such as environmental noise that caused large rigid body motions in the experimental set-up. The latter meant that electron holography was not consistently practiced in electron microscopy.

Electron holography was conceived to find a reliable way to correct the aberrations inherent in electron microscopes, mostly spherical aberrations, which meant that the viewed angstrom-sized images were out of focus. Gabor's main idea was to retrieve the phase difference, $\Delta\varphi(\boldsymbol{p})$ [3] to alleviate this aberration. At the time, the reference and object beams were combined using the Mollenstedt-type electron biprism [4] within the electron microscope. This device functions as a Fresnel optical biprism, i.e. it serves to split a beam equally into two wavefronts known as amplitude splitting interferometers. This splitting of the electron beam was used to record the holograms. EH in electron microscopy benefited both from the advent of the CCD and also from the introduction of the field emission electron gun (FEG) in the microscope, an instrument that provides a coherent, high-power electron beam, features that have proved indispensable for the development of EHI [5–7].

Many of the developments made for digital holographic interferometry can be applied to EHI [8]. The latter has acquired great relevance in microscopy and has recently demonstrated the potential of being a tool capable of experimentally extracting real information that can serve as a basis for correcting theoretical descriptions, both as a product of unproven and empirically formulated theories in nanosciences and the derived technologies.

15.1.2 Off-axis electron holographic interferometry

The electrostatic and magnetic Aharonov–Bohm phase shift experienced by an electron wave passing through a sample is recovered by off-axis electron holography. It can be linked to the object's electric field E and magnetic flux density B under kinematic imaging or scattering conditions. As an example, as it is transmitted by the object the electron beam phase introduces a type of curvature on the fringes recorded in the hologram, which we know is created by superimposing the electron beam passing through the object with an unperturbed reference beam (not being scattered by the sample). This phase is recovered using off-axis holography, with the retrieved phase map containing both fields.

An electron hologram is a fringe-modulated image comprising the amplitude and phase of an electron-transparent object, see figure 15.1 (which includes the

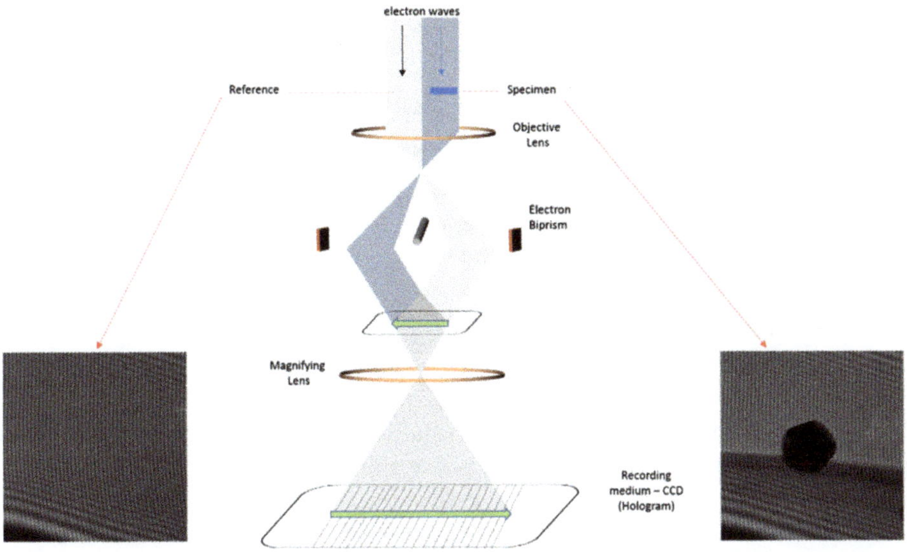

Figure 15.1. Off-axis EHI schematic view. Reproduced with permission from [8]. Copyright 2020 Springer.

Mollenstedt-type electron biprism). In this configuration, the object is placed in a position covering half of the electron wavefront, and the diffracted object beam is directed to the CCD. An object hologram is formed (right-hand side of the figure); the other half serves as the reference beam, and a reference hologram without a sample must be recorded separately for accurate phase reconstruction [9], see the left-hand side of the figure. After hologram acquisition, they are numerically reconstructed into complex images. It is possible to recover both the amplitude and phase images (the hologram reconstruction is done using the Fourier transform method, seen in chapters 3 and 6, i.e. the procedure is the same as that for holographic interferometry).

The intensity distribution at one point $\boldsymbol{p} = (x, y)$ in EHI can be given as

$$I(\boldsymbol{p}) = I_R(\boldsymbol{p}) + I_O(\boldsymbol{p}) + I_{\text{inel}}(\boldsymbol{p}) + 2\mu a(\boldsymbol{p})b(\boldsymbol{p})\cos\left[2\pi\Delta\boldsymbol{k}\boldsymbol{p} + \Delta\varphi(\boldsymbol{p}) + \varphi_0(\boldsymbol{p})\right], \quad (15.1)$$

where $I(\boldsymbol{p})$ is the intensity distribution on the image plane as recorded by the camera's sensor. $I_R(\boldsymbol{p})$ and $I_O(\boldsymbol{p})$ are the reference and object beam intensities, respectively. $\Delta\varphi(\boldsymbol{p})$ depicts the phase difference between the reference and object beams, where the reference beam is a hologram without an object, and the object hologram contains the information of the sample. $\varphi_0(\boldsymbol{p})$ denotes a phase term describing the location of the fringe pattern relative to the camera, also known as the lateral phase carrier/fringe, in which the dependence on $\boldsymbol{p}$ can describe camera distortions, i.e. fiber optic bundles for an electron microscopy type camera. $I_{\text{inel}}(\boldsymbol{p})$ stands for incoherent contributions coming from inelastic scattering electrons or spurious emission from a source other than the hologram, and μ is the interference fringe contrast. Contrast affects the accuracy of the reconstructed images, which is best measured with no object for transmission-type holography and with a perfect mirror for reflective holography and is denoted by

$$\mu = \frac{(I_{\max} - I_{\min})}{(I_{\max} + I_{\min})}. \tag{15.2}$$

EHI aims to retrieve the phase difference $\Delta\varphi(\boldsymbol{p})$ since the amplitude may be readily calculated by counting the number of resulting fringes, whose separation is given by the electron's wavelength λ. The phase term $\Delta\varphi(\boldsymbol{p})$ in electron microscopy can represent different variables from the object under study. In particular, this phase difference is due to several contributions [10]:

$$\Delta\varphi(\boldsymbol{p}) = \varphi_C + \varphi_M + \varphi_E + \varphi_G, \tag{15.3}$$

where the phase sub-indices stands for crystalline, magnetostatic, electrostatic, and geometric, respectively. The crystalline phase shift is expressed by [10, 11]

$$\varphi_C(x, y) = C_E \int V_E(x, y, z) \cdot \mathrm{d}z, \tag{15.4}$$

with

$$C_E = \left(\frac{\pi}{\lambda\varepsilon}\right) \equiv \frac{2\pi}{\lambda}\left(\frac{\varepsilon_k + \varepsilon_0}{\varepsilon_k(\varepsilon_k + 2\varepsilon_0)}\right), \tag{15.5}$$

where V_E is the electric potential, C_E is the electron interaction, ε is the total energy of the electron, which depends on ε_0 the energy of the electron at rest and ε_k the electron's kinetic energy, and λ is the electron wavelength.

The geometric phase shift is [11]

$$\varphi_G(x, y) = -2\pi \boldsymbol{g} \cdot \boldsymbol{p}, \tag{15.6}$$

where $\boldsymbol{g}$ represents the reciprocal lattice vectors of the perfect or 'reference' crystal, and $\boldsymbol{p}$ is a point on the (x, y) plane.

The principal component of the phase shift of a standard material, such as ferroelectric or magnetic, can be described generally as the sum of the electrostatic (φ_E) and magnetostatic (φ_M) phases as follows [11]:

$$\varphi_E(x, y) = C_E \int E(x, y, z) \cdot \mathrm{d}x \cdot \mathrm{d}z, \tag{15.7}$$

$$\varphi_M(x, y) = -\frac{e}{\hbar} \iint B_{\perp}(x, y) \cdot \mathrm{d}x \cdot \mathrm{d}z, \tag{15.8}$$

where e is the electron charge, E is the electric field associated with the electric potential $V_E(x, y, z)$, and B is the magnetic field. Here φ_E represents the electrostatic component that may include the mean inner potential and induced polarization. In EHI the phase retrieval is based on a Fourier procedure [12–14], discussed in chapters 3 and 6.

EH provides a unique phase imaging approach to characterize nanoscale magnetic and electrostatic fields. Quantitative full-field measurements can be directly linked to specific object characteristics from the relative phase shifts between the electron wave that passed through the object and the undisturbed reference electron beam.

15.1.3 EHI quantitative measurements

EHI can conduct measurements where the retrieved phase difference contains information such as in the following.

EHI can be used in the shape quantification of a gold (Au) decahedral nanoparticle. Figures 15.2(a) and (b) show the wrapped and unwrapped phase images, respectively, after hologram reconstruction. The particle shows a smooth contour moving from the center to the edge of the nanoparticle. In figure 15.2(b) the unwrapped phase image is calibrated using the average internal potential of the face-centered cubic (ffc) Au nanoparticle. The thickness profile can be calculated by considering the background signal as the reference plane. Figure 15.2(d) shows a thickness profile from the line shown in figure 15.2(b). Dedicated software written for TEM holography allows phase magnification, as shown in figure 15.2(c), where the phase obtained in figure 15.2(a) was amplified by 2.5 times [15].

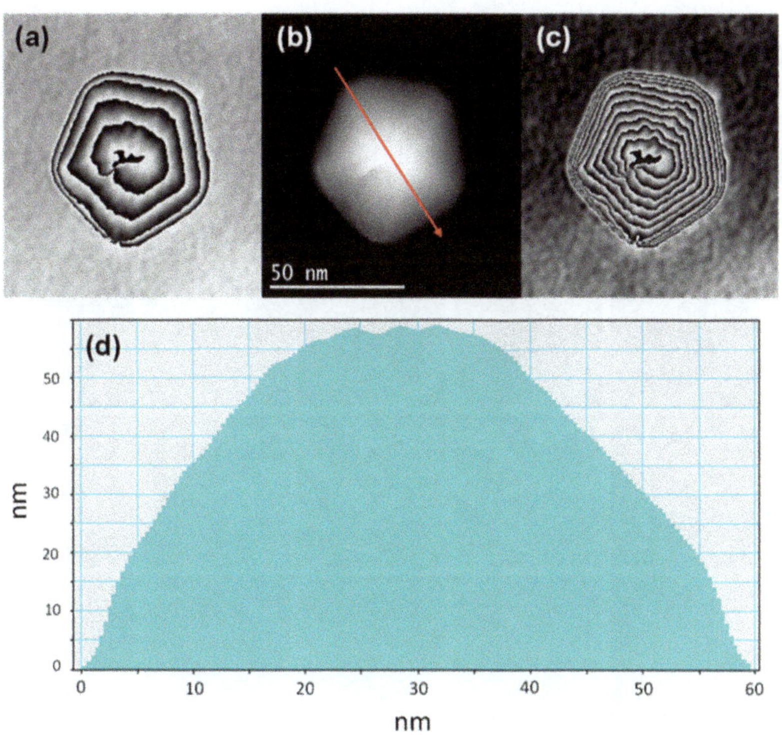

Figure 15.2. Au decahedral nanoparticle: (a) wrapped phase image; (b) unwrapped phase image; (c) 2.5 times magnification of the original phase in panel (a); and (d) computed thickness obtained using the Au crystalline potential. Reproduced with permission from [15]. Copyright 2013 Elsevier.

3D visualization of the *Staphylococcus aureus* (*S. aureus*) bacterium, which is a gram(+) bacterium of severe public health concern whose infections, including methicillin-resistant strains of *S. aureus* (MRSA), cause various diseases, including healthcare-associated infections and wound infections that may lead to pneumonia, bloodstream infections, and sepsis. *S. aureus* is usually found in the respiratory tract and skin.

In order to obtain 3D measurements, the bacteria were treated with avidin-streptavidin anti-*S. aureus* conjugated gold nanoparticles to confirm the specificity of antibody-functionalized (Ab-functionalized) Au nanoparticles. The location and effects of Ab-functionalized Au nanoparticles on *S. aureus* were inspected using off-axis EHI to determine the 3D sample structure and location of the particles attached to cell walls. The 3D reconstruction can be accomplished by acquiring several holograms from different viewing or electronic illumination positions [16]. A set of holograms was acquired by tilting the sample from −30° to +30° in 15° steps, as shown in figure 15.3(a). The resulting phase maps for each tilt are shown in figure 15.3(b), and their respective unwrapped phase maps are shown in figure 15.3(c). The series of unwrapped phase maps in figure 15.3(c) shows a phase variation from each other that is clearly due to the different orientations of the bacteria.

From the unwrapped phase maps, the surface contour may be extracted and represented as the real 3D image of the bacteria. Different views from the surface are shown in figure 15.4, from which it is possible to distinguish the surface morphology of the bacteria cell wall showing the prominent wave-like surface patterns. One advantage of EHI is acquiring surface information from a single object exposure and not by tilting the sample to reconstruct the object's surface (as done in optical DHI).

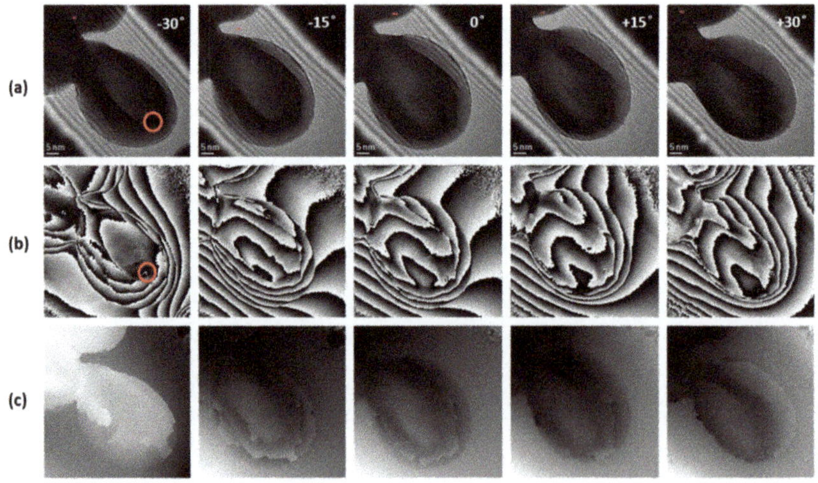

Figure 15.3. (a) Set of electron holograms at different tilting angles; (b) reconstructed phases; and (c) unwrapped phases that were taken using a biprism voltage of 20 V. The red circles show the location of a single Au nanoparticle. Reproduced with permission from [16]. Copyright 2017 Wiley.

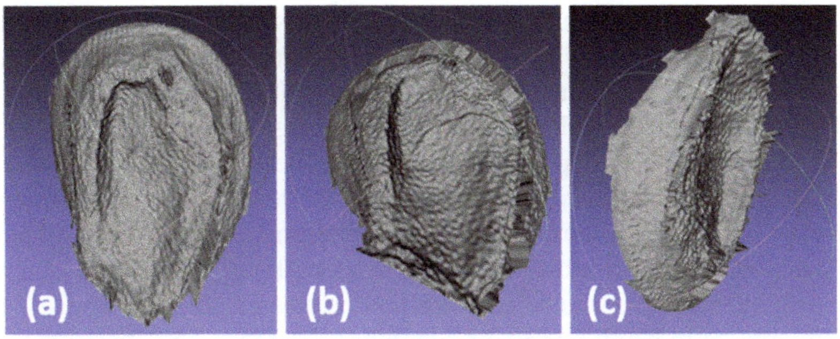

Figure 15.4. The surface contour of the reconstructed phase shows a 3D view of the bacteria obtained by EHI from three different views: (a) zero tilting, (b) vertical tilting, and (c) horizontal tilting. Reproduced with permission from [16]. Copyright 2017 Wiley.

Biological samples studied by conventional electron microscopy must be negatively stained or coated with metals in scanning and transmission modes to enhance their contrast. This procedure can affect cellular components, causing the interpretation–analysis process to form structural artifacts. Another limitation of conventional electron microscopy on biological samples is radiation damage, which represents an additional challenge in most cases. In comparison, electron holography has a considerable potential to retrieve information from the sample avoiding both problems, as its structural information can be extracted even at low voltages, under cryogenic conditions, and at low doses. In particular, electron holography is a highly sensitive imaging technique even for light atoms, which can discriminate chemical elements with a similar atomic number and identify structures that include different electric potentials. Having said this, EHI in TEM systems requires additional hardware for the latter, so efforts must be sought to construct dedicated and novel TEMs for EH. The potential for new research, to prove physical theories, and to reach discoveries is all present with EH; remember that the numerous techniques developed in optical DHI can be directly applied to electron holograms.

15.2 Quantum holography

15.2.1 Introduction

The invention of holography by Dennis Gabor (published at the end of the 1940s) provided its first fruitful applications during the 1960s, thanks to the invention of the first laser, a highly monochromatic and coherent light source. From then onwards, successful holography applications have grown constantly in a continuously increasing field of knowledge. Currently, many holography applications have reached such maturity that they started to be significant in industrial environments, such as developing security devices, product verification, rainbow holograms for seals, microscopic applications for medical devices, and biological characterization of samples, etc. Also, holography concepts have given birth to other widespread techniques such as coherence tomography, optical memory devices, and fully realistic 3D imaging displays. Many important advances in holography have to do with data handling and processing (the Big Data paradigm) and include algorithms based on neural networks, artificial intelligence, and deep learning, topics that need to be addressed explicitly and are outside the scope of this book. Also, in the literature, one can find many new topics that explain probable relations between relativity and holography or even the daring conceptual holographic nature of the Universe. Independently of these concepts and conceptions, a more practical application of holography lies in the development of quantum concepts giving way to notions and applications on quantum interferometry and quantum holography. Indeed, many colleagues working for many years in holography claim that we were working with quantum theories before knowing the meaning of such words. We do not pretend to write an in-depth review on quantum holography here, but aim to give the reader the basic information, including some basic equations and references to this topic in continuous and rapid progress within optical metrology.

15.2.2 Principles of quantum interferometry

Interferometry is a fully explained subject using the concepts of wavefronts since they are part of the electromagnetic (EM) spectrum. The photon is considered to be a wave or a particle. In the seminal theories on the quantum behavior of light, the double-slit experiment using single electrons plays a fundamental role. By 1927 Davisson and Germer [17] had observed the diffraction of electron beams demonstrating the wave-like behavior and properties of particles. Then, using a Mach–Zehnder interferometer set-up, Laszlo Marton, in the early 1950s, showed that electrons could interfere, again proving the wave-like behavior of these particles. Using an electron biprism (a very thin conducting wire placed at right angles with respect to an electron beam), Möllenstedt and Düker split an electron beam into two paths and combined them (overlapping), observing interference. Later, the electron biprism was used to develop off-axis electron holography (see the section in this chapter on EH). Finally, for the first time, in 1961 Jönsson performed the double-slit experiment with electrons demonstrating this particle's ability to interfere [18, 19]. This topic was followed by Tonomura, who observed electron interference using a 'very weak electron source' [20, 21]. Several interferometric experiments have been performed with neutrons, atoms, and molecules [22–25].

As a starting point in interferometry, the interference fringes are produced, i.e. in a classical Young's experiment, depending on the wavelength of the illumination source λ, the number of slits on the screen traversed by the traveling wave, and some other parameters such as the incidence angle or the diffraction index of the propagation medium. Meanwhile, considering the quantum version of Young's experiment, it is fundamental to consider the number m of detected photons since the de Broglie wavelength depends on the ratio λ/m [26]. Recall that the photons are conceptualized as massless particles from the quantum optics point of view. Furthermore, based on quantum concepts, the overlapping of wavefronts is not enough to explain interferometry; however, several analogies and comparisons have been made between classical and quantum interferometry [27].

The word photon was first coined by Gilbert Newton Lewis [28] in a letter to the journal *Nature* on 18 December 1926 to describe the unit of light that carries radiant energy. Einstein originally described such phenomena as a 'light quantum'. The photon in quantum optics represents the minimum unit of the electromagnetic field. In other words, it represents the quantization of the electromagnetic field, and it is defined by the probability of finding some photons (photon number function depending on n) inside a volume. Then, interferometry is a statistical process with a probabilistic description from the quantum point of view. The primary state of light defined by the photon can be described either as Fock or coherence states. The Fock state describes the photon's behavior as a particle within a fixed number of particles, and the phase is indeterminate. A coherent state describes the photon's behavior as a wave whose quantum mechanical uncertainty is distributed equally between the canonical conjugate coordinates, position, and momentum. The relative uncertainty in phase and amplitude are roughly equal. In a Fock state, the phase of the quantum state is indeterminate. Both conceptualizations of interferometry from Young's experiment perspective consider the interferometric pattern due to the arrival of many photons on a screen at different places and probabilities. Perhaps the main difference is that in the classical approximation, we consider the arrival of photons as an average in the shortest time possible that can be detected in the form of intensity. This average is a probabilistic function representation at different places on a screen in the quantum field. Any measurement made by the concept of quantum conceptualizations involves an irreversible loss of information, an issue that has been called the collapse of the wave-function. If we can reconstruct such a loss of information (wave-function reconstruction), we can achieve reversibility or be able to predict measurements.

For experiments conducted in quantum interferometry, the concept of entanglement of particles is fundamental. This concept is based on the coherent superposition of different states in a multi-particle system, meaning a non-local interaction or non-separable states between them. Theoretically, the use of entangled states will surpass the classical diffraction limit. Some other quantum interferometric experiments are based on the analysis of the correlations between detected states and position coincidences of entangled particles, giving rise to interference patterns [29].

Perhaps the Young experiment is an easy way of understanding the analogies or similarities in the interferometric principle between the classical and quantum points of view. In the classical conception, every slit or pinhole in Young's screen represents a secondary source of emerging spherical waves that share phase information. In the detection plane, a photodetector records the intensity of these overlapping waves. Then, each fringe on the interference pattern can be located and described promptly. The separation between maximum values depends on the light source wavelength used, and the minimum resolvable separation is equivalent to ($\lambda/2$) that is, the Rayleigh criteria or the resolution of the system. In the quantum version, it is required to consider the quantization of the electromagnetic field limited by a volume. This quantization represents a traveling wave with periodical boundary conditions, as can be expressed in terms of the quantum harmonic oscillator with eigenvalues

$$E_n = \hbar\omega\left(n + \frac{1}{2}\right). \tag{15.9}$$

Equation (15.9) means that the nth eigenvalue (state) has n quantum of energy (the photon) and frequency ω; $\hbar$ is the Planck constant and E_n the quantized electromagnetic field.

Also, the photon source is considered a 'photon gun', i.e. emission of photons with a ballistic trajectory until they reach the detector and emit s-propagation modes at the slits. The overall quantum system can then be conceptualized in two parts: the photon sources up to the Young's screen and a free propagation stage until the detection plane [27]. Up to this point, there are several assumptions of many quantum light properties and solutions for the quantized wavefront that the interested reader can find in the citations given.

The procedure to explain the quantum interference is similar to propagating a wavefront through Young's screen and then to the detection plane. In this respect, table 15.1 shows a set of equations that relate the two approaches, the classical and the quantum. Equations (15.10)–(15.12) on the left-hand side of this table represent the classical approximation of Young's experiment: $E(x)$ is a traveling wave (taken for simplicity as the electric field in the visible part of the EM spectrum), $k = 2\pi/\lambda$ it is the wave vector (λ is the wavelength), $|\bar{r}|$ is the magnitude of the distance to a point x on Young's screen, E_0 is the amplitude of the spherical wave, and i is the imaginary number. In the interaction with Young's screen, s represents the total number of slits, E_{0j} is the amplitude of the field from the jth slit, and $|\bar{r}_j|$ is the distance from the jth slit to the point x. The intensity $I(x)$ is the square of the total contribution of each secondary wave from the Young screen.

Table 15.1. Equations defining the basic relations for the classical and quantum cases in a basic Young's interferometer.

Classical	Quantum
Spherical wave emitted by a point source	Photon source
$E(x) = E_0 \frac{e^{-ik\|\bar{r}\|}}{\|\bar{r}\|}$ (15.10)	$\|\psi_0\rangle = \|n\rangle$ (15.13) $\|\psi_0\rangle = \|\alpha\rangle = e^{-\|\alpha\|^2/2} \sum_{n=0}^{\infty} \frac{\alpha^n}{\sqrt{n!}} \|n\rangle$ (15.14)
Interaction with the Young's screen	After Young's screen
$E(x) = \sum_{j=1}^{s} E_{0j} \frac{e^{-ik\|\bar{r}_j\|}}{\|\bar{r}_j\|}$ (15.11)	$\|\psi_Y\rangle = \frac{1}{\sqrt{s}} \sum_{j=1}^{s} \|\psi_{\alpha,j}\rangle$ (15.15)
Detection at the image screen detection (Intensity) $I(x) = \|E(x)\|^2$ (15.12)	Detection at the image screen (photon detection) $\|\psi_D\rangle = \hat{U}\|\psi_Y\rangle$ (15.16) $\langle M\rangle = \langle\psi_D\|\hat{A}^{\dagger m}\hat{A}^m\|\psi_D\rangle$ (15.17)

Equations (15.13)–(15.17) on the right-hand side of table 15.1 are the fundamental representations of the quantum case [30]. If the photon source represents a quantized monochromatic electromagnetic field in a single-mode cavity, then the solution is a series of eigenstates of coherent states. The most simple definition of a Fock state is the amplitude probability of finding a photon in a position x; in other words, the operator $\langle x,|, n\rangle$. The coherent state is a superposition of a Fock state with a Poisson distribution. In equation (15.13) ψ_0 is an eigenstate (from the quantified electromagnetic field solution) and n is an integer, and in equation (15.14) α is a coherent state. The latter is the most reliable representation of an ideal laser and thus, considering equation (15.14), after Young's screen, we have the contributions of all the entangled states given by equation (15.15), where $\psi_{\alpha,j}$ is the initial coherent state $|\,\alpha\rangle$ through the jth slit. Finally, equations (15.16) and (15.17) give an evolution operator $\hat{U}$ [31] and the detection of the mean m-photon counts $\langle M\rangle$, where $\hat{A}^{\dagger m}\hat{A}^{m}$ represent the annihilation and creation operators (quantum mechanics).

For the classical case presented in table 15.2, we refer the reader to chapter 1. In this case, $E_o$and E_r represent the object and reference waves with its wave vectors k, angular frequencies ω, and phases φ. For the optical hologram reconstruction, it is necessary to multiply the recorded intensity by the same reference used (see equation (1.30)). Thus, following the classic interferometric analysis where two slits in Young's screen are considered. The general expression for the interference pattern is defined by

$$I = I_1 + I_2 + 2\sqrt{I_1 I_2}\cos[(k_1 r_1 - k_2 r_2) + \Delta\varphi], \tag{15.18}$$

where the $(I_1 + I_2)$ is a background or bias intensity term and $2\sqrt{I_1 I_2}$ is the modulation intensity of the cosine function. In its argument, k_1, k_2 are the wave vectors from the secondary emission points of Young's screen, r_1, r_2 the distances to the observation plane, and $\Delta\varphi$ a random phase.

For the quantum interferometric case analysis, if a coherent state (equation (15.14)) is assumed, then the probability detection event at point x is [27]

$$< M > = \mathrm{e}^{-|\alpha|^2}\left\{\frac{m!}{s}\left[s + 2\sum_{j=1}^{s-1}(s-j)\cos(jm\Delta\theta)\right]\right\}. \tag{15.19}$$

In equation (15.19) the interference pattern is defined by the cosine function, which depends on the number of photons that are detected (m-photon detector) and a bias term, both defined in the rigged Hilbert space (RHS) that is the mathematical space underlying Dirac's notation in quantum mechanics. All the parameters involved in equation (15.19) were defined previously.

For both cases, the interferometric fringe pattern and the basic Young experiment are shown in figure 15.5.

Next, we briefly introduce quantum holography (QH) and describe what we believe is seminal experimental research. The conceptualization of quantum holograms will no doubt open the door to a series of novel applications impossible to

Table 15.2. Equations defining the basic relations for classical and quantum holography.

Classical Holography	Quantum Holography
Object and reference beams (1D) $E_o(x,t) = \|E_o\| \, e^{i(k_o x - \omega_o t + \varphi_o)}$ (15.18) $E_r(x,t) = \|E_r\| \, e^{i(k_r x - \omega_r t + \varphi_r)}$ (15.19)	Object and reference wave-function $\|\psi_o\rangle = e^{-i\omega_n(t)} \sum_n a_n \|n\rangle$ (15.22) $\|\psi_r\rangle = e^{i\varphi(\tau)} \sum_n b_n \|n\rangle$ (15.23)
The intensity of the beams' superposition $I(x,t) =$ $\|E_o\|^2 + \|E_r\|^2 +$ $2\,\Re e\left\{E_o\, E_r\, e^{i\left[\frac{(k_o-k_r)}{2}x - \frac{(\omega_o-\omega_r)}{2}t + \frac{(\varphi_o-\varphi_r)}{2}\right]}\right\}$ (15.20)	Coherent addition of the wave-function's $P_n =$ $\|a_n\|^2 + \|b_n\|^2 + 2\,\Re e\{a_n b_n^* e^{-i[\omega_n t + \varphi(\tau)]}\}$ (15.24)
	Incoherent fluorescence (time-integrated energy) $F_t = \sum_n(\|k_n\, a_n\|^2 + \|k_n\, b_n\|^2) +$ $2\,\Re e\{a_n b_n^* e^{-i[\omega_n t + \varphi(\tau)]}\}$ (15.25) with: $k_n = \left\{\{\sum_m \|\langle m\|n\rangle\|^2 \,(E_n - E_m)^4\}\right\}^{1/2}$ (15.26)
Hologram reconstruction with $(\omega_o = \omega_r)$ $I(x,y) = \|E_o + E_r\|^2\, E_r = (\|E_o\|^2 +$ $\|E_r\|^2\, E_r) + E_o^*\, E_r^2 + E_o \|E_r\|^2$ (15.21)	Recovery of the object wave-function $S(\tau) = \Re e\{a_n b_n^* e^{-i[\omega_n t + \varphi(\tau)]}\} +$ $\Im m\,\{a_n b_n^* e^{-i[\omega_n t + \varphi(\tau)]}\} =$ $\sum_n {k_n}^2 a_n b_n^* e^{-i(\omega_n + \omega_L)\tau}$ (15.27)

visualize clearly with today's knowledge, so we can only dare to dream of the enormous potential of QH scientifically.

15.2.3 Quantum holograms

The term quantum holography was probably first used by the beginning of the 1990s [32]. The initial works were closely linked to quantum mechanical concepts related to measuring a quantum event and its irreversible loss of information, also called the collapse of the wave-function. Then, quantum holograms were related to the

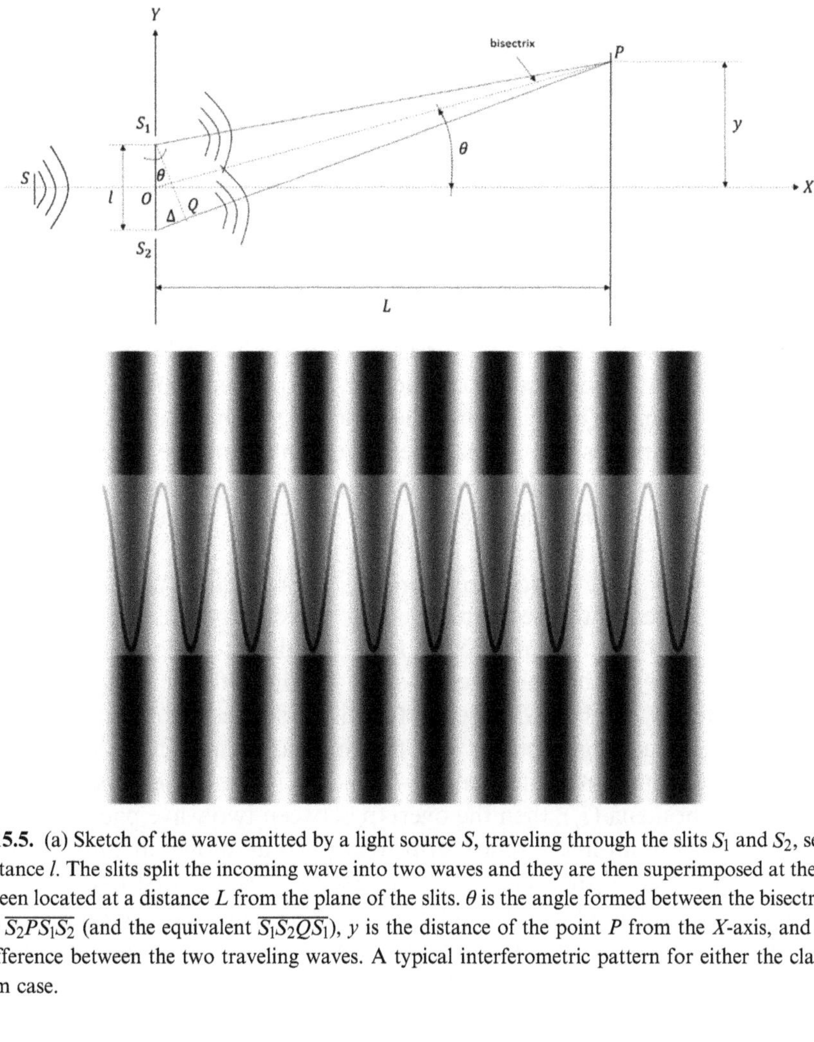

Figure 15.5. (a) Sketch of the wave emitted by a light source S, traveling through the slits S_1 and S_2, separated by a distance l. The slits split the incoming wave into two waves and they are then superimposed at the point P on a screen located at a distance L from the plane of the slits. θ is the angle formed between the bisectrix of the triangle $\overline{S_2PS_1S_2}$ (and the equivalent $\overline{S_1S_2QS_1}$), y is the distance of the point P from the X-axis, and Δ is the path difference between the two traveling waves. A typical interferometric pattern for either the classical or quantum case.

possibility of achieving reversibility, meaning the reconstruction of the wave-function. The core of such works was based on a systematic comparison of the holographic recording processes and the quantum measurement (wavefront recording and the collapsing problem, respectively).

The evolution of these concepts involves, at some point, the quantization of the electromagnetic wave encoding geometrical information. This paradigm encloses quantizing holographic representations of objects. In other words, the reconstruction of the wave-function for a recorded quantum mechanical surface. As in the case of quantum interferometry, in holography photon counting statistics is a must. All the basic equations implicated in the quantization of holographic representations can be found in [33].

As well as in the case of quantum interferometry, for the quantum holographic case, it is possible to follow similarities with classical holography since the fundamental problem in the classic case is to restore the loss of phase information. The concept of the phase in classical holography is fundamental since it provides information about the relative positions of the recorded object (i.e. before and after deformation) and its distance from the illumination source. The recovery of the phase information entails the complete description of the macroscopic geometry of the recorded object. This process implies the phase reconstruction of the object information, a variable loss in a conventional recording process such as photography. In order to reach this reconstruction of the wave-function, it is fundamental to have an undisturbed reference wavefront to be compared to that illuminating the object.

The role of the reference wave in quantum holography is also essential. In order to have an equivalent hologram reconstruction, a proposed method (quantum state holography) [34] consists in measuring the time delay τ between two emitted laser pulses. One of these pulses will be the object wave-function, the other will represent the reference wave, and τ is the path difference between the pulses. The recording hologram represents a time variation τ of a generated excited state (incoherent fluorescence).

In the quantum state holography model, a reference wave-function ψ_r is coherently overlapped with the corresponding object wave-function ψ_o. Then this operation results in a population function P_n of the nth state (photon count, equation (15.24)), which depends on the discretization of the reference and object wave-function, plus terms of phase difference functions depending on the time delay τ between the wave-function's pulses and the angular frequency of the source ω_L. Since the emitted time-integrated energy $E_n - E_m$ is a release process known as fluorescence phenomena (F_t), then the overlap between two wave packets, reference, and object, results in an interference term (real part) as shown in equation (15.25). Then the fluorescence of the object and reference waves, the first two terms of this equation, can be measured directly. The real part includes the known quantities b_n (coefficients of the reference wave-function) and the complex-value amplitudes a_n necessary to reconstruct the object wave information.

In quantum state holography, the recovery of the object wave-function can only be achieved numerically. The relative phase $\varphi(\tau) = \varphi_0 + \omega_L$ that is implicit in the wave packets represents the measured fluorescence, i.e. for example, for a fixed τ with known phases $\varphi_0 = 0$, $\varphi_0 = -\pi/2$, it is possible to calculate the coefficients a_n (equation (15.27)).

Since generating a quantum superposition state for a real physical surface is an impractical task [33], one approach for such realization is based on using an 'entangled pair of photons created by parametric down-conversion to construct entangled holographic representations for a surface'. Theoretically, quantum entanglement has been proposed for extracting holographic information from a remote 3D surface [35].

Recently, a method to record a hologram was proposed using a single photon to encode the complete information of the photon's shape (the reconstruction of the object wave-function), whose amplitude and phase were retrieved experimentally [36, 37]. In this experiment, a hologram of a single photon (HSP) was 'recorded' with the use of an intensified scientific complementary metal-oxide-semiconductor (I-sCMOS) camera system, made ad hoc for this purpose. It allows one to record the 'subtle variations of the detected photon positions and thus directly measure the empirical coincidence probability distribution'. With the reconstruction of the HSP, it is possible to numerically match the raw experimental data, measured independently, with the nearly identical amplitudes of the object and reference wave-functions. Figure 15.6 shows the reported measured and reconstructed HSP.

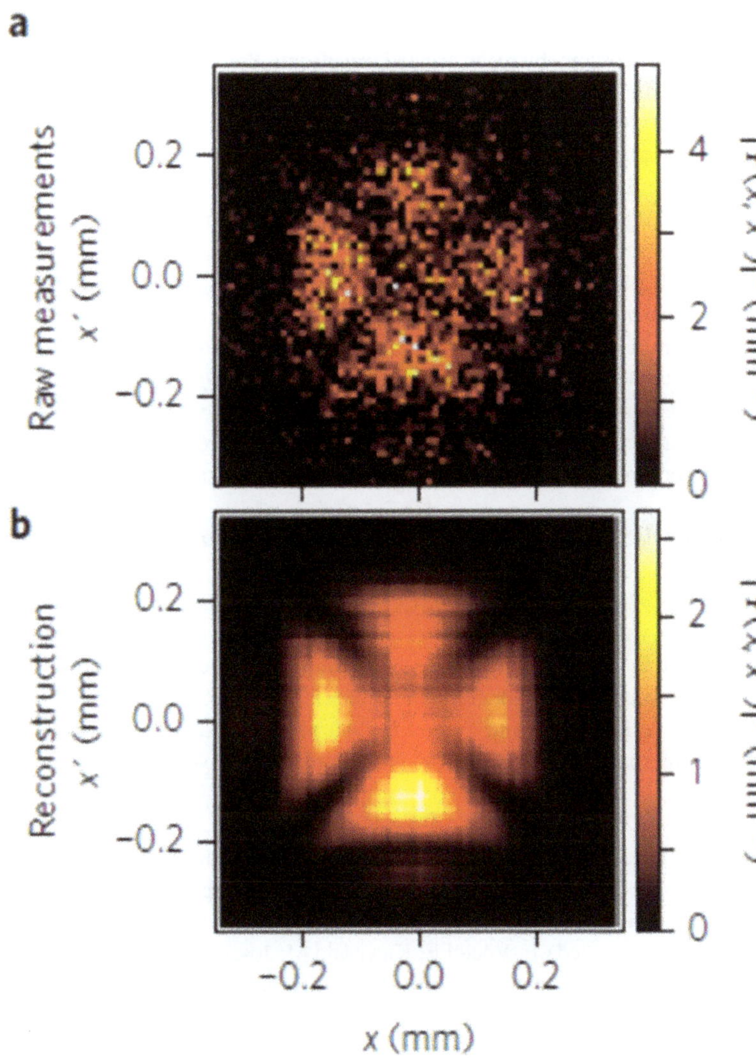

Figure 15.6. Measured and reconstructed hologram of a single photon: (a) a direct measured joint probability distribution that forms an empirical HSP; (b) its numerical reconstruction based on the matching of the raw experimental data measured and its coincidences with nearly identical amplitudes of the object and reference wave-functions. Reproduced with permission from [36]. Copyright 2016 Nature.

The field of quantum holography represents one of the most recent applications of holography. The research carries a series of challenges that will be solved depending on the ability to solve complex problems related to the need to apply the new ideas and tools that are continuously appearing—rapidly changing digital storage and processing technology provides new research tools. Quantum holography is still in the construction phase and will undoubtedly bring innovative developments of new paradigms to solve complex quantum problems, indeed, provide new tools in optical metrology in the future.

References

[1] Gabor D 1949 Microscopy by reconstructed wavefronts *Proc. Royal Soc. London* A **197** 454

[2] Gabor D 1951 Microscopy by reconstructed wavefronts: II *Proc. Phys. Soc.* B **64** 449

[3] Harada K 2020 Interference and interferometry in electron holography *Microscopy* **70** 3–16

[4] Möllenstedt G and Düker H 1956 Beobachtungen und Messungen an Briprisma-Interferenzen mit Elektronenwellen *Z. Phys.* **145** 377–97

[5] Tonomura A, Endo J and Matsuda T 1979 An application of electron holography to interference microscopy *Optik* **53** 143

[6] Tonomura A 1984 Application of electron holography using a field-emission electron microscope *J. Electron Microsc.* **33** 101

[7] Yatagai T, Ohmura K, Iwasaki S, Hasegawa S, Endo J and Tonomura A 1987 Quantitative phase analysis in electron holographic interferometry *Appl. Opt.* **26** 377–82

[8] del Socorro Hernández-Montes M, Mendoza-Santoyo F, Flores-Moreno M, de la Torre-Ibarra M, Silva Acosta L and Palacios-Ortega N 2020 Macro to nano specimen measurements using photons and electrons with digital holographic interferometry: a review *J. Eur. Opt. Soc.* **16** 16

[9] Völkl E, Allard L F and Joy D C 1999 *Introduction to Electron Holography* (New York: Springer)

[10] Hÿtch M J *et al* 2011 Dark-field electron holography for the measurement of geometric phase *Ultramicroscopy* **111** 1328–37

[11] Takeda M, Ina H I and Kobayashi S 1982 Fourier-transform method of fringe-pattern analysis for computer-based topography and interferometry *J. Opt. Soc. Am.* **72** 156–60

[12] Mendoza Santoyo F, Kerr D and Tyrer J R 1989 Manipulation of the Fourier components of speckle fringe patterns as part of an interferometric analysis process *J. Mod. Opt.* **36** 195–204

[13] Yamaguchi I 1984 Fringe formation in speckle photography *J. Opt. Soc. Am.* **A1** 81–6

[14] Cantu-Valle J, Ruiz-Zepeda F, Mendoza-Santoyo F, Jose-Yacaman M and Ponce A 2014 Calibration for medium resolution off-axis electron holography using a flexible dual-lens imaging system in a JEOL ARM 200F microscope *Ultramicroscopy* **147** 44–50

[15] Cantu-Valle J, Ruiz-Zepeda F, Voelkl E, Kawasaki M, Jose-Yacaman M and Ponce A 2013 Determination of the surface morphology of gold-decahedra nanoparticles using an off-axis electron holography dual-lens imaging system *Micron* **54–55** 82–6

[16] Ortega E, Cantu-Valle J, Plascencia-Villa G, Vergara S, Mendoza-Santoyo F, Londono-Calderon A, Santiago U and Ponce Pedraza A 2017 Morphology visualization of irregular shape bacteria by electron holography and tomography *Microsc. Res. Tech.* **80** 1249–55

[17] Rodgers P 2002 The double-slit experiment *Physics World* 1 Sept. https://physicsworld.com/a/the-double-slit-experiment/ (Accessed: February 2022)

[18] Jönsson C 1961 Elektroneninterferenzen an mehreren künstlich hergestellten Feinspalten *Z. Phys.* **161** 454–74
[19] Jönsson C 1974 Electron diffraction at multiple slits *Am. J. Phys.* **42** 4–11
[20] Tonomura A, Endo J, Matsuda T, Kawasaki T and Ezawa H 1989 Demonstration of single-electron build-up of an interference pattern *Am. J. Phys.* **57** 117–20
[21] Tonomura A 1993 *Electron Holography* (Berlin: Springer)
[22] Bernstein H J 1967 Spin precession during interferometry of fermions and the phase factor associated with rotations through 2π radians *Phys. Rev. Lett.* **18** 1102–3
[23] Marquardt F 2005 Fermionic Mach–Zehnder interferometer subject to a quantum bath *Europhys. Lett.* **72** 788–94
[24] Klepp J, Sponar S and Hasegawa Y 2014 Fundamental phenomena of quantum mechanics explored with neutron interferometers *Prog. Theor. Exp. Phys.* **2014** 082A01
[25] Danner A, Demirel B, Kersten W, Lemmel H, Wagner R, Sponar S and Hasegawa Y 2020 Spin-rotation coupling observed in neutron interferometry *J. Quantum Information* **6** 23
[26] Hecht K T 2005 *Quantum Mechanics* (Berlin: Springer)
[27] Sanabria H and Rodriguez-Lara B 2007 An introduction to quantum interferometry: Young's experiment with Fock and coherent states *Rev. Mex. Fis.* **23** 97–105
[28] Lewis G N 1926 The conservation of photons *Nature* **118** 874–5
[29] Bennink R, Bentley S and Boyd R 2002 'Two-photon' coincidence imaging with a classical source *Phys. Rev. Lett.* **89** 113601
[30] 6 ed Loudon R 1983 *The Quantum Theory of Light* 2nd edn (Oxford: Oxford University Press)
[31] Allen L and Eberly J 1975 *Optical Resonance and Two-Level Atoms* (New York: Wiley)
[32] Granik A 1990 Quantum holography *Proc. SPIE* **10308** 1030803
[33] Ludman J, Caulfield H J and Riccobono J 2011 *Holography for the New Millennium* (New York: Springer) ch 15
[34] Leichtle C, Schleich W P, Averbukh I and Shapiro M 1998 Quantum state holography *Phys. Rev. Lett.* **80** 1418
[35] Abouraddy A F, Saleh B E A, Sergienko A V and Teich M C 2001 Quantum holography *Opt. Express* **9** 498–505
[36] Chrapkiewicz R *et al* 2016 Hologram of a single photon *Nat. Photon.* **10** 576–9
[37] Szadowiak W, Kundu S, Szuniewicz J and Lapkiewicz R 2021 Self-referenced hologram of a single photon beam *Quantum* **5** 516

www.ingramcontent.com/pod-product-compliance
Ingram Content Group UK Ltd.
Pitfield, Milton Keynes, MK11 3LW, UK
UKHW051903100926
8556IPUK00005B/752

9 780750 33025